RENEWALS 458-4574

DATE DUE

GAYLORD			PRINTED IN U.S.A.

Wireless Mobile Networking with ANSI-41

Wireless Mobile Networking with ANSI-41

Second Edition

Randall A. Snyder
Michael D. Gallagher

McGraw-Hill
New York   Chicago   San Francisco   Lisbon   London
Madrid   Mexico City   Milan   New Delhi   San Juan
Seoul   Singapore   Sydney   Toronto

McGraw-Hill

A Division of The McGraw·Hill Companies

Copyright © 2001 by Randall A. Snyder and Michael D. Gallagher. All rights
reserved. Printed in the United States of America. Except as permitted
under the United States Copyright Act of 1976, no part of this publication
may be reproduced or distributed in any form or by any means, or
stored in a data base or retrieval system, without the prior written
permission of the publisher.

1 2 3 4 5 6 7 8 9 0 QWM QWM 9 8 7 6 5 4 3 2 1

ISBN 0-07-135231-7

The sponsoring editor for this book was Steve Chapman and
the production supervisor was Sherri Souffrance. It was set in
New Century Schoolbook by Patricia Wallenburg.

Printed and bound by Quebecor World Martinsburg.

McGraw-Hill books are available at special quantity discounts to use
as premiums and sales promotions, or for use in corporate training
programs. For more information, please write to the Director of Special
Sales, Professional Publishing, McGraw-Hill, Two Penn Plaza, New York,
NY 10121-2298. Or contact your local bookstore.

 This book is printed on recycled, acid-free paper containing a
minimum of 50% recycled, de-inked fiber.

DEDICATION

From Mike: I would like to dedicate this book to Halina, Samuel, Alexandra, and Kira, our most recent family addition. Thanks for the love.

From Randy: I would like to dedicate this book to my wife, Cecilia, whom I love dearly and to my dog Fred who kept me company while I typed away at the manuscript.

CONTENTS

FOREWORD

The wireless phone has gone from a gee-whiz gizmo to a commonplace fact of everyday life. Every one and a quarter seconds, 24 hours a day, 365 days a year, a new wireless subscriber signs on in the United States. This subscriber growth has been driven by two factors: competition and functionality. The wireless industry is the most competitive facility-based segment of all telecommunications. That competition, in turn, has driven technical innovation such as the amazing ability to have a call to your wireless phone number find you thousands of miles away—even if you are in another country.

Behind the tremendous growth in subscribership and functionality of the wireless phone is an underlying technology, known by the unlikely name of IS-41 (now known as ANSI-41). It is the *lingua franca* of the wireless age, the common protocol that underlies many wireless phone's features. The evolution of the ANSI-41 standard has been a critical element behind the majority of the enhancements in the current wireless environment. The fact that we, as consumers, take it for granted is a testament to its success.

This is a book for those who want to understand that *lingua franca* and, thus, understand the underpinning of the wireless revolution. The first decade of the 21st century will be the wireless decade, and a principal backbone technology for that reality is IS-41. Randy Snyder and Mike Gallagher have done a wonderful job in this book of helping us all—whether engineer or simply interested observer—understand one of the engines that will drive the decade.

Tom Wheeler, President & CEO,
Cellular Telecommunications & Internet Association (CTIA)

PREFACE

As the wireless industry continues to move forward at lightning speed, so has the development of new technologies and standards to support those technologies. Since the first edition of this book was published in 1997, many, many innovations have occurred. The basic technology standards of IS-41 are still intact; however, additions such as Over-the-Air Service Provisioning, new data services, new subscriber identifiers and a host of feature enhancements warrant an update to this text. Not only that, but since the first edition of this book was published, the IS-41 interim standard has been elevated to a full national standard, now known as *ANSI-41 Revision E*.

Who Should Read This Book

The subject matter of the ANSI-41 national standard for wireless telecommunications has typically fallen into the purview of telecommunications system engineers and equipment software developers only. This book attempts to serve many purposes, the primary one being to expand the understanding of ANSI-41 beyond this select group. For system engineers and software developers, this book can serve as a reference to the ANSI-41 standard, as well as explaining some of the more complex technical features of the standard. For wireless telecommunications technical and marketing managers, as well as executives, this book is intended to also be a learning tool. Although the ANSI-41 standard itself is the focus of the book, there is enough information included about wireless and cellular telecommunications networking to provide a better understanding of wireless network architecture and functionality.

This book is a direct result of our involvement as engineers in the wireless industry participating in the creation of the ANSI-41 standard, as well as many others. The bulk of the material is derived from the ANSI-41 standard itself, however there is valuable information provided to expand one's understanding beyond the standard. Information about the application of ANSI-41 in the real world, deployment strategies, insights into network architectures and the standards-making process are provided so that this book serves as more than just a reference for a single industry standard. The intent of this book is to provide a comprehensive introduction to ANSI-41, as well as a reference for those already familiar with the protocol.

All standards are open to interpretation and ANSI-41 is no different. Although we have taken part in the development of the ANSI-41 standard and the design of systems that use ANSI-41, the substance of this book reflects our understanding and interpretation of the standard. This understanding certainly should not be taken as a unique view. The ANSI-41 specification and related standards are the primary references for this book and we disclaim any responsibility for any usage of the interpretations described in this book.

Graphical Map of This Book

This book is divided into three parts. The subject matter of each part logically builds on the information presented in the previous part. Part 1 provides an introduction to the concept of wireless telecommunications networking, network architecture, and functions that enable mobility within the telecommunications network. Part 2 provides a full description of ANSI-41 including revisions, operational details and protocol usage with a focus on ANSI-41 Revision E. Part 3 provides ANSI-41 network implementations including network interconnections and practical information about network interoperability, roaming, and related technologies.

A Few Words about Terminology

Terminology can be a major stumbling block in the understanding of technical subject matter, and wireless telecommunications networks are no exception. Seemingly familiar terms develop new meanings and engineers begin to drown in a sea of acronyms. Also, informal usage of synonyms for many of the technical terms can lead to confusion and semantic arguments. This section presents an explanation of some commonly misunderstood and confusing terms that should help with the understanding of this book. Of course, anyone familiar with these terms can still find some ambiguity in our descriptions and definitions; however, it is our intention to clarify some of these semantics, provide distinctions between similar terms and provide a basis for consistent usage within this book.

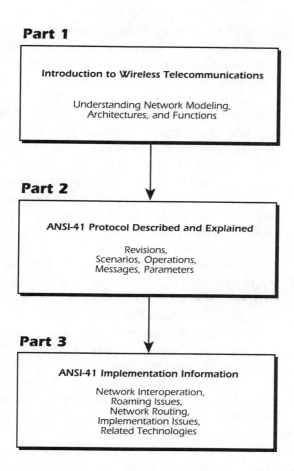

WIRELESS, CELLULAR, PCS, AND MOBILE

These terms can be very confusing based on the context in which they are used. Many standards bodies have vehemently argued the meaning of these terms from a technical, political and commercial perspective. The term *wireless* can encompass nearly any type of communications without the use of wires. Good examples are radio and infrared communications. Cellular systems, personal communications services (PCS), paging, wireless local area networks, and even remote control units are considered to be types of wireless communications.

The terms *cellular* and PCS are used in the U.S. to specifically refer to the current operating wireless telecommunications systems and are the type of systems referred to by ANSI-41. Cellular systems can be considered a subset of wireless communications employing certain unique characteristics, such as high extensible capacity, frequency reuse and mobility management, that controls communications to and from specific cells. The term cellular is sometimes used politically and commercially to differentiate first generation wireless telecommunications systems from next generation personal communications services systems. Technically, however, many systems considered to be PCS actually do employ cellular technology. This is an important concept. Generally, the terms cellular and PCS simply denote the differences in bandwidth allocated to wireless operating licenses (i.e, cellular in the 800 MHz–900 MHz range and PCS in the 1800 MHz–1900 MHz range). ANSI-41 can support any system based on these cellular technologies and the term wireless often times refers to both cellular and PCS networks.

The term *mobile* is generally used as a synonym to cellular and PCS technology systems. However, many wireless technologies exist that do not support *mobility*. Some of these technologies are known as *fixed wireless systems*. The terms *mobile* and *wireless* are used within this book to refer to systems employing both cellular and cellular-based PCS technologies.

INTERSYSTEM VERSUS INTERNETWORK

The ANSI-41 standard specifies cellular network *intersystem* operations. A system is generally defined as a single mobile switching center (MSC) system and its peripheral functionality (i.e., its associated HLR, VLR, etc.). The term *intersystem* in this context implies operations between MSC systems used to support wireless telecommunications. The term *internetwork* sometimes implies conversion of network messages between non-homogeneous networks.

Operation versus Message   An ANSI-41 operation refers to an entire operation process which includes all components of the information exchange used to effect that operation. An ANSI-41 message refers to each specific component of a given operation. For example, the Registration Notification operation consists of the entire information exchange process including the "Registration Notification Invoke" component as well as the "Registration Notification Return Result" component. Each of these components can be considered an individual ANSI-41 message.

Subscriber versus End-user   These terms are generally used synonymously. *Subscriber* is used throughout the ANSI-41 specification to mean the end-user of the wireless network services. The subscriber is the human using the mobile station (MS).

FEATURE VERSUS SERVICE

These terms are generally used synonymously. The term service is somewhat all encompassing and represents any function that can be supported in the wireless network. In the context of ANSI-41, the term feature specifically refers to a specific supplementary service provided to a subscriber, such as call waiting or call forwarding.

CDMA AND TDMA

Code division multiple access (CDMA) and time division multiple access (TDMA) are general categories of digital radio technology. There are many wireless radio standards (both national and international) that are based on these technologies. However, in reference to ANSI-41-based networks, the terms have specific meaning. In this book, CDMA specifically refers to the *ANSI/TIA/EIA-95* family of related standards and TDMA specifically refers to the *ANSI/TIA/EIA-136* family of related standards. The ANSI/TIA/EIA-136 TDMA family of standards is sometimes referred to as digital AMPS or D-AMPS (refer to the Bibliography for these standards).

CCITT VERSUS ITU-T

The original CCITT (International Telegraph and Telephone Consultative Committee) which published the well known *Blue Book* technical standards, officially changed its name to ITU-T (International Telecommunications Union–Telecommunications) in 1993. International standards published from this organization before 1993 carry the name CCITT, while those published after 1992 carry the name ITU-T. The appropriate term is used within this book depending on when the actual standard referred to was published.

CELL

An individual geographic area that is the topological component of a cellular or personal communications services system. The area is defined by the telecommunications coverage of the radio equipment located at the cell site. A *cell sector* is a geographic portion of a cell (typically a third) that is served by directional antennas dividing the coverage area of the original cell. The aggregate of the cell sectors that form an entire cell increases capacity by providing frequency reuse among the sectors. A *cell site* is the physical location of a cell's radio equipment and supporting systems. This term also refers to the equipment located at the cell site.

PROTOCOL

A protocol is simply a set of rules or conventions governing the interactions of processes, applications and components in a communications system. ANSI-41 is a standard protocol governing the communications among cellular telecommunications network elements.

ANSI-41 PARTS, CHAPTERS, AND SECTIONS

ANSI-41 is one of the few Telecommunications Industry Association (TIA) specifications that is divided into multiple parts (i.e., ANSI-41.1, ANSI-41.2, ANSI-41.3, ANSI-41.4, ANSI-41.5 and ANSI-41.6). These parts are sometimes referred to as chapters. Within each part of ANSI-41 is a Table of Contents where topics are divided into sections (e.g., Section 5.4.3 of Part ANSI-41.3).

REVISION 0

The term *Revision 0* is an informal term used to refer to the initial version of a standard or specification published by the TIA. This revision number is not officially listed as an actual revision number by the TIA. Subsequent revisions to an initial specification are officially denoted as *Revision A*, *Revision B*, *Revision C*, etc. Revision 0 documents carry no revision designator in their title.

1G, 2G, 2.5G, AND 3G

The term *3G* has become quite ubiquitous in the wireless industry. Generally, *1G* or first generation wireless technologies, refers to the original analog cellular telecommunications systems that are still in use today. *2G* or second generation wireless technologies, refers to the evolution of analog to digital cellular and PCS telecommunications systems. 2G systems include the GSM, TDMA, and CDMA systems operating today. These systems enable a variety of enhanced features and services such as short messages service (SMS), voice privacy, and a host of call features. *2.5G* wireless technologies refer to enhanced digital data capabilities that are interim technologies toward the evolution to full 3G wireless technology services. Examples of 2.5G services are the general packet radio service (GPRS) for GSM and TDMA systems, and enhanced high-speed data services for CDMA systems. *3G* or third generation wireless technologies, refers to the newest generation of digital wireless telecommunications services being standardized today. The International Telecommunications Union (ITU) is standardizing 3G technology as IMT-2000. Characteristics of these networks include true broadband data, multi-media messaging, and a new paradigm for the deployment of wireless data and telecommunications services.

Michael D. Gallagher
Randall A. Snyder

ACKNOWLEDGMENTS

We gratefully acknowledge the contributions of the following organizations:

- Telecommunications Industry Association, Inc. (TIA), for allowing us to use their published standards as source material for this book.
- *Cellular Networking Perspectives* Industry Newsletter, written by Mr. David Crowe, who continues to provide valuable insights into the wireless industry. David's ten years' of published information provided us with a great deal of source material and many outstanding ideas that added to the quality of this book.

We would also like to thank the following group of individuals—the contributors to the IS-41 Revision C standard: Jean Alphonse, Phil Audino, Cheryl Blum, Jane Bissonnette, Alain Boudreau, Terri Brooks, Arzu Calis, Kirk Carlson, Lynn Carlson, Sharat Chander, Ben Chen, Jeff Crollick, David Crowe, Erkin Cubukcu, Amy Dwyer, Robert Ephraim, A. Gains Gardner, Thomas Ginter, Anthony Golka, Ed Hall, Huel Halliburton, Alejandro Holcman, Michel Houde, Chuck Ishman, Terry Jacobson, Stephen Jones, Joe Juliano, Irfan Khan, Barry Kratz, Bill Krehl, Sunit Lohtia, Paxton J. Louis, John Marinho, Eileen McGrath-Hadwen, Charlene Meins, Robert Montgomery, David Munsinger, C. Peter Musgrove, Peter Oldfield, Gustavo Pavon, Gary Pellegrino, Tom Richter, Raul Rodriguez, Asim Roy, Todd Shingler, Jay Tang, Ed Tiedemann, Dennis Velte, Tim Verdonk, Terry Watts, John Willse, Mike Youngberg.

We also want to thank the additional contributors to the most recent version of ANSI-41: Susan Anctil, Arturo Arreaga, Brye Bonner, Wayne Bowen, Sam Broyles, Roger L. Bunting, Raul Cardenas, Louis S. Degni, Tim Dunn, Scott Easley, Andreas Flach , Gilberto Frederick, Leonel Garza, Jim Gately, Marcia Guza, Cathy Fitzpatrick, Eric Hill, Martin W. Hynes, Toni Johansson, Betsy Kidwell, Fred E. Kujawski, Ben Levitan, Sharon Lim, Louis Linneweh, Chad Loudon, Michael Marcovici, Stephen Meer, John Melcher, Jack Nasielski, Takeshi Nemoto, Masayoshi Ohashi, Hirofumi Okata, Jennie Ong, Marie Risser,

Doug Rollender, Greg Schumacher, Bob Slocum, Atul Thaper, Lee Valerius, Ross Wilken, Jake Wolski, Terry Yen, and Larry Young. Our apologies to those contributors that we missed, including the "behind the scenes" contributors who did not attend the TR45.2 meetings.

And special thanks to our sponsoring editor, Steve Chapman, for his infinite patience with us and for his ongoing support of our efforts to write this book.

Introduction to Wireless Telecommunications, Network Architecture, and Functions

CHAPTER 1

Basics of Wireless Telecommunications

To begin to understand the ANSI-41 signaling protocol, it is necessary to understand some basics about wireless telecommunications. This chapter provides a general overview of the most important concepts in wireless cellular telecommunications that apply to ANSI-41. For some readers, this will be a high-level review; for others, it may clarify some misconceptions and provide an overall understanding of cellular wireless technology that will be useful in understanding ANSI-41.

What Are Wireless Telecommunications?

The concept of wireless telecommunications can be viewed from two perspectives: the wireless subscriber's and the wireless network's. From the subscriber's perspective, *wireless telecommunications* is a service that allows telephone calls to be made or received while the telephone equipment moved from place to place or while it is in motion. From this perspective, the telephone handset (known as the *mobile station*) is wireless and affords the ability to be mobile.

From the network's perspective, *wireless telecommunications* is a service provided to end-users. Wireless telecommunications, within the context of ANSI-41, is a service based on a set of functions internal to the network known as *mobility management*. Mobility management functions enable the network to maintain location and subscriber status information so that end-users can make and receive calls while they move from place to place.

Origin of Advanced Mobile Phone System

Wireless telecommunications can be considered both a system and a service. The network equipment including antennas, radios, switches, databases, and all hardware and software within the network, represents a wireless telecommunications system that provides wireless telecommunications service to subscribers.

The first wireless telecommunications system based on cellular technology was (and is) known as *AMPS* (Advanced Mobile Phone System), a technology developed by Bell Laboratories in 1947. The term *cellular* refers to a network of small *cells* or radio transceivers, each providing a limited range of radio coverage, which are linked by a computer-controlled switching system that manages subscriber mobility and interfaces to the fixed wire-line telephone network. The technology is based on cellular frequency reuse (described later), providing a high-capacity system and allowing network access using low-power mobile stations (typically less than 6 W). The radio transceiver modulation methods used are based on analog frequency modulation (FM) signals similar to those used for commercial radio, but at a higher frequency range and lower bandwidth.

The first commercial cellular system in the United States became operational in Chicago in 1983. However, other countries around the world provided operational cellular systems several years earlier.

Today, cellular systems based on AMPS technology are implemented in more than 100 countries. It is interesting to note, however, that there is no single worldwide standard for the implementation of these cellular systems. The different systems deployed generally represent differing radio technologies, each based on the concepts of AMPS. The networking technologies used to link the cells together are also quite different. These technologies are considered supplemental to the defining characteristics of AMPS (i.e., cells, frequency reuse, etc.). In fact, AMPS can operate within a variety of networking schemes.

Some Basic Cellular Concepts

Basic Radio Technology

Cellular radio technology allows a subscriber to originate and receive telephone calls wherever compatible cellular radio coverage is provided. A *cell* is an individual radio coverage area controlled by a radio base station (BS) system. Individual calls within a single cell use different frequencies. These frequencies can be *reused* by other cells, provided that there is no interference with the other cells. The frequency reuse pattern of the cells is dependent on the distance between the cells and the radio transmission power.

First-generation radio technologies (AMPS-based) use signals based on analog FM for speech transmission. Subsequent generations of radio technology for wireless systems include NAMPS (narrowband AMPS), which is also based on analog FM, and a variety of sophisticated digital technologies based on TDMA (time division multiple access) and CDMA (code division multiple access).

The cellular radio spectrum (range of allowable and available radio frequencies) used for these cellular system technologies is regulated by government agencies in different countries. In the United States, cellular service providers are categorized by one of two sets of non-contiguous 25 MHz radio frequency bandwidths. PCS service providers are categorized by six sets of noncontiguous bandwidths: three sets of 30-MHz radio frequency bandwidths and three sets of 10-MHz radio frequency bandwidths.

Cellular A-band and B-band Carriers

The two sets of bandwidths licensed for cellular radio service are known as the *A-band* and the *B-band*. A-band carriers are cellular service providers originally termed the *nonwire* line licensees. These original licensees are companies that provide cellular service and are not associated with any local wire line telephone company.

B-band carriers are cellular service providers originally termed the *wire line* licensees. These licensees are companies that provide cellular service and are associated with the local wire line telephone company (i.e., the original Regional Bell Operating Company or RBOC) in the area where they provide cellular service.

The concept of A-band and B-band carriers was devised as part of the Modification of Final Judgment (MFJ) consent decree in 1982 that broke up the AT&T/Bell system monopoly in 1984. The AMPS technology originally developed by Bell Laboratories was given up to the seven RBOCs as part of the compromise to divest them from AT&T. The mandated provision to allow two cellular service provider licenses in a given geographic area was designed to provide competition between an independent cellular carrier and the cellular carrier owned by the local wire line carrier. Note that the cellular A- and B-band licenses are allowed to support analog or digital radio technologies. Figure 1.1 depicts the cellular radio spectrum licenses in use today.

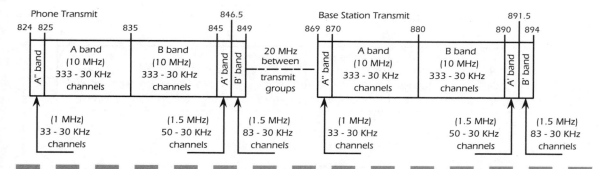

Figure 1.1 *Cellular licensed frequencies that are in use today. The A', B', and A" bands were originally set aside for control functions, but can be used for normal traffic.*

PCS A–F-band Carriers

In 1994, the U.S. government publicly auctioned six new sets of nation-wide wireless telecommunications licenses. These licenses had restrictions on who could own them different from those of the original A and B-band licenses. They were no longer limited to what type of carrier could own them (i.e., local wire line or independent wireless); rather, restrictions were put on the total number and types of wireless licenses that a single carrier could own in a given market.

In the U.S., PCS service providers are categorized by one of three sets of noncontiguous 30-MHz radio frequency bandwidths or one of three sets of noncontiguous 10 MHz radio frequency bandwidths. Note that the PCS A- through F-band licenses (see Figure 1.2) are allowed to support only digital radio technologies, typically operating among cell sizes of much smaller radii than analogous cellular systems. Figure 1.2 depicts the PCS radio spectrum licenses in use today.

Frequency Reuse

Cellular frequency licenses provide for each mobile station to occupy 60 kHz of bandwidth (30 kHz for transmission and 30 kHz for reception) within an entire radio frequency (RF) allocation of 25 MHz for each of the two cellular carriers (A and B) in a given area (i.e., 12.5 MHz for transmit and 12.5 MHz for receive for each carrier). PCS frequency licenses provide for each mobile station to occupy 60 kHz of bandwidth

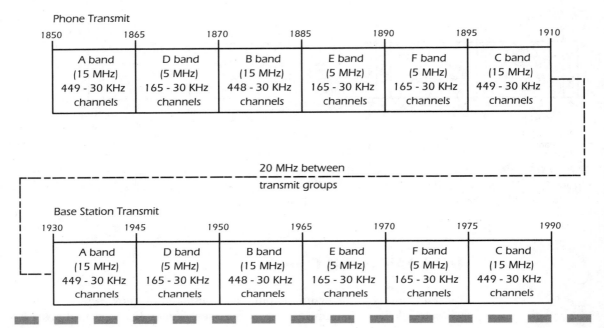

Figure 1.2   PCS licensed frequencies in use today.

(30 kHz for transmission and 30 kHz for reception) within an entire RF allocation of either 30 MHz (i.e., 15 MHz for transmit and 15 MHz for receive for each carrier) or 10 MHz (i.e., 5 MHz for transmit and 5 MHz for receive for each carrier). Note that although the terms cellular and PCS differentiate categories of frequency bandwidths, PCS systems employ the same basic technology of *cellular* systems.

Cellular systems use a technique known as *frequency reuse* (see Figure 1.3). A particular available channel frequency is transmitted from one base station at a power level that supports communications within a moderate cell radius around that base station (anywhere from a few hundred feet to about 50 miles!). Because this transmitted signal power is controlled to serve only a limited range, the same frequency can be transmitted simultaneously, or *reused*, by another base station, provided there is no interference between it and any other base station using that same frequency.

Figure 1.3 depicts a typical cellular frequency reuse model using a seven-cell pattern that provides uniform distances for channel reuse.

In the model, each base station is considered to be located at the center of a hexagon, with the hexagons (or cells) labeled A through G repre-

Figure 1.3
Seven-cell frequency
reuse pattern.

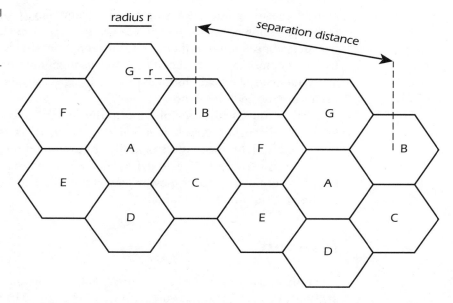

senting seven channel sets. The frequencies used for the channel sets in the A cells are the same, as are the frequencies in the B cells, C cells, etc. There are many possible frequency reuse patterns. Since the total number of channels for cellular and PCS is fixed, the selection of a reuse pattern and cell size determine how many subscribers can be supported in a given service area. Available capacity, however, is much greater than the actual number of channels accessible in a given cell. This is due to the nature of end-user calling behavior: not everyone wants to talk at the same time. A cell can typically serve 10 to 20 times more subscribers than the total number of channels supported.

Digital Radio

Digital radio technologies have been developed for use on the air (or radio) interface. These can dramatically increase the number of subscribers supported on the range of frequencies used for wireless telecommunications systems (note that the analog NAMPS technology also provides an increase in the number of subscribers supported). The two basic types of digital technology are time division multiple access (TDMA) and code division multiple access (CDMA). Many standards exist for the use of these basic technologies. TDMA is based on the use of time-interleaving of multiple signals to provide an apparently simultaneous transmission of those signals on a single radio frequency. CDMA is based on a technique known as *direct sequence spread spectrum* (DSSS) that digitally codes a base signal and employs different signal encoding patterns and frequency hopping to redistribute that base signal across a broad range of frequencies. Logical channels for each signal are created through the use of unique code sequences.

TDMA and CDMA can provide many advantages over analog-based systems. Examples are better voice quality, increased capacity, less noise and interference, and the ability to provide digital services such as data and messaging. The ANSI-41 networking protocol is designed to support versions of these newer-generation digital technologies as well as the original analog systems.

Handoff

Handoff encompasses a set of functions, supported between a mobile station (MS) and the network, which allows the MS to move from one cell to another (or one radio channel to another, within, or between cells) while a call is in progress. The handoff function requires sophisticated coordination between the network and the MS to transfer the MS smoothly from one radio channel to another during a call. There are two types of handoff: *intra*system and *inter*system.

Intrasystem handoff (see Figure 1.4) is a handoff between two cells or radio channels that subtend the same mobile switching center (MSC). In this case, no coordination is required between MSCs to support the movement of an MS between cells. Intersystem handoff (see Figure 1.5) is a handoff between two cells subtending two different MSCs. This type of handoff requires specialized signaling between the two MSCs to coordinate the movement of the MS between the cells. Since the ANSI-41 protocol is concerned with intersystem operations, it provides the operations necessary to support intersystem handoff. Intrasystem handoff is not within the scope of ANSI-41 and is handled via proprietary methods at the MSC.

There are three strategies for performing a handoff: MS controlled, network controlled, and MS assisted.

These strategies differ mainly in which side of the radio interface determines when to hand off the MS to another channel. MS-controlled handoff is a technique where the MS itself continuously monitors the radio signal's strength and quality. When predefined criteria are met, the MS checks the best candidate cell for an available traffic channel and requests that the handoff occur. Network-controlled handoff is a

Figure 1.4
Cells involved in
intrasystem handoff.

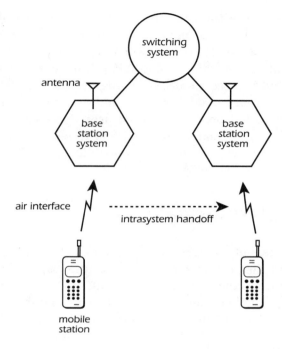

Figure 1.5
Cells involved in
intersystem handoff.

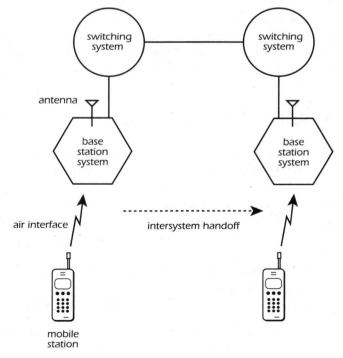

technique where the radio base station, MSC, or both, monitors the radio signal. When the signal's strength and quality deteriorate below a predefined threshold, the network arranges for a handoff to another channel. MS-assisted handoff (MAHO) is a variant of network-controlled handoff and is a technique where the network directs the MS to measure signals from surrounding cells and report those measurements back to the network. The network then uses these measurements to determine where a handoff is required and to which channel. The ANSI-41 protocol supports both network-controlled and MS-assisted strategies for intersystem handoff.[1]

Network Systems

In the context of wireless telecommunications systems, the network consists of entities not directly related to the radio interface between the mobile station and the radio base station. Network systems consist of switching functions, service logic functions, database functions, and all mobility management functions that enable subscribers to be mobile. These are the functions provided by the MSCs, control and database systems (known as *location registers*), and by other logical functional entities in the network. Network intersystem operations are required to provide communications between these entities to enable the mobility management functions. These intersystem operations, including intersystem handoff, are specified and standardized by ANSI-41.

Mobility Management

Mobility management is the primary set of functions supported by the network to enable subscriber mobility. Mobility management enables the network to keep track of the subscriber's status and location for the delivery of calls to that subscriber. It also enables the network to authorize a subscriber for service in a given cellular service area. The key component to mobility management is the control of the subscribers' *service profile*. The service profile is simply a database record in the network that contains information about each subscriber. This information includes temporary data such as current location and status of a subscriber as well as permanent data, such as the subscribed features (e.g., call waiting).

---

[1] Examples of MS-controlled handoff are in the Personal Access Communications System (PACS) and Digital European Cordless Telephone (DECT) protocols.

Basic Network System Architecture

Wireless telecommunications networks are primarily comprised of the following four basic network elements (see Figure 1.6):

- Radio systems
- Switching systems
- Data-based systems (i.e., location registers)
- Operations, administration, and maintenance (OA&M) systems

Figure 1.6
Basic wireless
telecommunications
system architecture.

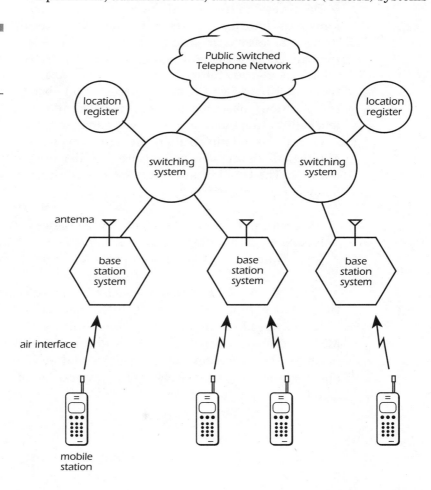

Radio systems consist of low-power mobile stations and base station systems. The radio system provides the communications path between

the mobile station and the cellular base station. The base station system includes the antennas, transceivers, and controller systems that provide radio access into the network.

Switching systems provide interfaces for subscriber traffic between the cellular network and other public switched networks and within the cellular network. The switching systems coordinate the establishment of calls to and from the wireless subscriber. These systems are directly responsible for managing transmission facilities, subscriber mobility, and call processing.

Data-based systems in the wireless telecommunications network are known as *location registers*. Location registers provide a database (hence, the term *data-based*) along with service logic to actively control the wireless services provided to the subscribers. The database provides information about subscribers to the network. This information includes the subscriber's identification, directory number (phone number), current location, subscribed features (such as call forwarding to voice mail), and call-routing information, as well as many other types of data.

Operations, administration, and maintenance (OA&M) make up a set of functions that enable the service provider to monitor and control the network. The OA&M functions allow the service provider to perform the following general actions:

- Observe and record operational characteristics of the network.
- Modify and configure the network equipment and functions.
- Identify and correct failures and defects within the network.

The ANSI-41 specification provides a standard protocol for the operations that enable subscriber mobility between MSC serving areas. ANSI-41 specifies the signaling communications that occur between MSCs, network location registers, and some specialized network nodes (such as short message service centers and authentication centers) to allow subscriber movement between networks based on the standard.

CHAPTER 2

Wireless Telecommunications Standards

ANSI-41 is a technical standard for wireless network signaling. This chapter explains the importance of standards and provides an overview of the organizations that govern the development of ANSI-41, and how those organizations relate to others that also define standards.

What is a Standard?

Technical standards are evident in everyday life all around us. There are basically two types of technical standards: those that are *prescribed* and those that are *de facto*. An example of a prescribed standard is the design and function of electrical appliance plugs and outlets in the United States. An example of a de facto standard is the placement of hot water faucets on the left side of a sink and cold water faucets on the right side. Standards serve two primary purposes: to make our lives easier and to save us money. Imagine the cost of installing many different kinds of electrical outlets to suit all the different types of appliance plugs that could be developed, or the inconvenience of trying both water faucets to discover which is cold and which is hot. Standards are supposed to prevent these problems and provide a commonly accepted authority for the design and function of all types of equipment.

Standards for the design and function of telecommunications equipment are no different in goal and purpose from those for any other type of technical standard. Telecommunications standards are prescribed for the design and function of equipment as simple as a telephone keypad or as complex and sophisticated as computer equipment and the services provided by a wireless telecommunications network.

In tangible terms, a telecommunications standard is a document that establishes engineering and technical requirements for processes, procedures, and methods that have been decreed by authority or adopted by consensus. The primary goal of the telecommunications standards process is to encourage the interconnectivity of telecommunications equipment and services by establishing and promoting technical recommendations in these areas. The telecommunications industry achieves this goal by creating and maintaining voluntary specifications that can optimize equipment compatibility. Standards are typically considered to be *recommendations*; that is, they are prescribed as purely voluntary. However, business needs usually show that it may not be very lucrative to stray from standardized designs and functions.

ANSI-41 (American National Standards Institute - 41) is the technical standard that prescribes the network model, functions, protocols, and services that provide wireless telecommunications network intersystem operations.

Scope of a Standard

Standards encompass many areas of engineering and technical requirements. They can be very brief and promote an existing engineering solution to a new problem (usually by reference), or they can be quite extensive (hundreds to thousands of pages) and establish the details for the processes and methods for developing an entire system. As stated before, the primary goal of telecommunications standards is to establish and promote technical recommendations that enable the interconnectivity of equipment and services. However, standards do not dictate an *implementation* or the methods for developing an individual solution to provide equipment or service. Thus, business competition among manufacturers is preserved while enabling interoperation of their equipment and services.

Telecommunications standards can include many aspects of the design and function of equipment. These include network models used for developing network architectures; descriptions of functions as perceived by an end-user and as provided within and between networks; descriptions of protocols or the methods of communication within and between functional and network entities; testing procedures for establishing equipment compatibility and standards compliance; and any procedures that assist in achieving the goals of the standardization process.

Who Makes and Uses Telecommunications Standards?

There are many organizations concerned with the development and use of telecommunications standards. The standards-making process requires cooperation at three basic levels: between and among industrial concerns (i.e., wireless service providers and equipment manufacturers), between industrial concerns and governmental concerns (e.g., service providers and regulatory agencies), and between nations.

Cooperation among all the concerned parties is not always possible, hence the existence of multiple standards (try taking your hair dryer to another country and plugging it in). A good example of multiple telecommunications standards is the transmission carrier standard T-1 used in North America (operating at 1.544 Mbits/s) and standard E-1 used in Europe (operating at 2.048 Mbits/s). To accommodate many conflicting interests, international standards-making organizations concentrate on producing what are known as *base* standards. Base standards contain variants, national options, and alternative methods for implementation-dependent needs. Adopting these variants means that an *implementation* is compatible with the standards, but there is no guarantee that equipment based on different variants can work together.

Standards Groups, Trade Groups, and User Groups

There are more than 250 organizations that prepare international standards. In fact, these organizations have developed about 20,000 technical standards. Ninety-six percent of all standards are developed by three international organizations: the International Telecommunications Union (ITU, formerly the CCITT), the International Organization for Standardization (ISO), and the International Electrotechnical Commission (IEC). Nearly half of these standards apply to telecommunications, information technology, and related fields.

The ITU is a treaty organization of the United Nations whose activities include standardizing telecommunications and spectrum management, regulating radio telecommunications, and managing frequency assignments that have international significance. The ITU also plays a key role in the evolution of seamless global telecommunications technology. The ITU membership consists of national delegations from more than 180 countries. The ISO is a voluntary nongovernment organization mainly providing standards for information technology. This group develops standards to facilitate international trade in goods and services. ISO membership comprises primarily national standards-making bodies including the American National Standards Institute (ANSI). More than 100 nations contribute to the ISO. The ISO and ITU work closely together in areas of common interest. The IEC is also a voluntary nongovernment organization primarily working in the area of electrical and electronic engineering. The IEC is a sister organization to the ISO, and its membership consists of about 50 contributing nations.

Functional standards are adapted from international base standards and contain only a limited subset of permissible variants. Adaptation from international base standards to national functional standards is provoked by the following types of organizations:

- Regional or national standards groups
- Trade groups
- User groups

Examples of regional or national standards groups are ANSI and the National Institute of Standards and Technology (NIST). Examples of trade groups are the Cellular Telecommunications & Internet Association (CTIA) and the Personal Communications Industry Association (PCIA). Examples of user groups are the Institute of Electrical and Electronic Engineers (IEEE) and the North American ISDN Users Forum (NIUF). These organizations usually provide well-defined requirements as input to the functional standards.

As part of the standards-making process, agreed-upon test specifications and methods are developed to ensure that equipment designed to the different variants (permitted within the functional standards) will work together. Independent test organizations can perform conformance tests and certify that telecommunications equipment and products comply with the standards. Figure 2.1 depicts a flow diagram of the overall standards-making process and the relationship among the contributors to standards development.

American National Standards and the TIA

There are many national organizations providing standards for North America. In this context, North America refers primarily to the United States and Canada. At the forefront of North American telecommunications standardization is ANSI. As a U.S. national standards-making body, ANSI is responsible for accrediting other U.S. standards-making bodies. Among these are the Alliance for Telecommunications Industry Solutions (ATIS), the Electronic Industries Alliance (EIA), and the

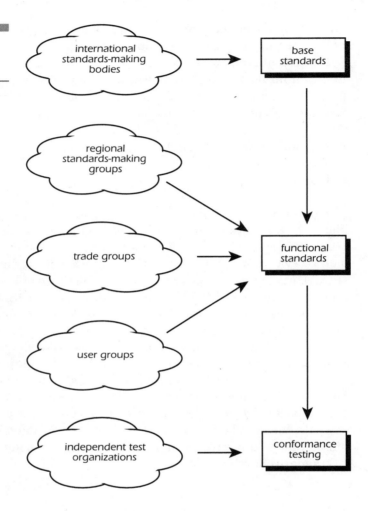

Figure 2.1
The standards making process.

Telecommunications Industry Association (TIA). The ATIS was formed during the impending Bell System divestiture to develop open industry standards to replace the closed, de facto Bell System specifications. An interesting note is that although most of these Bell System specifications are still prevalent throughout the telecommunications industry (currently sponsored by Telcordia—formerly Bell Communications Research or Bellcore), they are not considered prescribed standards (although they have become de facto standards in many cases). This is due to the fact that these specifications were developed in a closed forum, considering the business concerns of only the limited number of companies making up the Bell System. Even in the post-divestiture era, these specifications remain primarily focused on the interests of the former RBOCs and are not considered national standards.

The EIA was initially formed as a radio manufacturers' group and evolved to cover all areas of electronics information and communications technology. The TIA was formed in 1988 from the combination of the Information and Telecommunications Technology Group of the EIA and the U.S. Telecommunications Suppliers Association. The charter of the TIA is the formation of new land mobile telecommunications standards, although it does develop standards for technologies as diverse as wireless data, fiber optics, and satellite communications. The TIA is associated with the EIA and is an ANSI-accredited standards-making body. The TIA has developed most of the currently used standards for wireless telecommunications in the United States, including ANSI-41.

The TIA primarily develops what is known as *interim standards*, hence the former title *IS*-41 for the current ANSI-41 standard. These standards are considered interim because they have a limited life—originally five years and now three years. All interim standards developed by the TIA have the potential to eventually become full ANSI national standards, if they are agreed upon by the larger membership of ANSI, as ANSI-41 was.

The TIA is composed of many committees that develop wireless and other telecommunications standards. The committees concerned with developing wireless telecommunications standards are designated as *TR* committees; the TR designation being an artifact of the term *transmission*, which was the original technology being standardized in the early days of the EIA. There are currently nine separate TR committees within the TIA, as listed in Table 2.1.

TABLE 2.1

TIA TR Standards Committees

TIA TR Committees	TIA TR Committee Name
TR-8	Mobile and Personal Private Radio Standards
TR-14	Point-to-Point Communications Systems
TR-29	Facsimile Systems and Equipment
TR-30	Data Transmission Systems and Equipment
TR-32	Personal Radio Equipment
TR-34	Satellite Equipment and Systems
TR-41	User Premises Telephone Equipment Requirements
TR-45	Mobile and Personal Communications Systems
TR-46	Mobile and Personal Communications (primarily 1900-MHz GSM)

Standards projects begin with a contribution from an individual requesting the creation of a new standard in a specific technical area. A *contribution* is simply a written discourse that can be a request or proposition that an individual distributes at a TIA meeting. The scope and content of a contribution are very open, and contributions can contain any type of information. The contribution process is the primary method for furthering the development of standards in the TIA.

Aside from interim standards (ISs), the TIA also publishes other types of specifications. Among these are Telecommunications Systems Bulletins (TSBs). TSBs are not considered standards and do not carry the weight of standards; however, they do provide information that concerns existing standards or other issues of great importance to the industry.

After an interim standard has been published by the TIA, there is a three-year period within which one of three actions must be taken on the standard. An interim standard must be reaffirmed, revised, or rescinded.

Reaffirmation consists of a review that is intended to result in a decision that the technical content of an interim standard is still valid and does not require changes. A *revision* incorporates additional language into an interim standard that modifies its technical content or meaning. A *rescission* is the result of a review that determines that the technical content of an interim standard is no longer of value. Both the revision and rescission processes typically require the development of an official Standards Proposal in the same manner as required for a new interim standard. The TIA interim standard revision process was evident with IS-41. An initial version was published, followed by Revision A, Revision B, and Revision C. IS-41 Revision C was then elevated to a national standard, known as ANSI-41 Revision D, and subsequently to the current standard, ANSI-41 Revision E.

TIA Committee TR-45

TIA Committee TR-45 (Mobile and Personal Communications Systems) is the committee that maintains the ANSI-41 and related wireless telecommunications standards. The committee was formed in 1983 and has developed standards concerning interoperability, performance, and network operations for all areas of public wireless telecommunications in the cellular and PCS regions of radio frequencies. TR-45 is made up of subcommittees. Each subcommittee is responsible for a different area of wireless telecommunications technology. Table 2.2 shows a listing of these subcommittees.

TABLE 2.2

TIA TR-45
Subcommittees

TIA TR-45 Subcommittees	TR-45 Subcommittee Name
TR-45.1	Analog Technology
TR-45.2	Wireless Intersystem Technology
TR-45.3	Time Division Digital Technology (i.e., TDMA)
TR-45.4	Radio to Switching Technology
TR-45.5	Spread Spectrum Digital Technology (i.e., CDMA)
TR-45.6	Adjunct Wireless Packet Data Technology
TR-45.7	Wireless Network Management Technology

TIA Subcommittee TR-45.2 is the standards-formulating body that develops and maintains ANSI-41. Subcommittee TR-45.2 is composed of seven Working Groups (WGs). Each WG is responsible for a different area of wireless intersystem operations. Table 2.3 shows a listing of these working groups.

TABLE 2.3

TIA TR-45.2
Working Groups

TIA TR-45.2 Working Groups	TR-45.2 Working Group Responsibility
TR-45.2.1	Stage I Feature and Service Development
TR-45.2.2	Stage II Feature and Service Development
TR-45.2.3	Stage III Feature and Service Development
TR-45.2.4	Message Accounting
TR-45.2.5	Wireless Intelligent Networking
TR-45.2.6	International Applications
TR-45.2.7	Interfaces to Other Telecommunications Networks

WG1 (stage 1 feature and service development) is charged with developing standards that describe features and services from the subscriber's perspective. These standards encompass the basic description of a feature or service as well as the end-user interface to the network for using a particular feature or service. WG2 (stage 2 feature and service development) is chartered to develop standards that describe features and services from the network's perspective. These standards describe the messaging between nodes in the wireless telecommunications network. This standardized messaging is the basic mechanism that enables interoperation between equipment developed by different manufacturers. WG3 (stage 3 feature and service development) is charged with developing standards that describe and define the details of the messages and parameters that support a feature or service. Procedures for handling the messages and parameters are also standardized within this working group. WG4 (message accounting) works on developing standards that enable the wireless telecommunications network to disseminate and transport the call details that are used for billing and accounting purposes. WG5 (wireless intelligent networking) is chartered to develop standards that define and support additions to the wireless network that use additional intelligent networking technologies. These technologies include the standardization of call models and event triggers, which enable calls to be treated in a variety of ways and make possible many enhanced features and services. WG6 (international applications) is charged with developing standards that define and support radio and networking applications outside North America. WG7 (interfaces to other telecommunications networks) works on developing standards that define and describe the interfaces between the wireless telecommunications network and other types of networks. For example, interfaces between the wireless telecommunications network and the regular wire line network (i.e., the public switched telephone network, or PSTN) have been developed within this working group.

The ANSI-41 standard is primarily developed within Working Groups 2 and 3. However, since the standards developed in each WG are usually based upon or affect the standards developed in other WGs and subcommittees, portions of ANSI-41 have been addressed throughout TIA Committee TR-45. Note that the responsibilities of working groups can and do change over time as existing standards evolve and the need for new standards arises.

CHAPTER 3

Wireless Telecommunications Network Signaling

The ANSI-41 specification standardizes computer communications that provide signaling within and between wireless telecommunications networks. Signaling is the primary mechanism used to transfer control information through the network. This chapter provides an overview of wireless telecommunications networks and describes the different types of signaling used by the network to provide service to wireless subscribers.

What is a Wireless Telecommunications Network?

The wireless telecommunications network can be viewed from two perspectives:

- A logical view, where the network is represented by a generic functional model.
- A physical view, where the network is represented by the actual switches, specialized computers, and other equipment that comprise the nodes of the network.

The logical view of the network is a method of describing network topology to simplify discourse and study. The network topology is simply depicted as functional entities interconnected by branches. Each functional entity represents one or more logical network functions while the branches represent a relationship between those functions. In the logical view, there is no prescribed mapping of the functions and relationships to a physical implementation.

The physical view of the network is quite different. From a physical perspective, the network is simply an organization of computer-based systems that are capable of intersystem communications. These communications are accomplished by the interconnection of circuits between the specialized computing platforms.

So, what exactly is a wireless telecommunications network? It is a conglomerate of *physical* equipment and communications facilities consisting of electronic computer-controlled switches, location registers, and other processing centers along with transmission circuits. The circuits support efficient and intelligent communications among the pieces of equipment. The connectivity among all these systems by communications transmission circuits provide the wireless telecommunications service accessed by subscribers. These circuits represent the network branches between the equipment that supports the transmission of signals and user information.

Signals convey data that provide information or instructions to control the network. These control signals can be considered the primary mechanism that communicates the *intelligence* of the network. User information is the actual information content transferred through the wireless network. User information content is distinguished from signal information because it is information originating from a source end-user that is ultimately delivered to a destination end-user. The transmission of this user information through the wireless telecommunications network comprises the telecommunications service accessed by a network subscriber. The information content can be voice, data, facsimile, or any other type of information conveyed by the network as a service to subscribers.

Overview of Signaling

Signaling is the process of sending signals or signaling information. It is the transfer of special information to control communication. Signaling consists of a protocol or a specialized set of rules that govern the communications of a system. The signaling protocol is defined by three criteria:

- **Syntax**—How to construct the information
- **Semantics**—What the information means
- **Procedures**—What to do with the information

The protocol enables the effective use of the control information (i.e., signals) to provide meaningful communications within a network. Signaling is the mechanism used to operate, control, and manage the wireless telecommunications network. A good example of a signal is the common ringing alert signal we are all familiar with when someone is calling on a telephone. It is distinguished from the user information provided by the telephone network (i.e., voice), since it provides an indication that a party is calling, but it is not the information meant to be conveyed by the caller.

Signaling and signaling protocols have become very complex, especially when used to govern telecommunications and the sophisticated services provided today. These advanced signaling protocols provide for the transfer of information among network nodes, which enables what is known as *intelligent networking*. Intelligent networking is a method for providing and interpreting information within a *distributed network*. A distributed network is structured so that the network resources are distributed throughout the geographic area being served by the network. The network is considered to be intelligent if the service logic and functionality can occur at the distributed nodes in the network. The wireless telecommunications network is distributed and intelligent. Because intelligent networks require such sophisticated signaling, the signaling means has evolved from electrical pulses and tones into very complex messaging protocols.

Within the context of ANSI-41, signaling information consists of messages that contain parameters that support the function of mobility management throughout the network. This mobility management function is key to enabling subscriber mobility in wireless networks.

The transfer of *user information*—the traffic that is conveyed end-to-end between network end users—is controlled by the network signaling protocols. User information is a portion of what is commonly known as bearer information. Bearer information usually contains other information besides the user data. This information generally consists of message alignment, synchronization, or error-correction sequences. This added information is similar to signaling information, but is not considered signaling information content since it applies to each message individually and not to the functioning of the network as a whole.

Network Signaling and Access Signaling

Network signaling is used between network nodes to operate, manage, and control the network to support certain types of functionality (e.g., mobility, voice traffic, etc.). Network signaling is distinct from another type of signaling known as *access signaling*. Access signaling is used to manage communications between a network end-user and an access point in the network. The distinction between network and access signaling is one of perspective and function, since these two types of signaling relate to different portions of the network. A network end-user is sometimes considered part of the network—a confusing notion—but the signaling required between an end-user and the *access point* of the network is quite different from that required between network nodes within the network.

A good analogy to help distinguish access signaling from network signaling is the national highway system. On-ramps can be considered the mechanism supporting access onto the highway network. The highways and interchanges along the way can be considered "circuits" and "switches," respectively. On-ramps have characteristics distinctly different from the highways themselves. Highways can lead to many different places and points. On-ramps only provide a direct point-to-point connection between the entrance of the on-ramp and the entrance of the highway. This idea is also true of access signaling.

Network signaling is also distinguished from access signaling because of a characteristic of network signaling known as *adaptive routing*. Adaptive routing allows signaling messages to take alternate routes between points in the network in cases of failure or congestion. In other words, the traffic can adapt to new paths in the network, if for some reason the primary path has become inaccessible. Access signaling generally has no such capability (however, there are exceptions). Just as with the highway analogy, if an on-ramp is inaccessible, traffic cannot enter the network via that access point.

Note that in Figures 3.1 and 3.2, the communications between the radio base station and the mobile telecommunications network is designated as access signaling or network signaling. Generally, this connection is based on access signaling, since the radio base station can be considered the on-ramp onto the network of mobile switching centers (MSCs). However, some wireless telecommunications networks treat the base stations as part of the network. Although adaptive routing is not always employed, the same signaling message transport protocol can be used between the base stations and the MSC as between MSCs.

ANSI-41 is a network signaling protocol designed to provide mobility management signaling throughout the wireless telecommunications network. ANSI-41 signaling is provided among MSCs, location registers, and some specialized processing centers to support subscriber mobility within a single wireless service provider network and between many different wireless service provider networks.

In-band Signaling and Out-of-band Signaling

There are many different types of network and access signaling protocols. However, both network and access signaling protocols can be cate-

Figure 3.1
Access signaling and
network signaling
between networks.

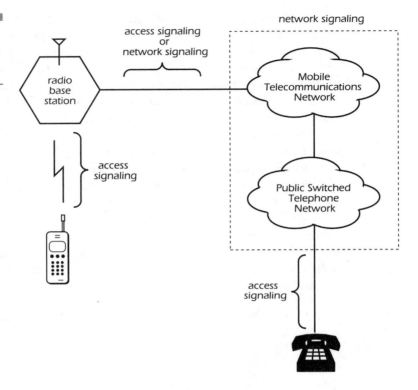

gorized into one of two signaling types: in-band signaling and out-of-band signaling.

In-band signaling is a type of signaling where the analog frequencies or digital time slots that carry the signals and signaling messages are within the bandwidth of the channel that carries the user information. In-band signaling uses a portion of the frequency band or bit stream that would otherwise be allocated for user information. An example of in-band signaling is dual-tone multifrequency (DTMF) access signaling—commonly known as Touchtone™ signaling—used to transmit dialed digits from a telephone to the network. The audio tones generated to inform the network of the party being called are transmitted as audio signals on the same channel as voice is transmitted during a conversation. These audio tones are generated at very precise frequencies. They consist of two specific voice frequencies that are combined to form the tone generated. Since they are very precise, it is difficult to duplicate exactly the tone with the human voice, minimizing the potential for inadvertent signaling errors.

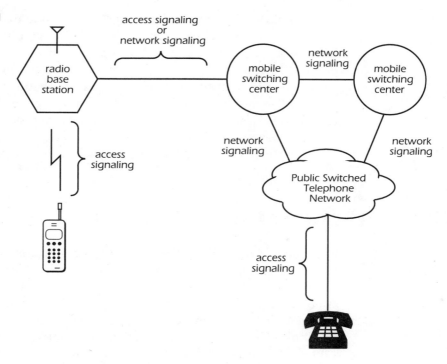

Figure 3.2
Access signaling and network signaling between network entities.

Out-of-band signaling is a type of signaling in which the analog frequencies or digital time slots that carry the signals and signaling messages are outside the bandwidth of the channel that carries the user information. Out-of-band signaling uses a frequency band or bit stream separate from the channels allocated for user information. Out-of-band signaling traffic is logically separate from the user traffic (i.e., on a different logical channel) within the same transmission line, or it may also be physically separate from the user traffic (i.e., on a completely different transmission line). An example of out-of-band signaling is ISDN, where the access signaling traffic is transmitted across the D-channel, which is separate from the B-channels used to transmit user information.

Out-of-band signaling is sometimes conveyed over a single digital channel separate from the user information channels. This single channel can carry signaling information for many bearer, or information, channels. That is, the signaling channel is used in *common* for many information channels. Because of this characteristic, the channel is called a *common channel*. Out-of-band signaling protocols that use a common digital channel are known as *common channel signaling protocols*. Common channel signaling allows signaling traffic to be consolidated and sent across a separate transmission link from the user traffic.

The primary advantage of sending this consolidated signaling traffic over separate transmission links is to prevent prohibited access by a user. Another advantage is the high-speed signaling transmission enabled by links that can be connected over a physically separate network, distinct from the network that carries the user traffic (e.g., voice traffic). The mechanism of out-of-band common channel signaling allows for high-performance distributed service logic across the telecommunications network as a whole, e.g., faster call setup times and the ability to efficiently support enhanced features such as 800-number service.

Signaling System No. 7

Signaling System No. 7 (SS#7) is a suite of common channel network signaling protocols defined by the ITU-T. ANSI has defined a national variant of this protocol, SS7 (without the "#"), specifically for North American telecommunications networks. SS7 is the primary building block on which enhanced intelligent telecommunications applications are built. These applications include call control and transaction capabilities that support database access as well as a variety of intelligent network functions and wireless telecommunications services. SS7 is designed to operate over a separate network distinct from the network that carries voice or user data (i.e., it is an out-of-band common channel signaling protocol). The scope of SS7 is extremely large, since it covers all aspects of control signaling for complex digital telecommunications networks.

SS7 is based upon *packet-switching* technology. SS7 packets (or messages) are used to convey signaling information from an originating point to a destination point through multiple switching nodes in the network. The SS7 messages contain addressing and control information used to select the routing of signaling information through the network, perform management functions (to provide high reliability), establish and maintain calls, and invoke transaction-based mechanisms in support of sophisticated applications. A transaction is a controlled exchange of information based on a query or command and the response to that query or command. The ANSI-41 signaling protocol provides operations for this type of transaction-based application using the SS7 protocol. The SS7 transaction operation mechanisms are used to query databases and invoke functions at remote points throughout the network. These mechanisms also support the delivery of status information and results to those database queries and invoked functions.

Since SS7 has so many advantages and provides a standard transaction-based protocol mechanism, it is ideal for performing the operations required to provide the mobility management function in the wireless telecommunications network. ANSI-41 is a signaling protocol that provides transaction-based operations to support subscriber mobility in the wireless telecommunications network.

Note that ANSI-41 signaling messages can be transported by communications protocols other than SS7 (i.e., using the CCITT X.25 packet switching protocol). However, SS7 is the more powerful and robust means for conveying the ANSI-41 signaling information through the mobile telecommunications network and is much preferred over other packet-switching protocols. More information on the use of the SS7 protocol is provided in Parts 2 and 3 of this book.

Overview of Intersystem Operations

The ANSI-41 signaling protocol provides *intersystem operations* in support of subscriber mobility management. ANSI-41 intersystem operations are operations performed between wireless systems that enable subscriber mobility in the following ways:

1. Subscribers can move between systems while a call is in progress.
2. Subscribers can originate calls while *roaming* (i.e., using their mobile stations in a system other than the home system where the wireless subscription was established).
3. Subscribers can receive calls while roaming.
4. Subscribers can activate and use supplementary call features while roaming (e.g., call forwarding).

In the context of ANSI-41 intersystem operations, a wireless system is defined as a single mobile switching center (MSC) along with its associated location registers, radio base stations, and processing centers. The term *intersystem* refers to subscriber movement between MSCs (or, more correctly, MSC serving areas), not between radio base stations served by a single MSC. The MSCs involved in the intersystem operations can belong to a single service provider network, or to different networks.

An intersystem operation consists of two elements: signaling messages (comprised of syntax and semantics) and functional procedures.

Intersystem signaling messages convey system and subscriber information between network nodes within a single wireless service provider network or between multiple wireless service provider networks. Functional procedures are the computer processes invoked at the network nodes when signaling messages are sent and received. The ANSI-41 intersystem operations support the following three basic mobility functions:

- Intersystem handoff
- Automatic roaming
- Intersystem operations, administration, and maintenance (OA&M)

Intersystem handoff is the set of functions that enables subscribers to move between wireless systems while a call is in progress. It is a more sophisticated procedure than *intra*system handoff, which is a handoff between radio base stations subtending a single MSC. Intrasystem handoff can be completely controlled (in a proprietary manner) within a single wireless system, and hence, there is no need for intersystem operations.

Automatic roaming is the set of functions that enables subscribers to originate calls, receive calls, and use supplementary call features *seamlessly* while roaming. The term seamless refers to the ability of wireless systems to provide these functions transparently to subscribers; i.e., subscribers do not need to take any special actions to use a mobile phone as they roam from system to system. The only indication provided to subscribers that they are roaming is usually the "ROAM" indicator displayed on most mobile stations.

Intersystem OA&M is the set of functions that provides trunk maintenance between MSCs. *Trunk maintenance* is a set of procedures required by telecommunications switches to manage the transmission circuits between switches that are exclusively used for intersystem handoff. This management consists of blocking, unblocking, resetting, and testing of those transmission circuits. Note that trunks between the PSTN and a wireless network use other trunk-maintenance procedures.

Origin of the ANSI-41 Solution

Standardized operations are necessary for the functioning of wireless systems, since subscriber mobility is supported between different wireless service provider networks that use equipment developed by different manufacturers. TIA Committee TR-45 was established in 1983 to develop the wireless technology standards, which continue to evolve. The initial version of the ANSI-41 specification (TIA IS-41 Revision 0), entitled *Cellular Radiotelecommunications Intersystem Operations*, was published as an interim standard by TIA Subcommittee TR-45.2 at the beginning of 1988. Contributors to this standard included wireless service provider companies and network equipment manufacturing companies. The relationship between these groups of companies is a simple one; service provider companies are customers of the equipment manufacturing companies from whom they purchase the equipment to deploy in their networks.

Without a standardized solution to intersystem operations, it would be difficult for wireless service providers to purchase equipment from different manufacturers and directly provide subscriber mobility between wireless systems. The ANSI-41 specification solved this problem and continues to evolve. There have been five subsequent revisions to the initial TIA IS-41 standard (Revisions A, B, C, D, and E). Each subsequent revision provides additional information to the previous version of the standard. Revisions are necessary for the following general reasons:

- To add new subscriber features to the standardized set
- To add functionality that supports new network requirements (e.g., over-the-air service provisioning)
- To fix errors found during the implementation of the standard
- To clarify text that was found to be open to many interpretations
- To remove functionality that was found to be unnecessary

Since the initial publication of ANSI-41 and the deployment of networks based upon the specification, the term *ANSI-41 network* was coined to describe wireless networks based on the standard. This term is generic and describes any network that uses ANSI-41 intersystem operations to provide mobility to subscribers. However, these networks also use other protocols, both standard and proprietary, to provide functions that are not within the scope of ANSI-41. Examples include OA&M

functions other than trunk maintenance and value-added features offered to subscribers that have not been addressed by ANSI-41 (i.e., intersystem support of proprietary features) or do not require intersystem operations.

CHAPTER 4

Wireless Telecommunications Network Reference Models

Network reference models are used to assist engineers in standardizing network functions and interfaces between functions. The model is an integral tool that can provide a representation of the physical network nodes, logical network functions, or both. In this chapter we define reference models, explain their use in the standards process, and describe the details of the ANSI-41 network reference model.

Purpose and Description of a Network Reference Model

A network reference model is a diagram that depicts the entities of a network and the interfaces between those entities. The model encompasses the definitions of the entities and the interfaces between them and depicts a graphical representation of the wireless telecommunications system as a whole. The network entities can represent physical network nodes that contain one or more functions or they can represent logical network functions only. The model is used to facilitate the definition and description of functions and protocols that can be standardized in the network. The model itself is usually not meant to depict a physical network implementation, rather only the basic interfaces between the minimum required functions or network nodes for the purpose of standardizing network services. A network reference model is used as the basis for a variety of network implementations, not as a description of a true physical network plan.

Physical Models versus Logical Models

A network reference model diagram consists of a representation of network entities connected by interfaces. Some models depict all network entities by a single shape (such as a square), while others depict them as different shapes with each shape representative of the kind of function provided by that network entity. An example of the latter type is the model generally used for North American SS7 networks, where circles represent end *signaling points* (SPs), squares with a diagonal through them represent *signaling transfer points* (STPs), and triangles represent *service control points* (SCPs). Each of these network entities provides different functions, and the depiction is meant to show external interfaces between physically distinct network nodes containing those functions. Figure 4.1 depicts the basic North American SS7 network reference model.

Figure 4.1
Example of a physical network reference model (basic SS7 network model).

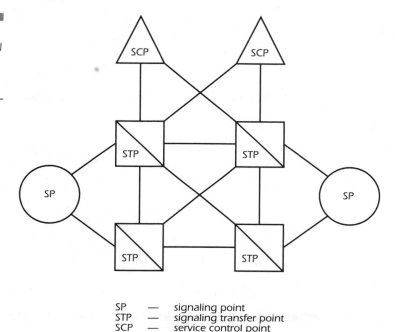

SP	—	signaling point
STP	—	signaling transfer point
SCP	—	service control point

Many network reference models are logical models; i.e., the network entities are representative of *logical functional entities* that may be implemented as physically separate network nodes or as functions combined within the same network node. The terms *logical* and *physical* can be confusing when we begin to discuss functionality and network reference models. A logical network reference model is based on logical functionality that is independent of implementation. Many physical implementations can be derived from a logical model. The logical functional entities within the model can be considered abstractions; implementations of those functional entities are left to system developers. Since a functional entity can represent one or more logical functions, the physical realization of the entity is an issue of network implementation dependency. The relationship between logical functional entities and physical network nodes can be one-to-one or many-to-one. Figure 4.2

shows the relationship of a logical view of the network to a physical view. The figure shows one possible one-to-one mapping of logical functional entities to physical network entities.

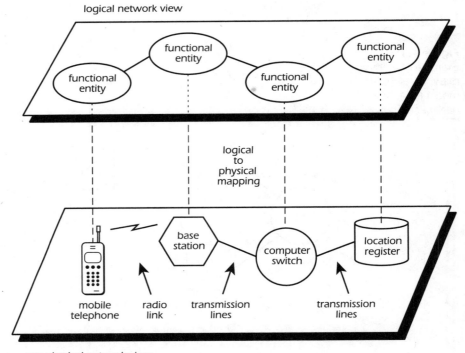

Figure 4.2
Logical and physical views of a network. Note that the logical view is mapped to only one of many possible physical views.

Logical models are used to represent *implementation-independent* functions and interfaces; that is, they usually depict functionality independent of the physical implementation of that function. The interfaces between the functions can be external or internal to the functional entity. Those that are implemented as internal do not necessarily require standardization since internal interfaces usually do not provide connections between network equipment from different manufacturers.

The depiction of the functional entities in the logical view of the network (Figure 4.3) is an example of a logical model. With reference to this model, there is no reason why a network location register could not be implemented as part of the computer switch. And, in fact, the point of this depiction is to allow that network implementation explicitly, if it is desired. Although the model is logical, for practical reasons some functional entities are truly representative of separate physical devices, such as the mobile telephone.

Figure 4.3
Example of a logical model.

Network Reference Models and the Three-stage Specification Process

Many telecommunications services and protocols are specified via a three-stage process, which was originally used to specify international Integrated Services Digital Network (ISDN) services, features, and capabilities, but is applicable to any telecommunications service (refer to *CCITT Recommendation I.130*). It is a generic method used to characterize services and facilitates a top-down engineering approach for designing them. The three stages are:

1. Stage 1 describes the service from an end-user's perspective.
2. Stage 2 describes the information flow between network entities (i.e., interfaces) to support the stage 1 service.
3. Stage 3 describes the protocol application of the stage 2 information flow.

(See Chapter 7 for a more detailed description of the three-stage specification process.)

The mobile telecommunications application services standardized by ANSI-41 are specified using this method, and there is a close relationship between this process and the use of a network reference model.

To provide a specific mobile telecommunications service to a subscriber, the service needs to be described in terms of what is provided to the subscriber and the actions, if any, the subscriber needs to take. This is specified in the stage 1 description of the service. The network reference model is used as a tool to help specify the stage 2 description of the service. Signaling information, user information, or both are transferred between network entities to provide the service to a subscriber. The service can be as basic as delivering a mobile-terminated call to the subscriber, or more involved, as in the call-waiting or conference-calling features. The information transferred to provide a service can be derived from a database (i.e., location register), from a processing center, through switched connections, by a radio system, or by any combination of these. The information can traverse these network entities within a single wireless service provider network or between different wireless service provider networks.

The network reference model is used to define the interface between two network entities that can be standardized to support a specific service. The information flow defined across a given interface for a specific service is part of the stage 2 description for that service. The stage 2 service description is sometimes referred to as the *network perspective* of a service, since it describes the service in terms of what is provided to each network entity and the actions, if any, those network entities need to take to provide the service. Without a network reference model, it would be difficult to establish a common point of reference for information transfer to provide a service supported by equipment developed by different manufacturers.

There are two basic philosophies for the interaction between a network reference model and the stage 2 description of a service:

1. The interface between network entities can be depicted first on the model to show the eventual need to provide a standardized stage 2 description for the interface.
2. The interface between network entities is added to the model only after the stage 2 description of the interface is specified, thereby justifying the presence of the interface on the model.

The ANSI-41 specification has been developed using both of these philosophies. Early in the development of ANSI-41, interfaces were depicted on the model to show the eventual need to provide a standardized stage 2 description for them. As the ANSI-41 standard evolved, it was more appropriate to add an interface to the model only after it was justified by a newly developed stage 2 description. The latter philosophy proved to be more appropriate. It is usually not very desirable to modify an existing network reference model since existing implementations are based upon earlier versions of the model. The problem with depicting functional entities and interfaces on the model before they are standardized is that sometimes the anticipated need for them never develops. Many network reference models contain extraneous network entities and interfaces that have not been standardized. They are simply left over as artifacts of an anticipated need.

Elements of a Wireless Telecommunications Network Reference Model

Regardless of whether a wireless telecommunications network reference model is logical or physical, it consist of the following five basic network entities:

- Radio systems
- Switching systems
- Location registers
- Processing centers
- Representations of external networks

These five network entities are principal elements of all wireless telecommunications networks. The network models also depict the interfaces, or *interface reference points*, between the network entities.

Radio Systems

Radio systems consist of the following three separate subsystems:

- Antenna systems
- Radio transceivers
- Radio transceiver controllers

Antenna systems convert electrical signals from a radio transmitter into electromagnetic waves that comprise the radio transmission signals sent to mobile stations. Conversely, antenna systems convert the electromagnetic radio waves into electrical signals that comprise the radio signals received from mobile stations. Antenna systems also manage radiated power to minimize interference.

Radio transceivers (sometimes called *base transceiver systems* or BTSs) consist of a combination of simplex radio transmitting and receiving equipment that employs common components for both transmitting and receiving. This equipment is often referred to as simply the *radio*.

Radio transceiver controllers (sometimes called *base station controllers*, or BSCs) are equipment that controls multiple radio transceivers. The controllers multiplex electrical signals from many radio channels into transmission signals that are sent to the wireless network. The controllers also send network signals to the appropriate radio transceivers for transmission to the mobile stations. Figure 4.4 shows the relationship of the entities that comprise the radio system.

Figure 4.4
Example of a radio system reference model. The A interface is shown between the switching system and the radio system. Within the radio system, the A-bis interface is between the BSC and BTS and the U-bis interface is between the BTS and the AS. Note that communications across these interfaces fall outside the scope of ANSI-41.

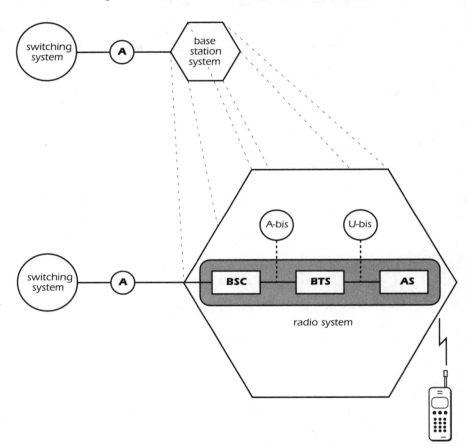

Switching Systems

Switching systems provide the function of transferring transmissions from one circuit to another in the network. The switching function controls the routing of signaling and user information to specific nodes in the wireless network. The switching systems consist of the transmission facilities and computing platforms that control the switch circuits to connect calls between users.

Location Registers

Location registers are data-based systems that control wireless subscriber services and contain the records and stored information related to wireless subscribers. These location registers are queried by other network functional entities to obtain the current status, location, and other information to support calls to and from mobile users. Location registers may also contain network address translation information to assist in the routing of calls to the appropriate network destination.

Processing Centers

Processing centers are peripheral network computing platforms that provide services to enhance the capabilities of the network. An example of a processing center is an *authentication center* (AC), which uses complex algorithms to authenticate the identity of wireless subscribers. Other examples are voice-announcement systems, user messaging systems, and interworking functions that convert circuit-switched data to packet-switched data and vice versa.

Representations of External Networks

External networks are integral elements in wireless telecommunications network models. They represent interconnections between the wireless network and the PSTN or other networks. Examples of these other networks are data networks, local area networks, the ISDN, and other dissimilar wireless networks. These representations are important since wireless networks usually need to interoperate with other networks to complete calls (e.g., wireless to wire-line calls).

Interface Reference Points

Interfaces depicted on network reference models are known as *interface reference points*. Each represents the point of connection between two adjacent physical or logical network entities. This point of connection is defined by functional and signaling characteristics and may define the operational responsibility of the interconnected network entities.

ANSI-41 Network Reference Model

The ANSI-41 network reference model is a logical and physical model consisting of logical and physical functional entities and interface reference points. This particular model is not meant to imply a physical implementation; however, as a practical matter, some functional entities are truly representative of separate physical devices. An example of this is the mobile station (MS), originally known as the *cellular subscriber station* (CSS).

Figure 4.5
The original ANSI-41 network reference model (from IS-41-A and IS-41-B). Note that the CSS is depicted as mobile telephone equipment and the PSTN and ISDN are depicted as network clouds. This is to distinguish them from the other logical functional entities that, from a practical perspective, do not necessarily imply physically separate equipment. The interfaces in **bold** represent the interfaces that are standardized in IS-41-A and IS-41-B. (Reproduced under written permission of the Telecommunications Industry Association.)

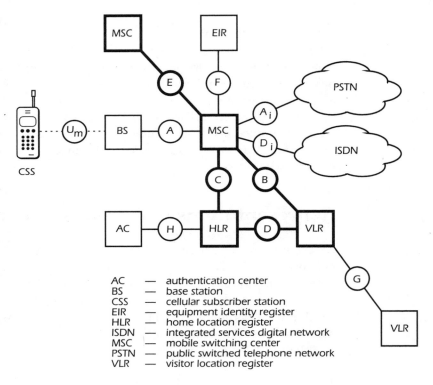

AC	— authentication center
BS	— base station
CSS	— cellular subscriber station
EIR	— equipment identity register
HLR	— home location register
ISDN	— integrated services digital network
MSC	— mobile switching center
PSTN	— public switched telephone network
VLR	— visitor location register

Figure 4.5 shows the first incarnation of the ANSI-41 network reference model (known as *IS-41* at the time). This model is purely logical and depicts the functional entities and interface reference points between those entities that may comprise an ANSI-41–based wireless

network. The model shows the interfaces within and between ANSI-41–based wireless networks as well as interfaces between the wireless network and the PSTN and ISDN. Since this IS-41 network reference model is a logical model, it is an abstraction on which physical implementations can be based. This means that more than one functional entity can be implemented on a single physical device. When this is the case, it is not necessary for the internal interface between those functional entities to comply with the standards.

Origin of the Network Reference Model

The first version of ANSI-41 (IS-41 Revision 0) did not explicitly define a network reference model. In fact, IS-41-0 considered only a single interface—that between a visited wireless system and the subscriber's home wireless system. No decomposition of each system into switches, location registers, and other functional entities was required in IS-41-0, since only basic intersystem handoff and subscriber validation functions were specified; there was no compelling need for a network reference model given these basic standardization requirements.

The first ANSI-41 network reference model (see Figure 4.5) was specified in IS-41 Revision A (IS-41-A) and was not modified until IS-41 Revision C (IS-41-C) several years later. It is interesting to note that the model was derived from the original version of the international CCITT Recommendation Q.1051, which specified the interfaces and application protocol for public land mobile networks. Only a few changes were made to the CCITT model for application to North American ANSI-41 networks.

Original ANSI-41 Functional Entities

Cellular Subscriber Station

The cellular subscriber station (CSS) is the functional entity that generally represents the wireless radio-telephone equipment. In the CCITT Q.1051 model, the CSS is known as the *mobile station* (MS). The CSS was named differently to distinguish it from MS, which was used in European wireless networks. Since the CSS is a logical functional entity, it may not necessarily be a mobile telephone. In fact, it is defined in the ANSI-41 standard as "the interface equipment that terminates the radio path on the user side of the network." Many equipment implementations can comply with this definition; however, the most common implementation today is certainly a mobile telephone. The CSS incorporates user interface functions, radio functions, and other service logic functionality.

Base Station

The base station (BS) is the functional entity that represents all the functions that terminate radio communications at the network side of the CSS interface to the wireless network. It controls radio resources and manages network information required to provide telecommunications services to the CSS. The BS essentially consists of the radio transceivers and radio transceiver controller functions serving one or more cells. A cell is the geographic area defined by the telecommunications coverage of the radio equipment located at a given cell site. The cell site is the physical location of a cell's radio equipment (i.e., the BS) and supporting systems. The BS incorporates radio functions and radio resource control functions.

Mobile Switching Center

The mobile switching center (MSC) is the functional entity that represents an automatic switching system. This switching system constitutes the interfaces for user traffic between the wireless network and other public switched networks, or other MSCs in the same or other networks. The MSC provides the basic switching functions and coordinates the establishment of calls to and from the wireless subscribers. The MSC is directly responsible for transmission facilities management, mobility management, and call processing functions. The MSC is in direct contact with one or more BSs on one side and with external networks and functions on the other side. The MSC incorporates switching functions, mobile application functions, and other service logic functions.

The term *MSC* refers to the logical functionality of the wireless telecommunications switch including the signaling protocols, mobility management application, and functional interfaces. Sometimes MSC is confused with *mobile telephone switching office* (MTSO). The term *MTSO* is analogous to the wire-line term *central office*. MTSO refers more to the physical architecture of the wireless switching office including the switching equipment, hardware interfaces, and the real estate where the switching system resides.

Home Location Register

The home location register (HLR) is the functional entity that represents the primary database repository of subscriber information used to provide control and intelligence wireless networks. The term *register* denotes control and processing center functions as well as the database functions. The HLR is managed by the wireless service provider and represents the "home" database for subscribers who have subscribed to service in that home area. The HLR contains a record for each home subscriber that includes location information, subscriber status, subscribed features, and directory numbers. Supplementary services or features that are provided to a subscriber are ultimately controlled by the HLR. An HLR may serve one or more MSCs. The HLR incorporates database functions, mobile application functions, and other service logic functions.

Visitor Location Register

The visitor location register (VLR) is the functional entity that represents the local database, control, and processing functions that maintain temporary records associated with individual network subscribers. The VLR is managed by the wireless service provider and represents a "visitor's" database for subscribers served in a defined local area. A visitor can be a wireless subscriber being served by one of many systems in the home service area, or a subscriber who is roaming in a nonhome, or visited, service area (i.e., another service provider area). The VLR contains subscriber location, status, and service information derived from the HLR. The local network MSC accesses the VLR to retrieve information for the handling of calls to and from visiting subscribers. The VLR may serve one or more MSCs. The VLR incorporates database functions, mobile application functions, and other service logic functions.

Authentication Center

The authentication center (AC) is the functional entity that represents the authentication functions used to verify and validate a mobile station's identity. The AC manages and processes the authentication information related to a mobile station. This information consists of encryption and authentication keys as well as complex mathematical algorithms used to hinder fraudulent use of network services. The AC incorporates database functions used for the authentication keys and authentication algorithm functions.

Equipment Identity Register

The equipment identity register (EIR) is the functional entity that represents the database repository for mobile equipment-related data. An example would be a database of the electronic serial numbers (ESNs) of mobile equipment along with the status of that equipment. Such a database could assist in preventing stolen or fraudulent equipment from being used to access network services. However, the functions of the EIR are not defined in any TIA standard.

Public Switched Telephone Network

The public switched telephone network (PSTN) is the functional entity that represents a network completely separate from the wireless telecommunications network. The PSTN refers to the regular wire-line telephone network that provides service to the general public. The PSTN is commonly accessed by ordinary telephones, key telephone systems, private branch exchange (PBX) trunks, and data transmission equipment. The interface between the MSC and the PSTN represents the capabilities to originate calls from wireless phones to wire-line phones and to terminate calls from wire-line phones to wireless phones.

Integrated Services Digital Network

The Integrated Services Digital Network (ISDN) is the functional entity that represents a network completely separate from the wireless telecommunications network. It refers to the wire line network that provides enhanced digital services over special digital transmission lines to special digital terminals. The ISDN is commonly accessed by computer terminals and switching equipment employing digital adapters that interpret the ISDN digital signals to provide high-speed and broadband services. The interface between the MSC and the ISDN represents the capabilities to originate calls from wireless phones to destinations within the ISDN (e.g., voice-mail systems) and to terminate calls from the ISDN to wireless phones.

Original ANSI-41 Interface Reference Points

Table 4.1 shows the interface reference points specified in the original ANSI-41 network reference model (i.e., in IS-41-A and IS-41-B).

There are more interfaces and functional entities included in the model than are actually addressed by the IS-41 intersystem operations. IS-41-A and IS-41-B addressed intersystem operations for the B, C, D, and E interfaces only. The other interfaces were depicted on the model for the following reasons:

- To show the need for future standardization in subsequent revisions of IS-41.
- To show the need for future standardization in other specifications related to IS-41.
- To show the relationship of other standard interfaces to the IS-41 protocol.

TABLE 4.1

The Original Interface Reference Points in IS-41-A and IS-41-B (Interfaces in **Bold** Represent Interfaces Standardized in Both IS-41-A and IS-41-B)

IS-41-A/IS-41-B Interface	Functional Entities Using the Interface
A interface	BS—MSC
A_i interface	MSC—PSTN
B interface	**MSC—VLR**
C interface	**MSC—HLR**
D interface	**HLR—VLR**
D_i interface	MSC—ISDN
E interface	**MSC—MSC**
F interface	MSC—EIR
G interface	VLR—VLR
H interface	HLR—AC
S_m interface	within CSS
U_m interface	CSS—BS

The interfaces that have since been standardized (i.e., in ANSI-41 or related standards) are the A, A_i, D_i, and H interfaces. The interfaces that have not been addressed are the F, G, and S_m interfaces. The functionality that can be provided by the F and G interfaces (i.e., the MSC-EIR and VLR-VLR interfaces, respectively) is not yet a necessity for wireless telecommunications. The S_m interface defines functionality within the CSS only, and was determined to be beyond the scope of network standardization.

The other standard interfaces that relate to ANSI-41, but are not within its scope, are the radio system interfaces. The interfaces related to the radio systems—the A, S_m, and U_m interfaces, are included in the model for the following reasons:

- ANSI-41 is designed to support a variety of radio-system standards separate from, but related to, the network.
- ANSI-41 supports intersystem handoff functions affected by the radio systems.

Original incarnations of mobile switching and radio systems were implemented and deployed as bundled systems obtained from a single equipment vendor. Therefore, the A interface evolved as proprietary and there was no need to standardize it. The U_m interface represents the radio, or air, interface between the mobile telephone and the radio base station. This interface was originally specified and standardized by the TIA as *EIA/TIA IS-3*, and subsequently *ANSI/TIA/EIA-553*, as the AMPS protocol. The latest version of this standard is *ANSI/TIA/EIA-691*. Currently, there exist many standards for the radio interface such as Narrowband AMPS (NAMPS), TDMA, and CDMA.

Current standards do exist for the A interface, supporting the necessary operations to communicate from the MSC to the radio systems (i.e., in *ANSI/TIA/EIA-634* and *TIA/EIA/IS-653*). These standards support SS7, TCP/IP, and frame relay as transport protocols on this point-to-point communications link. However, it is very rare to see one vendor's MSC on an ANSI-41 network and another vendor's radio systems subtending from that MSC. These A interface standards are typically used to enable new technology trials and for attempts to demonstrate inter-technology networking, but are generally not used to support commercial deployment of multivendor equipment.

Second-generation ANSI-41 Functional Entities (IS-41-C)

The second generation of the ANSI-41 network reference model is specified in IS-41 Revision C (IS-41-C). The following changes were made to the original reference model:

- The name of the CSS was changed to mobile station (MS).
- The S_m interface was removed.
- Short message service (SMS) functional entities and interfaces were added.

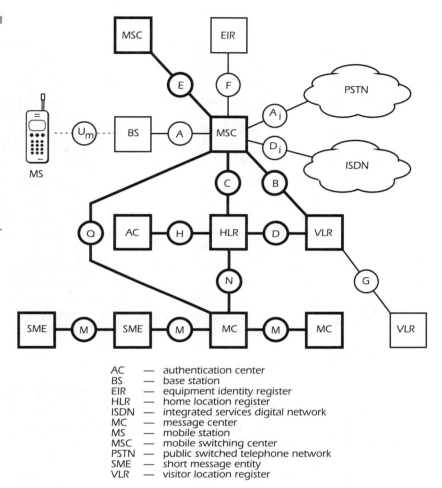

Figure 4.6
The second generation ANSI-41 network reference model (from IS-41-C). The interfaces in **bold** represent the interfaces that are standardized in IS-41-C. (Reproduced under written permission of the Telecommunications Industry Association.)

AC	—	authentication center
BS	—	base station
EIR	—	equipment identity register
HLR	—	home location register
ISDN	—	integrated services digital network
MC	—	message center
MS	—	mobile station
MSC	—	mobile switching center
PSTN	—	public switched telephone network
SME	—	short message entity
VLR	—	visitor location register

Figure 4.6 shows the second-generation ANSI-41 network reference model. The most prominent change from the previous model is the addition of functional entities supporting short message services (SMS). SMS is a set of services that supports the storage and transfer of short text messages (200 bytes or less) through the wireless network. The SMS functional entities were added to the model after the stage 2 description of SMS justified their presence. The name change from CSS to MS implies no change of functionality. The need to show a distinction between this functional entity and other similar and analogous entities specified in other network standards was no longer an important consideration. The change was implemented to make the terminology of the mobile telephone more consistent with common industry usage.

Message Center

The message center (MC) is a functional entity that stores and forwards short messages for SMS. The store-and-forward function provides a method of forwarding short messages to their destination recipient or storing those messages if the recipient is unavailable to receive them. The store-and-forward function can be distinguished from the real-time delivery requirements of voice calls. Short messages can be stored in a database until it is convenient for them to be sent to their specified destinations. The MC can store messages that are either sent from a mobile station (i.e., mobile-originated messages) or destined for a mobile station (i.e., mobile-terminated messages). Besides storing and delivering short messages, the MC also performs signaling functions to support the other delivery functions, such as MS location and status queries and mapping of destination addresses.

Short messages can be sent to the MC from any functional entity that includes the function to support SMS message originations. Conversely, short messages can be received by any functional entity that includes the function to support SMS message terminations. These functional entities are known as *short message entities* (SMEs).

Short Message Entity

The SME is a functional entity that can originate short messages, terminate short messages, or do both. Basically, *SME* is a generic term for any entity that can send or receive short messages via the ANSI-41 SMS. An SME may be associated with an ANSI-41 functional entity (e.g., an HLR, MC, or MSC) or associated with an entity external to ANSI-41. A mobile station also requires SME functionality for it to support the transmission of mobile-originated short messages and the reception of mobile-terminated short messages.

Second-generation ANSI-41 Interface Reference Points

Table 4.2 shows the interface reference points specified in the IS-41-C network reference model.

The M, N, and Q interfaces were added to support the MC and SMEs. These interfaces represent signaling and bearer service communications required to provide the SMS functions.

IS-41-C and other related standards evolved to address many of the interfaces that were included, but not yet standardized, in the original network reference model (IS-41-A and IS-41-B). The standardization of the A interface is a relatively new system requirement.

TABLE 4.2

Current Interface Reference Points in IS-41-C and the Most Current Standard Where They Are Addressed (Interfaces in **Bold** Represent Interfaces Standardized in IS-41-C)

IS-41-C Interface	Functional Entities Using the Interface
A interface	BS—MSC
A_i interface	MSC—PSTN
B interface	**MSC—VLR**
C interface	**MSC—HLR**
D interface	**HLR—VLR**
D_i interface	MSC—ISDN
E interface	**MSC—MSC**
F interface	MSC—EIR
G interface	VLR—VLR
H interface	**HLR—AC**
M interface	**SME—SME, SME—MC, MC—MC**
N interface	**MC—HLR**
Q interface	**MC—MSC**
U_m interface	MS—BS

The A_i and D_i interfaces are specified in TIA specification *ANSI/TIA/EIA-93*. The *A* and *D* in these interface names stand for *analog* and *digital*, respectively. The subscript *i* stands for *interface*. A_i and D_i represent the interfaces between the mobile network and the analog PSTN and ISDN, respectively. These interfaces, however, are treated as a single interface in ANSI/TIA/EIA-93, supporting both analog and digital network interconnection protocols. This single specified interface can be mapped onto either the A_i or D_i interface of the model. The reason is that the distinction between the use of different network signaling protocols for the PSTN and ISDN does not exist networks deployment today.

The H interface is standardized in IS-41-C. It represents the interface between the HLR and the AC supporting the authentication functions. This interface was originally standardized in TIA specification TSB51, as an addendum to the already published IS-41-B. The information in TSB51 was subsequently refined, revised, and incorporated into IS-41-C and ANSI-41-E.

Current ANSI-41 Functional Entities

The current ANSI-41 network reference model is specified in *TIA/EIA TSB100*. The following changes were made to the second-generation (IS-41-C) network reference model:

- The generic M interface supporting the message center (MC) and short message entity (SME) for short message service was divided into three distinct interfaces, M_1, M_2, and M_3.
- Over-the-air service provisioning (OTASP) functional entities and interfaces were added.
- The number portability database (NPDB) functional entity and Z interface were added.
- The public packet data network (PPDN) and associated interworking function (IWF), along with the P_i interface, were added.

Figure 4.7 shows the current ANSI-41 network reference model. The most prominent changes from the previous model are the addition of functional entities supporting wireless number portability (WNP), over-the-air service provisioning (OTASP), and support for interconnection from the wireless network to a public packet-switched data network (PPDN).

Number portability is a set of functions that supports the ability of a wireless subscriber to retain a directory number (i.e., the MIN) in the event that that subscriber changes wireless carriers within the same local region. This portability can also extend to wire-line directory numbers. In essence, it allows the subscriber and not the service provider to own the local directory number so that this number can be used regardless of which local wireless or wire-line service provider is used.

Over-the-air service provisioning is a set of functions that supports the initial and subsequent programming of the MS over the air (i.e., the radio interface). Typically, mobile stations require manual programming of MINs, SIDs, and other activation parameters to enable network access and "turn-on" the subscription. OTASP enables this programming to be performed by the wireless service provider remotely and "over the air," without the manual intervention of a qualified dealer.

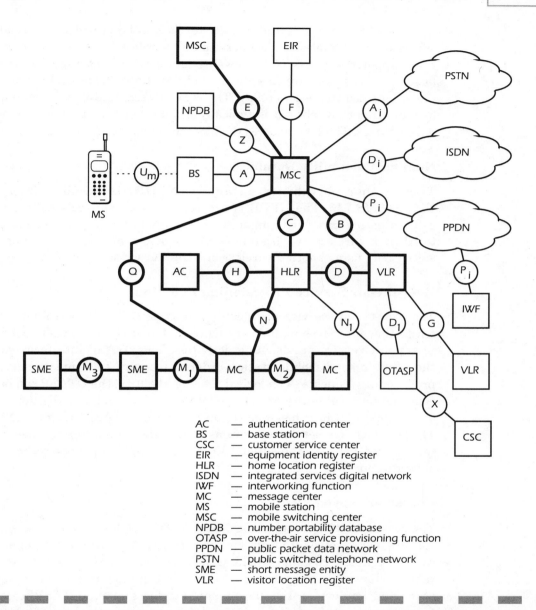

Figure 4.7 The current ANSI-41 network reference model (from TIA Subcommittee TR45.2). The interfaces in **bold** represent the interfaces that are standardized in ANSI-41.

Public packet-switched data network interconnectivity is designed to support a circuit-switched to packet-switched interworking function

(IWF) enabling packet data protocol services to be provided to wireless subscribers. Examples of these services are wireless Internet connectivity that supports wireless Web browsing and a variety of other Internet-based services that are more commonly available from desktop wireline–connected computers today. The MSC is essentially a telecommunications circuit switch and is typically not designed for packet-switched data services. The IWF enables the MSC to act as a packet switch.

Number Portability Database

The number portability database (NPDB) is standardized in *TIA/EIA/IS-756*. The NPDB is a repository that supports delivery of calls to *ported* directory numbers, ones originally owned and administered by a wireless or wire-line service provider and now belonging to subscribers using another wireless or wire-line service provider.

Over-the-air Service Provisioning Function

The over-the-air service provisioning function (OTASP) is a functional entity that interfaces with one or more customer service centers (CSCs) to support service provisioning activities. The OTASP interfaces with the MSC to send mobile station orders necessary to complete service provisioning requests. As a logical functional entity, the OTASP can be a separate physical entity or may be located within and be indistinguishable from an HLR or message center (MC). The OTASP employs ANSI-41 short message service operations to transfer subscriber provisioning information between the serving VLR and the HLR to support the new subscription.

Customer Service Center

The customer service center (CSC) is a functional entity where the service provider receives information (typically in the form of a telephone call) from a new customer wishing to subscribe to wireless service or to request changes in the existing service. The CSC uses completely proprietary and implementation-dependent interfaces to the OTASP to perform network and mobile station related changes necessary to accomplish the service provisioning process.

Public Packet Data Network

The public packet data network (PPDN) is the functional entity that represents a network completely separate from the wireless telecommunications network. It refers to a packet-switched data network that provides packet-data services over special data links to digital terminals. The PPDN is commonly accessed by digital data terminals, for example, a mobile station (e.g., TDMA, CDMA, or other digital packet data protocol), a wireless desktop client (e.g., a personal computer), a personal data assistant, or any other device capable of sending and receiving packet-switched data. The most common realization of a PPDN today is the Internet, whose underlying network protocol is the packet-switched Internet protocol (IP). The interface between the MSC and the PPDN represents the capabilities to originate packet-switched data transmission from wireless phones to destinations within the PPDN (e.g., the Internet) and to terminate packet-switched data transmission from the PPDN to wireless phones.

Interworking Function

The interworking function (IWF) is the functional entity that provides protocol conversion between any two wireless network functional entities. The definition in ANSI-41 is broad, but typically the IWF supports conversion between circuit-switched transmission protocols and packet-switched transmission protocols. The IWF is sometimes necessary to support transmission between a packet-switched network entity (e.g., an IP router) and a circuit-switched network entity (e.g., an MSC). One of the primary functions of an IWF can be to enable the support of packet data to circuit-based devices (such as mobile stations designed primarily for voice telephony services) and circuit data (such as voice telephony) to packet data-based devices (such as voice-over-IP terminals).

Current ANSI-41 Interface Reference Points

Table 4.3 shows the interface reference points specified in the ANSI-41 network reference model. The Z interface was added to support the NPDB. This interface represents the signaling communications required to provide the number portability functions.

The D_1, N_1, and X interfaces were added to support the OTASP. These interfaces represent the signaling communications required to provide the OTASP functions.

The P_i interface was added to support signaling between the wireless network and a PPDN to support wireless packet-switched data services. In reality, PPDN interconnectivity typically requires an IWF, although it may not be necessary to support packet-switched functions. This is why the IWF uses the same P_i interface as direct connectivity between the PPDN and MSC. The IWF is shown with a separate connection to the PPDN to enable it to connect to other network elements beside the MSC. In fact, the IWF is defined to enable connectivity to any wireless network element. The P_i interface is not standardized, but was added to show the potential future need to standardize this interface and to acknowledge existing vendor solutions that provide implementation-dependent solutions to this function today.

TABLE 4.3

Current Interface Reference Points in ANSI-41 and the Most Current Standard Where They Are Addressed (Interfaces in **Bold** Represent Interfaces Standardized in ANSI-41)

Interface	Functional Entities Using the Interface	Where Addressed (Most Current Standard)
A interface	BS—MSC	ANSI/TIA/EIA-634
A_i interface	MSC—PSTN	ANSI/TIA/EIA-93
B interface	**MSC—VLR**	**ANSI-41-E**
C interface	**MSC—HLR**	**ANSI-41-E**
D interface	**HLR—VLR**	**ANSI-41-E**
D_i interface	MSC—ISDN	ANSI/TIA/EIA-93
D_1 interface	VLR—OTAF	TIA/EIA/IS-725
E interface	**MSC—MSC**	**ANSI-41-E**
F interface	MSC—EIR	*not standardized*
G interface	VLR—VLR	*not standardized*
H interface	**HLR—AC**	**ANSI-41-E**
M_1 interface	**MC—SME**	**ANSI-41-E**
M_2 interface	**MC—MC**	**ANSI-41-E**
M_3 interface	**SME—SME**	**ANSI-41-E**
N interface	**MC—HLR**	**ANSI-41-E**
N_1 interface	HLR—OTAF	TIA/EIA/IS-725
P_i interface	MSC—PPDN, PPDN—IWF	*not standardized*
Q interface	**MC—MSC**	**ANSI-41-E**
U_m interface[1]	MS—BS	ANSI/TIA/EIA-691, ANSI/TIA/EIA-95, ANSI/TIA/EIA-136
X interface	OTAF—CSC	*not standardized*
Z interface	MSC— NPDB	TIA/EIA/IS-756

[1] Note that other radio interface standards exist, but these specifications are most representative of the technology supported by ANSI-41.

ANSI-41 Functional Entities Supporting the Wireless Intelligent Network

The current ANSI-41 network reference model also supports wireless intelligent network (WIN) functionality and is specified in *TIA/EIA TSB100*. The following functional entities were added to support WIN:

- Intelligent peripheral (IP, not to be confused with *Internet protocol*)
- Service control point (SCP)
- Service node (SN)

Note that these functional entities were originally specified in the Bellcore Advanced Intelligent Network (AIN) specification for wire-line intelligent network (IN) functionality in the late 1980s. These functional entities provide the distributed network intelligence required to support the IN call models, trigger points, and call treatment to provide new and efficient services to subscribers. Since WIN functionality requires many of the same characteristics of the wire-line IN network reference model, it made sense to include these same functional entities in the ANSI-41 network reference model.

Figure 4.8 shows the additions to the ANSI-41 network reference model to support WIN capabilities. Although WIN functionality is explicitly specified in *TIA/EIA/IS-771*, and not ANSI-41, it is worth mentioning due to its close relationship to, and involvement in, the ANSI-41-based MSC and HLR.

Intelligent Peripheral

The intelligent peripheral (IP) is a functional entity that performs special *resource management* functions required to treat a call in a specific way. For example, a voice-mail system can be considered an IP, as are systems that provide voice announcements, interactive voice and response systems, and digit-collection systems. IP is a very broad term, so a wide variety of services and devices, used to manage the resources to treat certain calls in special ways, may be considered intelligent peripherals in this model.

Service Control Point

The service control point (SCP), in the broadest sense, is a functional entity that performs real-time database and transaction processing

Figure 4.8
The current network reference model for the ANSI-41-based wireless intelligent network (WIN). This model is an extension to the basic ANSI-41 network reference model that includes additional functional entities required by WIN.

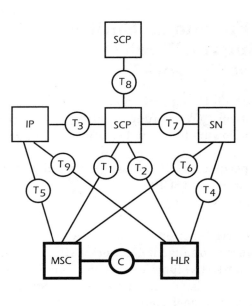

HLR	— home location register
IP	— intelligent peripheral
MSC	— mobile switching center
SCP	— service control point
SN	— service node

functions to provide service control and service functionality in the wireless intelligent network. However, the SCP is an entity most commonly implemented as a physically separate SS7 network node that provides the *service logic* and *transaction capabilities application part* (TCAP) protocol to perform database operations to treat special calls in special ways. The best example of SCP functionality is digit translation performed in a database to provide 800-number toll-free calling services, where the 800 number is translated to the actual called-party number to deliver a call.

Service Node

The service node (SN) is a functional entity that provides service control, service data, special resources, and call control functions specifically to support intelligent network services. The SN can be considered a variant, or combination of both an IP and SCP, providing both resource management and service logic functions to treat special calls in special ways.

ANSI-41 Interface Reference Points Supporting the Wireless Intelligent Network

Table 4.4 shows the interface reference points for WIN specified in TIA/EIA TSB100. The reference points are labeled T_1 through T_9 to explicitly separate all of the T interfaces to represent WIN interfaces. Note that both the MSC and HLR can communicate to each of the WIN functional entities.

TABLE 4.4

ANSI-41 Interface Reference Points for WIN

Interface	Functional Entities Using the Interface
T_1 interface	MSC—SCP
T_2 interface	HLR—SCP
T_3 interface	IP—SCP
T_4 interface	HLR—SN
T_5 interface	MSC—IP
T_6 interface	MSC—SN
T_7 interface	SCP—SN
T_8 interface	SCP—SCP
T_9 interface	HLR—IP

Mapping the Logical Model to the Real World

The ANSI-41 network reference model is a *logical* model depicting logical functional entities and interfaces that may comprise a wireless telecommunications network. But how does this "logic" relate to real implementations of networks? How is the model representative of the many network implementation variations that can and do exist?

Wireless telecommunications network standards enable any wireless service provider, large or small, conveniently and efficiently to offer a standard set of wireless telecommunications services to subscribers. From a technical perspective, differences in the number, quality, and options of the standard services, as well as the deployment of additional nonstandard services, are what differentiate one service provider from another. The ANSI-41 standard is designed to allow many diverse implementations; it has been developed, and continues to be enhanced, with these implementation considerations in mind.

A wide variety of differing ANSI-41 network implementations exists today and all can claim compliance to the ANSI-41 standard. While the model is logical and intangible, the actual network is physical and very tangible.

An individual ANSI-41 network is usually comprised of many MSCs, each deployed as a separate physical switching platform. The number of MSCs supported within a single wireless service provider network is a function of the traffic capacity and geographic service area supported. Although the ANSI-41 standard allows multiple MSCs to be served by a single distinct VLR, there is usually a one-to-one relationship between MSCs and VLRs. Each MSC usually employs the VLR functionality internally; that is, the VLR is not normally deployed as a physically separate computing platform distinct from the MSC. This is due to the very close relationship between the MSC and VLR functions as defined by ANSI-41.

HLRs are deployed as both separate network platforms and internal to MSC platforms. An HLR usually provides service to many MSCs in a single network. If an HLR is deployed internal to an MSC, other MSCs in the network communicate with that MSC as if it were a standalone HLR. A network can support multiple distinct HLR platforms, serving specified geographic areas, or based upon some other division of subscribers. The HLR, though, is still considered to be a single logical entity in a given wireless service provider network; i.e., a single subscriber profile record cannot exist in more than one *active* HLR. The term *active* refers to the concept of physical network node replication. HLRs may be deployed as an *active/standby* replicated pair, since a failure will interrupt service to subscribers (see Figure 4.9).

Many functional entities can be deployed as redundant systems. Redundancy, in the form of node *replication*, allows for very high service availability. Replicated platforms are usually separated geographically. This avoids service interruptions based on catastrophic failure at a given site. The most common catastrophic examples used by engineers are hurricanes, earthquakes, and fires. There are two configurations for deploying replicated network nodes: load-sharing pair and active/standby pair.

A *load-sharing pair* requires both of the replicated platforms to be active and serve the network equally, with no distinction as to which platform sends, receives, or processes information. In case of failure of one of the platforms, the other can handle all the network traffic. An

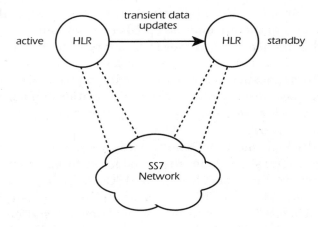

Figure 4.9
Active and standby HLR configuration. A special data link is used between the HLRs so that transient activity affecting the active database can be "mirrored" on the standby database. Each HLR of the replicated pair supports multiple links to the SS7 network. The SS7 network manages the rerouting of traffic to the standby (newly active) HLR when the previously active HLR fails.

active/standby pair requires one platform to be active, and the other to be in a standby mode. Only the active platform handles network traffic and processing. In case of failure of the active platform, the standby platform takes over and becomes the new active platform. The active/standby configuration is usually more appropriate for many mobile telecommunications network nodes (e.g., MSCs, HLRs, VLRs, ACs, and MCs). This is due to the nature of the real-time transaction processing required by the mobile telecommunications network nodes. Multiple messages related to one or more transactions for a single subscriber need to be handled by the same platform for data consistency. In general, the transient information communicated by the transactions cannot be efficiently shared in real time between two active platforms to keep them synchronized.

In the active/standby scenario, the active platform usually updates database changes on the standby platform across a data link in *near* real time. When a failure does occur, some data are inevitably lost, but this loss is minimized by continually updating the standby database.

Some network implementations are based upon both replicated platforms being ready at the same time, with each one able to support the entire traffic load in case the other fails. Note that this type of redundancy is distinct from fault-tolerant capabilities *within* a single computing platform, such as a switch. Fault tolerance enables a computing platform to operate uninterrupted or recover quickly in the case of faults within the platform. However, this type of fault tolerance does not protect against catastrophic events at a single location.

Authentication centers are usually implemented as part of the HLR. This is because there is a very close relationship between the HLR and the AC. However, ACs are also deployed as separate external platforms in the network.

Although short-message services are maturing in the network, message centers are generally considered to be separate physical platforms, distinct from HLRs and MSCs. Short message entities are usually considered to be software applications. An SME is normally implemented within an SMS-capable mobile station. The physical MC can support SME applications as well. An SME can also be a separate physical device that can even be external to the ANSI-41 network, communicating with the MC via ANSI-41 or any other data communications protocol.

Several considerations for network implementation options are given in the ANSI-41 standard. Two considerations provide the most freedom to network developers:

- ANSI-41 specifies the required intersystem operations only—mechanisms internal to each functional entity are left open to developers.
- The specified ANSI-41 intersystem operations explicitly defined between functional entities do not preclude the participation of other entities in the operations.

These considerations are very important and powerful. They emphasize the need for standardized intersystem operations between the defined functional entities, while allowing network engineers to implement optimal solutions to their specific internal requirements.

The first consideration limits the scope of the ANSI-41 standard. The internal mechanisms developed for a given functional entity affect many aspects of service and performance, of which ANSI-41 signaling is only one part. The second consideration allows network designers and equipment manufacturers to employ a variety of schemes to implement the standard functions in ANSI-41, as well as a multitude of other nonstandard features and services.

Relationships Between Functional Entities

The association between any two functional entities communicating with each other via ANSI-41 can be defined in the following logical ways:

- A one-to-many relationship
- A one-to-zero-or-more relationship
- A many-to-many relationship

These relationships only apply to the association between two functional entities at a single point in time. For example, an MS has a relationship with a single MSC at any point in time (from an ANSI-41 network perspective), even though it can have a relationship with many different MSCs at different times.

Defining the relationships between two functional entities in this way facilitates the understanding of the scope and purpose of the ANSI-41 intersystem operations specified between those entities. While it can be argued that defining these relationships limits the possible implementations of ANSI-41–based networks, the definitions are actually intended to limit ambiguity and interpretations of the ANSI-41 standard, while allowing freedom of implementation. In fact, it can even be argued that each functional entity has a many-to-many relationship (either direct or indirect) with any other functional entity! However, depicting a more confusing perspective of the reference model is not likely to be helpful. For reasons of simplicity, the depiction of many-to-many relationships was left out of the ANSI-41 standard.

Other Network Interfaces

Besides the functional entities and interfaces specified in the ANSI-41 network reference model, and the functions they represent, there are many other interfaces and functional entities that need to be implemented and deployed for the wireless network to be operational. The following list provides some of the basic functions not directly addressed by functional entities and interfaces specified in the ANSI-41 network reference model:

- Billing and accounting functions
- Customer service functions
- Operations, administration, and maintenance (OA&M) functions
- Provisioning functions

Multiple Interpretations of the Network Reference Model

As a general consequence of the protocol standardization process, multiple interpretations of standards are inevitable. In fact, many contributors to a given standard may insist on a degree of ambiguity so that they are free to interpret the standard in an *individual* way. This concept tends to be in complete opposition to the purpose of standardization; however, it does happen.

Many efforts were made to avoid ambiguity in the ANSI-41 specification. This is especially true of the network reference model. The emphasis on *logical* functional entities rather than physical nodes in the network allows for many diverse implementations, but can be confusing to those used to applying more physical models. From a purely academic-standards perspective, the model represents functions only; from a pragmatic perspective, mobile stations are telephone handsets, base stations are radios, MSCs are computer-controlled switches, HLRs and VLRs are data-based location registers, MCs are store-and-forward processing centers, and SMEs can be anything that originates or terminates a short message. Registers and processing centers can be combined with each other or with switches. The idea is to allow freedom of implementation of the ANSI-41 protocol while providing standard unambiguous descriptions. While there is no ambiguity in that goal, multiple interpretations can and do exist, and the resolution of those interpretations is continuing work for the standards committees.

CHAPTER 5

Introduction to Wireless Functionality

Before the details of the ANSI-41 protocol are described and explained, it is important to have an overall view of all the basic functions of the wireless network, even those not addressed by the ANSI-41 standard. This will allow a better understanding of the scope of ANSI-41 and also place the wireless telecommunications network into an ANSI-41–based perspective. In this chapter we describe, in general, all the categories of wireless network functionality and identify which of these categories and general functions are standardized by ANSI-41.

General Scope of Wireless Functionality

Wireless telecommunications networks are similar in many ways to other communications networks. They incorporate network nodes and communications interfaces between those nodes. Generally, the communications supported by any type of network can be defined by the following characteristics:

- *Physical and electromagnetic parameters*. The physical and electromagnetic parameters define the means of propagation of signals across an interface between network nodes.
- *Channel structures*. The channel structures define the *transmission paths* that the signals take, with each path being separated by some physical means (e.g., frequency or time).
- A *protocol*. The syntax, semantics, and procedures enable the *functions* that can be supported by communications across the network interfaces.

Although these characteristics are very closely related—due to the nature of systems based on layered protocols—ANSI-41 is primarily concerned with the intersystem operations of a wireless telecommunications network. These intersystem operations comprise the protocol necessary to enable functions that provide intelligent communications in the wireless network. The standard also, to some extent, addresses the other characteristics of the network; however, these other characteristics (i.e., physical and electromagnetic parameters and channel structures) are standardized in other ANSI and TIA specifications.

Wireless telecommunications networks (including ANSI-41–based networks) differ from other types of communications networks through the *functions* that they support. These functions are enabled by the signaling protocol. Five generic categories of functions enable subscriber mobility in the wireless telecommunications network:

- Mobility management
- Radio system management
- Call processing and service management
- Operations, administration, and maintenance (OA&M)
- Terrestrial transmission facilities management

The five categories of functions named are common terminology in the industry. They are used here for convenience and understanding of the wireless functions as they relate to the entire wireless telecommunications network system. Many individual functions and subfunctions can be considered in more than one category, and good arguments can be made for placing some functions in one category or another. Although it is convenient to categorize the wireless functions in this way, there is not always a clear one-to-one mapping between each individual wireless network function and one of these five broad categories of wireless network functionality.

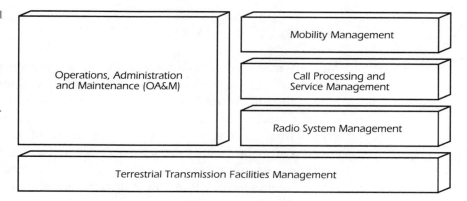

Figure 5.1
Model showing the general relationships among the five categories of wireless functionality.

Figure 5.1 depicts a model of the five categories of functions. Terrestrial transmission facilities management supports the means for conveying communications in the network. Radio system management, call processing and service management, and mobility management provide the basic functions to enable mobile service to subscribers. OA&M provides control and management of all network functions as well as the equipment deployed in the network.

Many of the functions in these categories are not supported by the ANSI-41 intersystem operations; some of the functions are performed within a single switching system, and others are performed across non-standard interfaces. Some of the functions are defined by other standards and specifications. In fact, many functions in these categories are not standardized or specified anywhere yet! The ANSI-41 standard provides the intersystem operations primarily for the *mobility management* category of functions performed between systems; however, there are many ANSI-41 operations that also support functions considered to be in the other categories and within a single system only.

This may seem confusing at first, but many wireless network functions can be divided into subfunctions, each in a different functional category. For instance, the group of functions needed to perform intersystem handoff for a mobile subscriber can be considered to be in the radio system management category. The mobility management functions support certain radio-specific parameters passed between switching systems to determine and control the radio access. The radio system management functions control radio resources and manage radio channels to enable the handoff between the cells.

Mobility Management

Mobility management is a broad category of functions that enables wireless subscribers to be mobile. Mobility can be defined in a very broad sense. For instance, does a portable FM radio provide mobility? In a sense it does, since a user of the radio can receive service while she or he moves from place to place. But in cellular-based wireless systems, the concept of mobility has a much narrower definition. Cellular networks are based on radio transmissions, similar to FM radio, and if a subscriber moves only within the coverage area of a single cell, he or she can obtain the same degree of mobility as the user of the portable radio. Cellular systems, however, support two-way communications and frequency reuse (see Chapter 1), which allow the system to support more subscribers and to provide individual radio service within and between individual cells. From both a network and subscriber perspective, mobility in cellular systems refers to the ability of the network to track a subscriber's status and location and continue to provide service in the following situations:

1. While the subscriber is being served in a cell associated with a mobile switching center (MSC) in the home service area.
2. While the subscriber is being served in a cell associated with a visited MSC.
3. While the subscriber is moving between contiguous cells served by the same MSC (home or visited).
4. While the subscriber is moving between contiguous cells served by different MSCs (home or visited).

These situations introduce the mobility concepts known as *automatic roaming* and *handoff*. These capabilities are the cornerstone of mobility management in wireless cellular telecommunications. *Automatic roaming* is the ability of a subscriber to obtain automatically mobile telecommunications service in areas other than the home service provider area, i.e., without additional actions being taken by the subscriber or a party calling the subscriber.

In this sense, mobile telecommunications *service* refers to the ability of a subscriber to make calls, receive calls, and apply any additional subscribed features to calls, such as call waiting, within the home serving system or in a visited serving system.

Handoff is a combination of mobile telephone and network functions that provides the capability for a subscriber to receive continuous service while a call is in progress, without interruption, when the subscriber is moving between contiguous cells.

Assuming that the cellular geographic coverage area has no gaps, a subscriber can theoretically receive *seamless* and *ubiquitous* service anywhere within the coverage area, even if the coverage area is the entire land area of the Earth! Unfortunately, wireless cellular systems today are still lacking, so that even within a defined coverage area, service may be interrupted due to radio transmission problems and system capacity limits.

Seamless Roaming

The idea of a seamless and ubiquitous cellular network is a goal that many service providers try to attain—but it is very elusive. *Seamlessness* (a good industry buzzword) refers to the ability of a subscriber to obtain wireless telecommunications services while roaming in the exact same way that they are provided in the home service area (i.e., no apparent *seams* between network service areas). For instance, the sequence of dialed digits that a subscriber enters when making a call would always have the same effect anywhere in North America, such as dialing 411 to obtain directory assistance or 911 for emergency services. Even today, these services are not seamless. Dialing 411 for directory assistance or 911 for emergency service does not work everywhere. Many wireless networks still do not provide equal access to long-distance service (i.e., choice of long-distance carrier) when originating a call, while some do.

Ubiquitous Wireless Service

Ubiquity refers to the ability of a subscriber to obtain wireless telecommunications services *everywhere*. This does not necessarily imply seamlessness, but it does imply that a subscriber could make and receive calls anywhere, even if the *call treatment* is different. Call treatment refers to the methods that the network uses to handle a call in a certain way. Ubiquity is also a characteristic of radio coverage. It implies that there are no gaps or holes in a radio coverage area.

The terms *seamless* and *ubiquitous* are often used in conjunction to imply consistent service capabilities, accessible from anywhere and everywhere.

The MSC and VLR as Combined and Separate Systems

Throughout this book we consider the serving system as a single entity encompassing the MSC and VLR functional entities. This simplifies the descriptions of ANSI-41 application processes and also is representative of nearly all of the ANSI-41 implementations currently in service.

However, ANSI-41's definition of the MSC-to-VLR interface makes the standalone visitor location register a viable network entity. For example, ANSI-41 does not assume that the MSC is "just a switch" without a database or feature control intelligence; if this were the model for ANSI-41, the MSC-to-VLR interface would be burdened with signaling, limiting the potential for separate functional entities. On the other hand, the MSC is not given full control of the operation of the serving system; authentication, for example, assigns specific processing duties to the VLR, which makes the idea of sharing a VLR among multiple MSCs particularly attractive. So, while we describe automatic roaming processes in terms of a single *MSC/VLR* entity, the reader should keep in mind that the potential for separation exists.

Subscriber Identification

Subscriber identification is a set of processes to uniquely identify a wireless subscriber. In ANSI-41 networks this identification also pertains to the subscriber's *subscription and service profile record* maintained at the home location register. That is, the identification of a subscriber essentially refers to the identification of a service profile as well.

A unique subscriber ID is used to register and qualify a subscriber for service; however, the unique subscriber ID is also used to enable *all network functions pertaining to an individual subscriber*. The full unique subscriber ID in most ANSI-41 networks is a concatenation of two unique parameters: the mobile identification number (MIN) and the electronic serial number (ESN). Note that in many ANSI-41 operations, only the MIN is used to identify a subscriber. The MIN-ESN combination is used primarily for the registration and authentication functions. The international mobile station identity (IMSI) is gaining ground as the successor to the MIN. The IMSI offers an improved address space and has particular advantages for international applications of ANSI-41.

Mobile Identification Number

The MIN is the single most important parameter used in ANSI-41 networks. It is a unique 10-digit decimal number programmed into the MS. This unique value is assigned to the mobile subscriber by the wireless service provider. This MIN is transmitted over the air interface (no matter which radio standard is used) during registration to inform the network of the identity of the mobile station (MS) accessing the network. The MIN is also used as a key field to access the wireless subscriber service profile record stored in the HLR (this is why we say that the subscriber ID also identifies a subscription). The service profile record contains the current location and status of the subscriber as well as other information pertaining to the subscription, such as subscribed features and credit status. Note that the format of the 10-digit MIN parameter used in the ANSI-41 network is a 40-bit value, with each digit encoded as 4 bits. The same 10-digit MIN value transmitted over the air interface (AMPS, TDMA, and CDMA) is specially encoded as two fields: the 24-bit *MIN1* and the 10-bit *MIN2*. The translation between these encoding formats can be performed at either the base station or the serving

MSC. The value transmitted over the air is fewer bits because of the importance of conserving bandwidth over the air—bandwidth is the single most limited resource in wireless cellular systems. Figure 5.2 shows the two formats for the MIN.

Figure 5.2
The two formats of
the MIN.

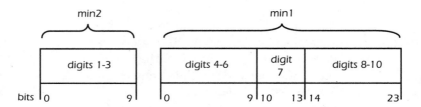

digit 1	digit 2	digit 3	digit 4	digit 5	digit 6	digit 7	digit 8	digit 9	digit 10

bits |0   3|4   7|8   11|12 15|16 19|20 23|24 27|28 31|32 35|36 39|

ANSI-41 mobile identification number (MIN)

min2 — digits 1-3
min1 — digits 4-6 | digit 7 | digits 8-10

bits |0   9| |0   9|10   13|14   23|

air interface mobile identification number (MIN)

The MIN was originally designed to contain a North American Numbering Plan (NANP) 10-digit number. This is because the MIN was meant to be the subscriber's directory number (i.e., the dialable telephone number) as well as the subscriber ID for network signaling. This is due to early cellular systems' being designed for North America only, incorporating wire-line system concepts.

From an ANSI-41 protocol perspective, the MIN is purely the unique identifier of a mobile subscriber and subscription. In most implementations, the MIN is usually the dialable directory number of the MS as well. However, this is not a requirement, and the MIN is not specified as the subscriber's directory number in ANSI-41. The Mobile Directory Number (MDN) serves this purpose.

Electronic Serial Number

The ESN is a unique 32-bit serial number permanently stored in the MS equipment by the manufacturer. It identifies an MS, rather than a subscriber or subscription. The internal circuitry that provides the ESN is usually isolated from fraudulent contact and tampering. Many mobile stations are manufactured so that attempts to alter the ESN render the MS inoperative. Figure 5.3 depicts the format of the ESN.

Figure 5.3
The two formats of
the ESN.

Manufacturer's Code 8 bits	Serial Number 24 bits

or

Manufacturer's Code 14 bits	Serial Number 18 bits

electronic serial number (ESN)

The ESN consists of a manufacturer's code and a serial number. There are currently two formats for the ESN. The first and most common uses a unique 8-bit manufacturer's code and a unique 24-bit serial number for that manufacturer's code. Due to the increasing number of manufacturers' codes (many individual manufacturers retain multiple codes for different models of MSs), the manufacturer's code was increased as an option to a 14-bit code with an 18-bit serial number. The Federal Communications Commission (FCC) has mandated the presence of the ESN in the MS, so that radio transmissions can be identified, if necessary. Since the MIN and ESN are used in conjunction to identify a subscriber uniquely for automatic roaming functions, the physical MS terminal is always tied to a wireless subscription in ANSI-41. This is a problem for subscribers who need to replace the MS, since the subscription has to be modified for the new MS with a new ESN.

International Mobile Station Identity

The international mobile station identity (IMSI) is a unique identifier originally standardized in *CCITT Recommendation E.212*. The IMSI (Figure 5.4) consists of up to 15 digits (0–9) whose structure is specified by CCITT E.212. The first three digits of the IMSI are the mobile country code (MCC), which is the country where the wireless service is provided. The remaining digits are the national mobile station identity (NMSI). The NMSI consists of two fields: the mobile network code (MNC) and the mobile station identification number (MSIN). The MNC is a 2- or 3-digit code that uniquely identifies the network to which the subscriber belongs. The MSIN is a 9-digit field that uniquely identifies a subscriber within a network within a country. One can immediately see how the IMSI improves on the MIN by providing a larger address space and a clear identification of the subscriber's home country and network.

Figure 5.4
The format of the
IMSI.

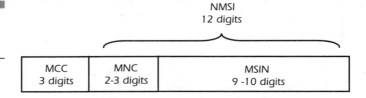

international mobile station identity (IMSI)

The IMSI is used in the Global System for Mobile communications (GSM) wireless technology. However, the IMSI in GSM stands for international mobile *subscriber* identity, to connote the identification of a subscriber, separate from the mobile station. The IMSI is not used for dialing or call routing purposes through the PSTN, ISDN, or PPDN. It is used to identify a subscriber for the GSM signaling functions.

The IMSI is an important and powerful mechanism for identifying wireless subscribers. In an MS that employs a *smart card* (known as the subscriber identity module or SIM in GSM), the IMSI is normally contained in the card. In MSs that are not smart-card–operated, the IMSI is contained in the physical mobile station. When used as the unique identifier in a smart card, it enables the separation of the subscriber identity from the mobile station identity. In this case, the subscription is not tied to the MS. The subscriber can use other compatible mobile station equipment without any changes to the subscription.

The IMSI was first standardized for use in ANSI-41 technology in the radio interface specifications ANSI/TIA/EIA-136 (TDMA) and ANSI/TIA/EIA-95 (CDMA) to identify the subscriber over the radio link to the mobile station. The IMSI has been added to the ANSI-41 standard, and its use is specified in *TIA/EIA/IS-751*, which introduces the *mobile station identity* (MSID) concept. The MSID can take on the value of either a MIN or an IMSI. IS-751 specifies that, in all messages containing a mandatory MIN parameter, the MIN is replaced by the MSID. Unfortunately, from a backward-compatibility standpoint, this changes the MIN from a mandatory parameter to an optional parameter.

The use of the IMSI gives the ANSI-41 network an alternative method for identifying a subscriber instead of the MIN-ESN pair. When the IMSI is used, it replaces the MIN-ESN pair to identify a subscriber uniquely throughout the ANSI-41 network (although the ESN is still a mandatory parameter in a number of ANSI-41 messages and is necessary for functions like authentication). Replacing the unique mandatory identifier for a subscriber requires modifications to the MSCs, VLRs, and HLRs in a wireless service provider's network as well as all billing, provisioning, and OA&M systems, a daunting and expensive task to say the least, especially when one considers that during the transition to IMSI use, ANSI-41 networks will be required to support both the MIN-ESN pair and the IMSI as identifiers of a wireless subscriber!

Of course, the use of the IMSI also allows the ANSI-41 network to take advantage of smart-card technology within the MS. As of December 2000, the only smart-card–operated phones deployed that use the ANSI-41 network are those used for GSM/AMPS dual-mode roaming purposes (see Chapter 17). However, standards have been developed that support the use of a removable user identity module (R-UIM) in ANSI-41 networks. These standards are

- TIA/EIA/IS-820, for CDMA mobile stations
- ANSI/TIA/EIA-136, a series of standards for TDMA mobile stations
- TIA/EIA/IS-808, the basic ANSI-41 enhancements to support R-UIM

Temporary Mobile Station Identity

The temporary mobile station identity (TMSI) is an alias for the international mobile station identity (IMSI). This alias is used for security purposes and is assigned dynamically by the VLR, which stores the IMSI value and maps a random TMSI to the IMSI. When implemented, the TMSI is used as the MSID between the MS and the MSC/VLR instead of the IMSI. This mechanism prevents the IMSI from being obtained from the air interface and can be changed by the network and sent to the mobile station (MS) based on a number of events. These events include MS power-off, deregistrations, or any implementation-dependent events (such as timer-based) determined by the VLR.

An added benefit of the TMSI is that it is shorter than either the IMSI or the MIN; e.g., in TDMA systems, the IMSI can be either 20 or 24 bits in length, versus the 34-bit MIN and the 50-bit IMSI. This offers practical paging efficiencies.

MSC Identification

From an ANSI-41 perspective, the current location of a subscriber is recorded in the HLR as the mobile switching center identification (MSC ID) of the system currently serving the subscriber. The MSC ID is a unique three-octet number that identifies each individual MSC in the network and the group of cells associated with that MSC. The structure of the MSC ID is shown in Figure 5.5.

Figure 5.5
The format of the MSC ID.

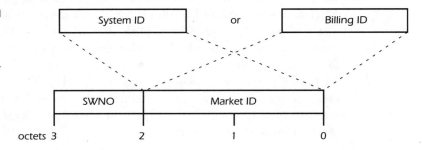

mobile switching center identification

The MSC ID is a concatenation of two parameters, the two-octet market identification (market ID) and the one-octet switch number (SWNO). The SWNO identifies an individual MSC within a defined market area. SWNO values are assigned for each MSC by the service provider. The market ID is split into two types of ranges of values: System ID (SID) and Billing ID (BID). The SWNO parameter is allocated by the service provider to uniquely identify each MSC within a given market identified by a SID or a BID.

SIDs

SIDs are 15-bit values assigned by the FCC, which are allocated to each wireless service provider.

For wireless networks, SIDs are uniquely allocated for each *cellular geographic service area* (CGSA) covered by a licensed operator. CGSAs represent the 733 U.S. government-defined geographic areas that are individually licensed to the two 800 MHz cellular service providers in that market area (i.e., the A-band and B-band). A CGSA is further designated as either a *metropolitan statistical area* (MSA) or a *rural service area* (RSA). An MSA denotes one of the 305 largest urban population cellular markets. An RSA denotes one of the 428 less-populated areas defined for the country. SID values 0 to 2175 have been allocated for North American wireless cellular service providers.

For personal communications services networks, SIDs are allocated for each *trading area* covered by a licensed operator. Trading areas represent the 544 Rand McNally-defined geographic areas that are individually licensed to the six 1900-MHz PCS service providers in that market area (i.e., the A–F bands). A trading area is further designated as either a *major trading area* (MTA) or a *basic trading area* (BTA). An MTA denotes one of 51 large PCS markets. A BTA denotes one of 493 small PCS markets. SID values 4096 to 7679 have been allocated for North American PCS service providers.

A parameter known as the *home SID* is programmed into each valid mobile station. This home SID represents the system ID of the home service area of the subscriber. When a subscriber is roaming, the home SID stored in the MS will not match the SID broadcast by the serving system cell site. This causes the "ROAM" indicator on the MS to activate, indicating that the subscriber is no longer being served by the home system. This home SID is never transmitted from the MS to the network.

BIDs

BIDs are 15- or 16-bit values assigned by CIBERNET, a corporate organization that is a wholly owned subsidiary of the *Cellular Telecommunications & Internet Association* (CTIA). The BID values can be allocated to each cellular and PCS service provider, upon request. BIDs are assigned to any market, which can be any geographic territory, desired by the service provider.

The purpose of a BID is to identify a portion of a service area separately (i.e., a subset of the entire area represented by a SID). Some service providers choose to operate all their markets using the same SID, thus preventing the "ROAM" indicator from activating on the mobile station while the subscriber is in the service provider's territory. However, a distinction among markets within a service provider's territory is still desirable for business reasons, and BIDs serve this purpose. Some of these business reasons are the generation of marketing statistics, roaming agreement restrictions, and to provide a distinction among recorded call detail information.

Markets identified as BIDs are still identified by a SID for broadcast over the air by cell sites. The BID overrides a SID when more specific location information is required. When BIDs are used to separately identify portions of a service area, the SID must still be used for one portion of that area. A service provider can request and use multiple BIDs for a single network to provide greater resolution of information for any purpose. BID values are derived from the same pool of market ID numbers as SID values. Market ID values 26112 to 31103 and 32768 to 40960 have been allocated as BIDs for North American cellular and PCS service providers.

Note that BIDs are never broadcast by a cell site; rather the associated SID is broadcast, and the BID is used for call detail recording and billing purposes.

Cell Site Identification

The identification of the cell currently serving a subscriber is managed by the serving MSC only. This more-specific location information is not normally stored at the HLR; however, it can be passed to the HLR for statistical purposes or for providing subscriber location information to another network entity. From an ANSI-41 perspective, the MSC ID is adequate location information for an HLR. The ANSI-41 intersystem operations also pass a *serving cell ID* parameter between MSCs to support the intersystem handoff function.

Mobility Management Functions

Mobility management provides the functions supporting the mobility of wireless subscribers. Mobility management generally encompasses the following functional subsets:

- Automatic roaming (including the component functions of mobile station service qualification, mobile station location management, and mobile station state management)
- Authentication
- Intersystem handoff

Conspicuously missing from this list of functions is a process known as *registration*. The reason that registration is not explicitly included as a mobility management function here is that the term is really representative of a subset of automatic roaming functions. The registration process primarily consists of mobile station service qualification and mobile station location management. These functions enable a subscriber to "register" with the network; that is, to indicate to the network location register functional entities the location and status of the mobile station.

The registration-related functions generally must be performed before the network can provide services to the subscriber; however, there are exceptions. (For example, access to emergency services by dialing 911 may not require registration—because in some cases, a subscription may not be necessary.) Figure 5.6 shows the network registration process.

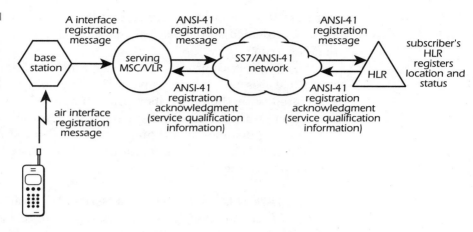

Figure 5.6
The MS sends a registration message to the serving system. The message is propagated to the home system (the HLR) to notify it of the subscriber's location and status. The HLR then responds to the serving system with service qualification information.

The process of registering a mobile subscriber is unique to wireless networks. In wire-line networks, there is no need for a telephone user to register with the network, since the user is not mobile and the user's location and status, as far as the network is concerned, are constant and static. That is, a wire-line user has no need to make the network aware of the location and status of the telephone being used to make or receive calls since it is always directly connected to the network via a physical line.

Automatic Roaming

The entire process of automatic roaming includes those functions performed between the mobile station and the network that allow the MS to inform the network of its current location, status, identification, and other characteristics. These functions allow a subscriber to obtain mobile telecommunications service outside the home service provider area without requiring special subscriber actions. The mobile station location and status information is recorded in the network so that services can be provided to the subscriber (based on the MS identification) at that location.

MOBILE STATION SERVICE QUALIFICATION   MS service qualification is the set of functions that qualifies the MS (i.e., the subscriber) to use the wireless telecommunications network services. This service qualification includes functions to validate the subscriber and manage the subscriber's service profile information.

Validation consists of examining the identification, location, and status information provided by the MS and comparing it to information stored in the subscriber's service profile. Network access can be either allowed or denied on the basis of service or financial criteria defined for that subscriber.

Managing service profile information consists of handling the specific set of features and service capabilities, including service restrictions, associated with the subscriber. An example of this information is a restriction on the types of calls a subscriber can make, such as limitations on long-distance or international calls. A subscriber's HLR service profile record is the network repository for the subscriber's service qualification information.

MOBILE STATION LOCATION MANAGEMENT   MS location management is the set of functions that stores, updates, and cancels the MS's (i.e., subscriber's) location. When the subscriber initially registers with the network, the current location is stored in the service profile record. As the subscriber moves from one serving area to another, the location is updated. The subscriber's location is canceled in the network when the subscriber is no longer considered active (for example, when the MS is turned off).

When a VLR serves multiple MSCs, or when a system serves multiple markets, many levels of the MS's location resolution are possible (see Figure 5.7):

- *The serving market ID*. If the MS is authorized for service for a particular market, the HLR will be provided a location update if the market changes, even if the serving VLR remains constant.
- *The serving MSC ID*. If the MS is authorized for service at the serving MSC's MSC ID, the HLR will be provided a location update if the MSC ID changes or if the serving VLR changes (these two are the same for a serving system [MSC/VLR] that has a single MSC ID).
- *The serving location area ID*. If the MS is authorized for service in the location area of the serving MSC, the HLR will be provided a location update if the location area changes or if the serving VLR changes.

MOBILE STATION STATE MANAGEMENT   MS state management is the set of functions that manages the activity status of the MS (i.e., the subscriber). The subscriber's state is considered by the network to be either *active* or *inactive*. When the state is active, the subscriber is considered available to receive calls. When the state is inactive, calls are generally not sent to the subscriber by the network. The subscriber's state is considered inactive when the subscriber is not registered, outside the radio coverage area, or is intentionally inaccessible based upon the modes of certain features or capabilities.

Figure 5.7
In ANSI-41, the location of a mobile station can be resolved to the SID only, MSC ID (SWNO and SID), or to a location area representing a group of cells.

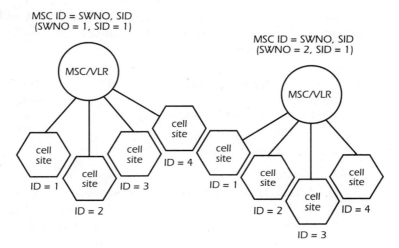

Authentication

Authentication is a mobility management process by which information is exchanged between an MS and the network for the purpose of *authenticating* (i.e., verifying with a strong degree of confidence) the identity of the MS. Authentication provides robust security against unauthorized or fraudulent access to the network above and beyond simple subscriber validation. Note that the authentication process requires an authentication-capable MS as well as the network support to perform the ANSI-41 authentication intersystem operations and calculations.

The authentication process is designed to be performed while a subscriber is in the home system or roaming. This means that if a subscriber with an authentication-capable MS roams to a network that is authentication capable, the network can challenge the MS to prove authenticity.

BACKGROUND AND NEED FOR AUTHENTICATION   Fraud has become a widespread problem in the wireless industry. The most well-known type of fraud in wireless networks is *cloning* fraud. Cloning is accomplished by reprogramming an MS with the identification values of a valid MS (i.e., the MIN and ESN). The valid identification values are typically obtained by illegally scanning the control channels on the air interface. It is relatively easy to obtain the identification data of legitimate subscribers. These data are transmitted without any form of encryption from the MS to the base station over the air whenever registration occurs. Special radio scanners can easily capture these values during the registration process of a valid MS. Any fraudulent MS programmed with a valid subscriber identification can then make calls as if it were the original legitimate MS.

Authentication-capable systems attempt to prevent fraud and cloning by supporting special algorithms that are performed in both the network and the MS. These algorithms use the mobile identification values and additional secret-key values that are never transmitted over the air.

Intersystem Handoff

Intersystem handoff is a mobility management process that enables the handoff of an MS from one cell to another where the two cells involved are served by two different systems (i.e., two different MSCs). Intersystem handoff transfers a call in progress from a radio channel in one cell to a radio channel in another cell in a *neighboring* system. In general, handoff functions are usually considered to be in the category of radio system management rather than mobility management. This is because the handoff functions involve the management of radio signal measurements, radio traffic channel allocation, and continual monitoring and control of radio resources at the base stations. However, since intersystem handoff enables the wireless subscriber to be mobile while involved in a call, it can also be considered part of mobility management as we have defined it.

Intersystem handoff provides for continuous terminal mobility management while a call is in progress. It is designed so that not only calls in progress, but calls in progress involving special features, such as three-way calling, can be handed off between MSCs. Note that the two MSCs involved in the intersystem handoff process can belong to a single wireless service provider or to two different wireless service providers.

Radio System Management

Radio system management is a broad category of functions that manage the radio resources, connections, and radio transmission paths between the mobile station and the network, before, during, and after a call. Radio system management can be defined in a very broad sense. It encompasses the functions to establish and release radio connections, maintain those connections, and control all resources associated with the type of radio technology supported by the wireless system (i.e., analog AMPS, digital TDMA, and digital CDMA). These functions are usually specific to the interfaces between the base station systems and the MSCs where radio transmission paths are *transcoded* (converted from one digital coding plan to another) for terrestrial transmission in the network. Individual radio system management functions are generally beyond the scope of the ANSI-41 standard except for the intersystem handoff functions.

The radio system consists of the subscriber's MS and the network radio equipment—antenna systems, base transceiver systems (BTSs), and base station controllers. The MSC can also be considered part of the radio system if it contains functions that control the operations of the base station (e.g., network-controlled handoff) (see Figure 5.8). Radio system management provides the functions that manage the radio and signaling transmission paths between the MS and the base station (i.e., the U_m interface) and the radio control signaling between the base station and the MSC (i.e., the A interface).

Figure 5.8
Radio system management controls aspects of the radio system including the base station controller (BSC), base transceiver system (BTS), and the antenna system (AS). The MSCs are involved in the coordination of radio resources for handoff.

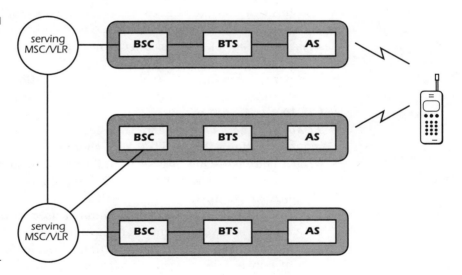

Radio System Management Functions

Radio system management generally encompasses the following functional subsets:

- Radio channel management
- Transmission mode management
- Power control
- Handoff control

Note that these are only the most general functional subsets and that there are other functions performed by the radio systems in the wireless telecommunications network.

Radio Channel Management

Radio channel management is a set of functions that manages the set of radio channel resources within each cell site. Radio channel management generally consists of the following functions:

1. Channel allocation and initialization
2. Channel configuration
3. Traffic channel management

The term *channel* in this context can be very confusing since it has many meanings in wireless systems. The following terms are used to describe channels for wireless telecommunications:

- *Radio frequency (RF) channel*—A single allocation of contiguous spectrum (i.e., 30 kHz in AMPS, 30 kHz in TDMA, and 1.25 MHz in CDMA).
- *TDMA channel*—One or more timeslots of a set of six timeslots carrying information within a TDMA RF channel.
- *CDMA channel*—One of a set of information channels distinguished from other information channels by a unique encoding scheme within a CDMA RF channel.
- *Traffic channel*—An information channel used to carry voice or user data and, in some cases, channel-associated signaling.
- *Control channel*—An information channel dedicated to the transmission of nonchannel-associated signaling data (e.g., mobile registrations and page responses) and supplementary service data. The control channel is generally split into a one-way channel to the MS and a one-way channel from the MS.

Channel allocation and initialization is a function that controls the assignment of radio channels to wireless subscribers to support calls and signaling. These radio channels provide the communications across the air interface between the base station and the MS.

Channel configuration is a function that manages the set of channels defined for use within a cell. Channels can be configured for different use and for a given set of frequencies supported by a cell.

Traffic channel management is responsible for dynamically modifying the set of traffic channels to meet real-time traffic demands on the radio system.

Transmission Mode Management

Transmission mode management is a set of functions that manage the transmission methods and characteristics of the circuits used between the radio system and the MSC. The function determines which communication modes can operate on a particular circuit or channel. Examples of these modes are signaling only, signaling and speech, and data only. These modes are dependent on the type of air interface supported (i.e., AMPS, TDMA, or CDMA) and the types of channels supported. The function also manages the transcoding and rate adaptation methods for the circuits supported.

Power Control

Power control is a set of functions that manages the real-time transmission power of the MS and the BTSs within the cell. The transmission power is defined within a certain range of values. The power can be dynamically modified based on interference levels and signal fading to provide adequate service to subscribers.

Handoff Control

Handoff control is a set of functions that allows the MS to move from one cell to another (or one radio channel to another, within or between cells) while a call is in progress. *Intrasystem* handoff (between two cells or radio channels that subtend the same MSC) is outside the scope of ANSI-41 since no coordination is required between MSCs to manage mobility of the MS. *Intersystem* handoff is a handoff between two cells subtending two different MSCs. This type of handoff combines functions from both radio system management and mobility management (i.e., ANSI-41 intersystem operations) to coordinate the movement of the MS between the cells.

Handoff control includes handoff preparation methods and measurements to determine the need for an MS to change channels or cells and support the communications switch between those channels or cells.

Call Processing

Call processing (also known as *call control*) is a broad category of functions that establishes, maintains, and releases calls to and from the wireless subscriber. It consists of the operations performed from the initial reception of an incoming call through the final disposition of the call. A call can be defined as a temporary communication between telecommunications end-users for the purpose of exchanging information. A call includes the sequence of events that allocates and assigns resources and signaling channels required to establish a communications connection. A call can be a conventional telephone call or another type of communications connection, such as data transmission (see Figure 5.9).

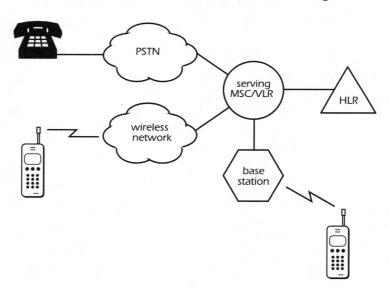

Figure 5.9
Call processing involves all network signaling, management, and connectivity to complete calls between a wireless subscriber and a wire-line phone or another wireless subscriber. The phones involved may be served by different networks.

The call processing functions establish and release terrestrial transmission paths for calls as well as invoke and manage call features that provide a specific variation on the way a call is treated. Many call processing functions are beyond the scope of the ANSI-41 standard; however, many aspects of call processing involve the ANSI-41 mobility management functions. These aspects include call delivery to a mobile subscriber and the application of call features for a mobile subscriber.

Basics of Call Processing

The basic capabilities to support mobile-originated calls and mobile-terminated calls involve the coordination of many functions, most of which are transparent to the parties involved in the call. These functions become much more complex when a subscriber is roaming and during an intersystem handoff. The mobile-originated call scenarios are simpler than the mobile-terminated call scenarios, since MS location and status information does not need to be obtained. Also, terminating calls to a roaming MS involves special routing to deliver the call to the system currently serving the subscriber. Note that the MS must be registered with the serving system prior to originating a call.

Dialing Plan

Mobile-originated calls are any calls initiated by the MS. The dialing plan refers to the format of numbers that a mobile subscriber dials to reach a called party. These dialed digits may include special routing information, such as choice of long-distance carrier, special digits for operator access (i.e., dialing a zero), and the actual directory number address digits of the called party. The wireless dialing plan is standardized and specified in *ANSI/TIA/EIA-660*.

The subscriber enters (or, euphemistically, *dials*) the number to be called. In wireless systems today, the subscriber performs *pre-origination dialing*, meaning that the numbers are dialed and sent to the system before any connection is requested. This is why mobile subscribers need to press a SEND key to transmit the dialed digits. This is in contrast to wire-line systems where *post-origination dialing* is used. In wire-line systems, the off-hook signal from the receiver informs the network that a call is about to be made, and the network responds with a dial tone. The called party's digits can then be dialed.

Numbering Plan

The numbering plan for North America is specified by the North American Numbering Plan (NANP), which defines the format of all dialable numbers in North America. Wireless service providers obtain ranges of dialable directory numbers from the North American Numbering Plan

Administration (NANPA) that follow this format. These are the directory numbers given to subscribers when they become initially authorized for service.

The numbering plan defines the format of directory numbers (i.e., addresses) that represent a party that can be called. It differs from the dialing plan, since it only defines the address format and does not include digits that are *dialed* for special access or to provide additional information. Directory numbers for wireless systems in North America generally use the same numbering plan and dialing plan as wire line systems.

In wire line networks, a directory number indicates an area (i.e., a three-digit *area code*), an exchange, and a line (see Figure 5.10). An *exchange code* (or office code) is used to indicate a specific switch that serves lines directly connected to telephones. A *station number* is used to indicate a specific line directly connected between the switch and the telephone.

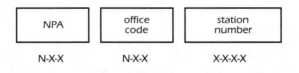

Figure 5.10 The basic NANP directory number format for wire-line and wireless networks. N represents any digit within the range 2 to 9. X represents any digit within the range 0 to 9. Note that the formats are identical although the last four digits represent a different type of user.

In wireless networks, a directory number indicates an area (i.e., the same three-digit area code), an exchange, and a *subscriber*. The exchange code indicates a specific MSC that serves the home subscriber. The subscriber number indicates a specific MS served by that home system. Since a mobile station's location can change, the method of routing calls geographically is contradictory to the concept of mobility. ANSI-41 provides signaling that allows the network to locate the MS and route calls to the subscriber in any location.

Call Processing Functions

Call processing generally encompasses the following functional subsets:

- Call establishment
- Call release
- Call features

Note that these are only the most general call processing functional subsets and there are other call-processing functions performed in the wireless telecommunications network.

Call Establishment

Call establishment (also known as *call setup*) is a set of functions that arranges for the connection of wireless calls. There are essentially two types of wireless calls:

- *Mobile-originated calls*—Calls that are placed from a mobile station.
- *Mobile-terminated calls*—Calls that are made to a mobile station.

Mobile-originated calls are established from an MS (the calling party) to a telecommunications termination point (the called party), which can be either within or outside the wireless telecommunications network. Mobile-terminated calls are established from a telecommunications terminal (the calling party), which can be either within or outside the wireless telecommunications network, to an MS (the called party). The method for delivering a mobile-terminated call to a roaming mobile subscriber is an ANSI-41 feature known as *call delivery*. Note that a mobile-to-mobile call is usually treated as the combination of a mobile-originated call scenario and a mobile-terminated call scenario. Both mobile-originated and mobile-terminated calls typically make use of the public switched telephone network (PSTN) to create the connections between the calling and called parties. This is due to the extensive infrastructure and routing capabilities that exist in the PSTN. There is no need to deploy new switching capabilities for calls that can be handled by existing networks. However, connection through direct trunks between MSCs in close proximity is sometimes used for mobile-to-mobile calls. This architecture is typically used in high-traffic areas where there are many mobile subscribers and many mobile-to-mobile calls.

The establishment of a wireless call involves two types of signaling to properly connect the call between parties: ANSI-41 call processing signaling and traditional call control signaling.

ANSI-41 call processing signaling is used to obtain the location, status, routing, and any special call treatment information about a mobile subscriber to complete mobile-terminated calls properly. It is also used to obtain call treatment and routing information for mobile-originated calls. This signaling is provided by the ANSI-41 intersystem operations. For intersystem handoff, ANSI-41 signaling is also used to control the inter-MSC trunks.

Traditional call-control signaling is used to establish trunks and propagate the call, as well as information about the call, from the calling party to the called party. Three types of call-control signals are used to complete calls: *supervisory signals*, to transmit the busy and idle states of lines and trunks; *address signals*, to transmit the digits of the called party's number and other information through the network; and *call progress signals*, to inform the calling party about the status of the connection that is progressing through the network (e.g., alerting, busy, routing failure).

Examples of other information provided with the address signals are the calling party's number, charge number information (also known as *automatic number identification* or ANI) used for billing, originating line or number information, network routing information, and special call-treatment information. Call-control signaling is generally provided by the out-of-band SS7 ISDN User Part (ISUP) protocol or an in-band multifrequency (MF) signaling protocol.

Special call-treatment information includes *call features*, also known as *supplementary services*. These features can apply to mobile-originated calls (e.g., three-way calling) and mobile-terminated calls (e.g., call waiting). Depending on the feature, the signaling can be provided by ANSI-41 call processing signaling, traditional call control signaling, or a combination of both.

MOBILE-ORIGINATED CALLS   Mobile-originated calls are calls a subscriber makes to any other party. The subscriber dials the digits to initiate the call processing functions with the serving system. When the serving MSC receives the dialed digits, the following procedures are performed to complete the call (see Figure 5.11):

1. *Authentication.* The authentication procedure can be performed locally or with the home system to verify the identity of the MS. This procedure may be skipped if the subscriber dials an emergency number such as 911.
2. *Mobile station service qualification.* The serving MSC queries the VLR to get service profile information and validate the subscriber's ability to make a call defined by the dialed digits (e.g., an interna-

tional or long-distance call). In many cases, the HLR is queried for this function for greater security.

3. *Digit analysis*. The serving MSC analyzes the digits for format and routing purposes.

4. *Obtain routing information for the call*. The serving MSC routes the call to its destination based on internal routing tables using the called party's dialed digits, special instructions obtained from the VLR or HLR, or origination triggers that can be used to request special call origination treatment from anywhere in the network.

5. *Application of call features*. The serving MSC uses the service profile to apply appropriate subscribed mobile-originated features to the call.

Once the call is authorized, routing information obtained, and call features applied, the call can be established using call-control signaling techniques (e.g., SS7 ISUP or MF) appropriate to the selected network.

Figure 5.11
Basic MS call origination. The MS dials the called party's number. The serving MSC validates the subscriber and then routes the call to its destination. The HLR may also need to be accessed if the VLR does not hold the subscriber's service profile, or if service qualification is requested. Authentication may be performed if it is supported by the MS and the serving system.

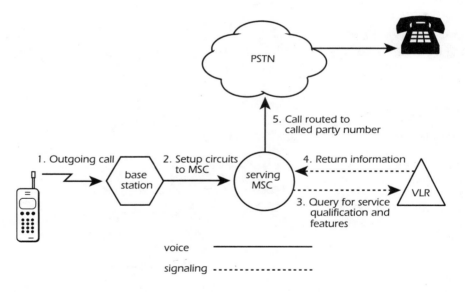

MOBILE-TERMINATED CALLS   Mobile-terminated calls are calls made to a subscriber's MS. The subscriber can be in the home area or roaming. The delivery of mobile-terminated calls to a roaming subscriber is a call feature known as *call delivery* in ANSI-41. If the subscriber is being served by the same MSC as the one where the incoming call arrives, call termination is a relatively simple process since no location information or special signaling or routing is required. The MSC where calls are initially delivered is known as the *originating MSC*. The originating MSC may be the home MSC or a gateway MSC that may not be controlled by the home service provider.

When a subscriber is being served by the same MSC as the one where the incoming call arrived, the following procedures are performed to complete the call (see Figure 5.12):

1. *Obtain MS registration status.* The MSC queries the HLR to determine if the MS is available to receive a call.
2. *Application of terminating call features.* The MSC queries the HLR to apply appropriate subscribed mobile-terminating features to the call.
3. *Page the MS.* The serving MSC alerts the MS and waits for it to answer.

Figure 5.12
Basic call delivery to the home system. The serving/originating MSC queries the HLR for registration status and features. The HLR returns the status and feature information to the MSC, which then establishes the call to the MS.

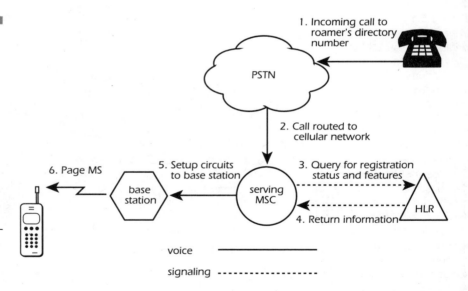

When a subscriber is roaming, and the originating MSC receives the dialed digits, the following procedures are performed to complete the call (see Figure 5.13):

1. *Obtain MS registration status.* The MSC queries the HLR to determine if the MS is available to receive a call.

2. *Obtain serving system routing information.* The MSC queries the HLR (which subsequently queries the current serving system) to obtain routing information to deliver the call to the location where the subscriber is being served.

3. *Apply terminating call features.* The MSC queries the HLR to apply appropriate subscribed mobile-terminating features to the call.

4. *Deliver the call to the serving system.* The originating MSC establishes the call to the serving system using call control signaling techniques (e.g., SS7 ISUP or MF) appropriate to the selected network.

5. *Page the MS.* The serving MSC alerts the MS and waits for it to answer.

Figure 5.13
Basic call delivery to a visited system. The originating MSC queries the HLR for registration, location, and routing information to deliver the call. The HLR queries the serving system for routing information in the form of a TLDN. The serving MSC assigns a TLDN and returns the digits to the HLR, who sends them to the originating MSC. The originating MSC establishes a call using the TLDN to the serving MSC, which then establishes the call to the MS.

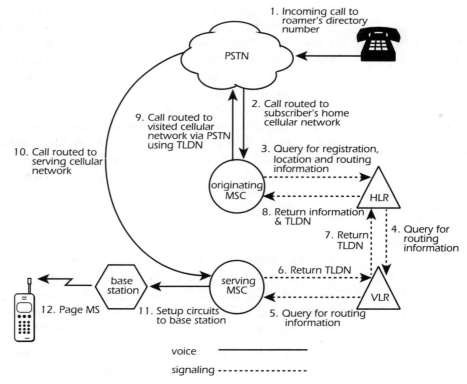

Of course, a call to a roaming subscriber can be redirected based upon subscribed features such as call forwarding. This can be done unconditionally, or when the call cannot be completed. There are many reasons why a call may not be completed to a subscriber:

■ The subscriber does not answer the call after a period of time.

- The subscriber is already engaged in a call (i.e., busy).
- The MS does not respond to paging (e.g., the subscriber has moved beyond the wireless coverage area or the subscriber has turned off the MS and the registration has not yet been canceled).
- The subscriber's location is unknown.
- The subscriber is reported as inactive.

TEMPORARY LOCAL DIRECTORY NUMBERS (TLDNS)   Delivering a call to a roaming subscriber requires a *temporary local directory number* (TLDN), a real NANP directory number assigned by the serving system to roaming subscribers for a brief period (about 20 seconds) to support call delivery. The use of TLDNs is necessary because calls to a roaming subscriber must be delivered somewhere other than the subscriber's home system. This is due to the fact that the subscriber's directory number is a geographic number, implying the subscriber's home system. Ranges of TLDN values are reserved by the wireless service provider from the entire set of directory numbers obtained from the NANPA. These numbers are associated with a serving MSC and are never assigned to mobile subscribers.

The home system begins a series of ANSI-41 signaling queries to locate the subscriber. The serving MSC assigns and maps a TLDN to the subscriber's identity at the serving system. The incoming call is redirected to the serving MSC, via the PSTN, using the TLDN as the called party's number. The call can then be established to the subscriber at the serving system.

TLDNs are dynamically allocated and released when call delivery is completed.

THE TROMBONING PROBLEM   *Tromboning* is a problem that occurs because of the use of TLDNs and the fact that directory numbers are geographically based. Tromboning refers to the shape that a call path takes during call delivery to a roaming subscriber (see Figure 5.14). The following scenario clearly represents the problem:

Randy has a mobile subscription and lives in San Jose, California. His number is 408-221-6567. Randy travels to New York with his MS. Mike lives in New York and calls Randy on his mobile phone. The call is established from New York to San Jose, since it is a 408-number. Randy's home MSC makes the appropriate signaling queries and receives a TLDN from the serving MSC in New York, which is a 212-number. The call is then redirected from San Jose to New York, where the call is completed when Randy answers the phone.

The actual call path in this scenario traverses the United States twice to complete the call. Call setup time is increased, and there are two long-distance legs to the call: one is charged to the calling party, the other is charged to the roaming subscriber.

This problem exists even in the ANSI-41 standard. The solution to the problem requires additional signaling and changes to network implementations. Note, however, that solving this problem may not be very desirable to long distance carriers who make money from its existence.

Figure 5.14
The tromboning problem. A call made to a roaming MS in New York is routed to the MS's home MSC in San Jose and then redirected back to New York via a TLDN to the serving MSC.

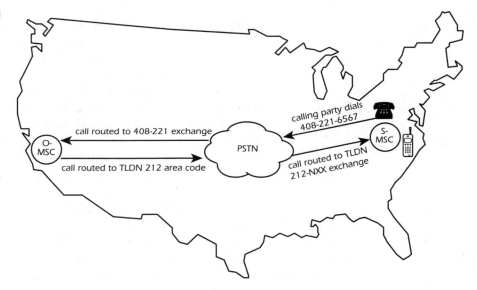

call routed to 408-221 exchange

call routed to TLDN 212 area code

PSTN

calling party dials 408-221-6567

call routed to TLDN 212-NXX exchange

O-MSC

S-MSC

Call Release

Call release (also known as *call disconnect*) is a set of functions that releases wireless calls. Calls can be disconnected by one of the following three entities:

- Calling party
- Called party
- Network

The wireless calling or called party disconnects by hanging up, e.g., pressing the END key. The network can disconnect a call for a variety of reasons, most due to a direct failure or to avoid the failure of a system involved in the call. Wireless calls can be disconnected by entities in the network regardless of whether the call is mobile originated or mobile terminated. In wire-line networks the disconnection is normally under control of the calling party.

Call release involves the release of resources used to establish and maintain the call so that they can be available for other calls. These resources include lines, trunks, and stored-program control switch resources such as timers, memory, and real-time software processes.

Call Features

Call features (also known as *supplementary services*) are services provided to wireless subscribers that offer variations on the *basic* service. Basic service refers to the ability of a subscriber to make or receive calls. In fact, in ANSI-41 networks, the ability of a subscriber to receive calls while roaming is considered to be a feature, known as *call delivery*. The ANSI-41 protocol provides intersystem operations to support the application of many call features for wireless subscribers during roaming and intersystem handoff.

Call features typically enhance the way a call is treated and can apply to mobile-originated and mobile-terminated calls. Some features affect only the wireless subscriber, and others affect the other party, or parties, involved in the call. Examples of the more conventional call features are call forwarding, call waiting, three-way calling, and calling party number identification. An example of a mobile-terminated feature is call forwarding, since the calls originally intended for the wireless subscriber are to be redirected to another termination point. An example of a mobile-originated feature is three-way calling since it allows a mobile subscriber to originate a new call while already engaged in a call.

Operations, Administration, and Maintenance

OA&M is a broad category of functions that enables the wireless service provider to monitor and control the wireless telecommunications network. OA&M tends to be a catchall term describing all activities involved in managing and operating a telecommunications network.

Operations generally include coordinating and operating the network on a day-to-day basis. Personnel in a network operations center monitor alarms and traffic throughout the network.

Administration generally includes the functions to ensure that the network operates at a high quality of service. It involves correct network sizing for users. The difference between operations and administration is that operations involves collecting data, whereas administration involves analyzing those data.

Maintenance generally includes troubleshooting, reporting, testing, and repair. It incorporates all functions that ensure that the network does not fail.

OA&M functions are primarily used to observe, record, and analyze operational characteristics of the network. OA&M functions are generally beyond the scope of the ANSI-41 standard; however, there are a few OA&M intersystem operations specified in the standard. These operations are used for a very small subset of the OA&M functions necessary to operate a wireless network.

The typical scope of OA&M functionality includes both unsolicited methods and inquiry and response methods for obtaining operational information about the wireless network. These statuses can be obtained on demand, on a priority basis, or on a time-interval basis. The most crucial information obtained includes the following wireless network measurement and statistical data:

- Network and service performance
- Network and service availability
- Network and service usage

These data are used to optimize the resources, performance, and quality of service of the network. The data are also used as a marketing tool to determine the types of network services accessed by subscribers.

Most wireless service providers have implemented a centralized *operations center* as part of the network to perform the OA&M functions (see Figure 5.15). The operations center is customized to support and manage the equipment and services specific to the network. Communications are supported between the operations center and the other network elements either directly or indirectly through mediating elements.

OA&M generally encompasses the following functions:

1. *Configuration management*—To control the provisioning, configuration, and modification of the network elements and the functionality within those elements.
2. *Network engineering*—To manage the parameters and statistics used to evaluate the efficiency of the wireless system.

Figure 5.15
Typically, a centralized operations center controls the telecommunications network elements.

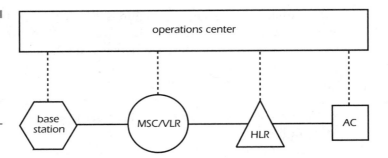

3. *Network change control*—To manage changes in network elements while the network is operating.

4. *Circuit management*—To control the inter-MSC trunks (i.e., trunks directly between adjacent MSCs) in the network to support intersystem handoffs.

5. *Fault management*—To reduce the number of network element failures as well as minimize the effects of a particular failure.

6. *Performance management*—To evaluate and report on the behavior of equipment and the effectiveness and efficiency of network systems, services, and elements.

7. *Overload control*—To manage traffic overload conditions at the network elements.

Note that OA&M functions are almost always implementation dependent. The functions listed here are only the most general categories of functions; many other OA&M functions are performed within the wireless telecommunications network.

Terrestrial Transmission Facilities Management

Terrestrial transmission facilities management is a broad category of functions that manages the physical means of providing voice and data communications services to wireless subscribers. These functions manage the wire-based trunk and line types and support the physical rates and encoding of the trunks and lines associated with the switching system (i.e., the MSC). The management of these facilities and resources is beyond the scope of the ANSI-41 standard except in the case of inter-MSC trunks supporting intersystem handoff.

The typical scope of transmission facilities management functionality includes digital transmission functions for both bearer and signaling channels. This is true even for analog AMPS-based systems since the analog voice can be digitally encoded for transmission across T-1 facilities. The primary characteristics for digital transmission channels include the following:

1. Selection of speech vocoding and data interworking techniques (since transmission rates and encoding schemes on radio channels are different from terrestrial transmission channels).
2. Selection of transmission modes (e.g., voice, data, etc.).
3. Companding techniques.

These characteristics apply to particular channels and for particular call types. They are dependent on the physical wire-line facilities and the types of bearer and signaling communications supported by the MSC.

Summary of ANSI-41 Functional Scope

This chapter presented a comprehensive overview of the basic functions required to operate a wireless telecommunications network. The ANSI-41 standard provides signaling in the form of intersystem operations to support a subset of these functions. This section provided an outline of the functions supported by ANSI-41 only. The protocol details of the ANSI-41 mobile application part (MAP) are described in subsequent chapters.

Wireless Network Functions Addressed by ANSI-41

As discussed previously, there are five generic categories of communications functions that provide, support, and enable subscriber mobility in the wireless telecommunications network. ANSI-41 provides many, but not all, of the functions in each of these categories. Table 5.1 shows the functions supported by ANSI-41.

TABLE 5.1

Functions
Supported by
ANSI-41

Functional Category	Functions Supported by ANSI-41
Mobility management	MS service qualification
	MS location management
	MS state management
	authentication
	intersystem handoff
Radio system management	handoff control (intersystem handoff only)
Call processing	mobile-originated calls
	mobile-terminated calls
	call release (for intersystem handoff only)
	call features and control
OA&M	circuit management (for inter-MSC trunks only)
Terrestrial transmission facilities management	none

Areas of ANSI-41 Application

The ANSI-41 standard specification is a large and complex document. It is sometimes difficult to grasp the power that the protocol provides. ANSI-41 can be applied to a variety of network architectures and services to enable the mobility of wireless subscribers. ANSI-41 was written to support the features specified in the *ANSI/TIA/EIA-664* cellular features standard. ANSI-41 can support additional features and services; careful study of the operations and parameters provided reveals a multitude of tools to support numerous services. Keep in mind that basic wireless telephony may be the most prevalent application of ANSI-41, but it is not the only application.

Intersystem Signaling and Intrasystem Signaling

ANSI-41 provides intersystem signaling operations, with a *system* defined by the standard as a single wireless telecommunications switch and related functional entities (i.e., location registers and processing centers). In deployed networks, a system is sometimes viewed as a single service provider network, consisting of many MSCs, VLRs, and at least one HLR. *Inter*system operations are usually viewed as the operations performed between one wireless service provider's functional entity and another wireless service provider's functional entity, as in the scenario of a roaming subscriber (i.e., between a *serving* MSC and the subscriber's service provider HLR). But ANSI-41 can be, and is, used just as appropriately for *intra*system operations. The ANSI-41 operations work just as well between functional entities within a single service provider network.

Some deployed networks use all proprietary protocols for operations within their own networks and use protocol conversion techniques to perform the ANSI-41 operations for intersystem signaling only. These networks can offer additional proprietary features. The major drawback of not using ANSI-41 operations for intrasystem operations is that there is no standard allowing the interconnection of different manufacturer's equipment within that system. These systems are usually composed of equipment manufactured by a single vendor, which limits the ability of the wireless service provider to deploy another vendor's equipment.

ANSI-41 Capabilities

The ANSI-41 signaling operations transfer information and invoke commands between functional entities. Taken from this simplistic perspective, ANSI-41 can be applied to a variety of wireless applications. Although the original impetus for ANSI-41 was public 800-MHz cellular telecommunications, the standard can be used to support many capabilities:

- Personal communications services (PCS)
- Network interfaces to wireless PBXs
- Personal base stations and related applications
- Rudimentary international roaming
- Enhanced mobile station functions (e.g., short-message services)
- Voice-mail systems
- Interactive voice response systems

These capabilities are only a sample of the application of the ANSI-41 protocol. As more capabilities are desired by wireless service providers, the desire for equipment manufactured by different vendors increases, and the need for standardized intersystem operations becomes greater.

PART **2**

ANSI-41
Explained

CHAPTER 6

The Road to ANSI-41

Since the mid-1980s, the TIA has been developing the ANSI-41 standard for cellular intersystem operations. The standard was always intended to evolve with the growing demands of the wireless marketplace. Initially, only basic capabilities to provide mobility were addressed. These capabilities eventually became the building blocks that now allow wireless service providers to offer a rich set of services to subscribers. This chapter follows the evolutionary path of ANSI-41, which leads to the sophisticated mechanisms that support truly advanced wireless services.

IS-41 Revision 0 (IS-41-0)

In the beginning, there was Interim Standard-41, named simply because 41 was the next available number for a new TIA standard. In 1984, the development of the first version of a cellular internetworking standard had begun. IS-41-0 was eventually published in February 1988 and addressed the following basic wireless intersystem functions:

- Basic intersystem handoff
- Basic service qualification
- Basic OA&M (circuit management)

Basic intersystem handoff included the core functions still used today. For example, handoff measurements, handoff forward, handoff back, and inter-MSC trunk release were all supported. Multiple intersystem handoffs were allowed as a propagation of the basic handoff-forward procedures.

Basic service qualification addressed the operations to validate a mobile subscriber. Validation was a function to address the *financial accountability* of a roaming subscriber. This validation could only be performed when the roaming subscriber was visiting a neighboring system connected by direct point-to-point signaling links to the subscriber's home system. These operations were specified to occur between a visited MSC and a home MSC across a generic data network. There were no HLRs, VLRs, or other functional entities specified to manage more sophisticated functionality.

Basic OA&M included the same functions specified today in ANSI-41 (i.e., inter-MSC circuit maintenance).

IS-41-0 did not specify a network reference model. In fact, there was no need, since there were only two functional entities involved, connected by a data network to transport the IS-41 application signaling. The communications specified included the CCITT recommendations for the Transaction Capability Application Part (TCAP) as an application service element. This application protocol was specified for use over X.25, as the primary choice, or North American ANSI SS7 (1984 version), as a secondary choice. The X.25 protocol was used in all the original field trials for IS-41.

IS-41 Revision A

IS-41-0 was a good start, but it obviously did not satisfy the true wireless needs of either subscribers or the original cellular service providers. Within three years of the publication of IS-41-0, IS-41-A was published (in January 1991). IS-41-A was the first significant step to providing elaborate wireless capabilities and services while improving on the basic functions specified in IS-41-0.

The first network reference model appeared in IS-41-A. This model introduced the concept of providing standard interfaces between particular functional entities with defined responsibilities. The HLR, VLR, and AC, along with interfaces to the PSTN, were the start to defining a true distributed intelligent network supporting wireless telecommunications.

IS-41-A addressed the following wireless intersystem functions:

- Intersystem handoff
- Enhanced service qualification
- Location management
- Mobile station state management
- Call delivery
- Cellular feature support

Intersystem handoff was essentially the same as in IS-41-0, except for minor modifications to the handoff procedures and the use of ANSI TCAP as the transaction protocol. Enhanced service qualification included subscriber validation from any serving system that had a roaming agreement with the home system and subscriber profile management. These functions were born of the introduction of an HLR.

The concept of an HLR also supported the MS location management and MS state management procedures. Call delivery to a roaming subscriber via a temporary local directory number (TLDN) was introduced. This same mechanism is used as the basis for call delivery today.

IS-41-A supported the first group of cellular features to enhance subscriber service. The following features were specified in the initial version of *TIA/EIA/IS-53* (Revision 0), now known as *ANSI/TIA/EIA-664*.

- Call forwarding—unconditional
- Call forwarding—busy
- Call forwarding—no answer
- Call waiting
- Three-way calling
- Feature control

One of the most important changes specified in IS-41-A was the use of ANSI SS7. ANSI TCAP was specified instead of CCITT TCAP or the other ANSI SS7 standards (1988 version). X.25 continued to be specified for network transport, but it was no longer favored over SS7.

IS-41 Revision B

IS-41-B was published in December 1991, only a short time after IS-41-A. Although there were many significant changes in IS-41-B, they were clearly anticipated before IS-41-A was even published. Changes were made primarily to intersystem handoff and network routing mechanisms.

IS-41-B addressed the following wireless intersystem functions:

- Path minimization
- Flash feature support after handoff
- Support for TDMA handoff parameters
- Use of SS7 global title translation

Although these functions are significant, they are quite subtle. Networks implementing IS-41-A could provide many enhanced services without these functions. Path minimization and intersystem handoff to a third MSC are fairly rare occurrences. Although not as efficient and flexible as global title translation, SS7 point-code routing works just fine. Because of these reasons, IS-41-B was not implemented for quite a while, and in many cases not at all. Networks that support IS-41-A may have no need for many of the enhancements that IS-41-B specifies. Many networks skipped IS-41-B implementations in favor of the feature-rich IS-41-C standard.

Not long after IS-41-B was published, implementation and deployment issues began to arise concerning IS-41-A. Most of the IS-41-A issues also applied to IS-41-B, since the primary functions remained unchanged. Ambiguities were also discovered in the standards. This is not surprising since these standards addressed some very complex issues and the industry was growing rapidly. Thus, the development of Telecommunications Systems Bulletins (TSBs) began. These publications do not carry the weight of interim standards; however, they are provided as addenda to the published standards and to solve particular problems specific to one or more of the standards.

Within three years of the publication of IS-41-B, several TSBs were published to add enhancements or correct problems with both IS-41-A and IS-41-B. The following TSBs directly pertain to IS-41 functionality:

- *TIA/EIA TSB41*—IS-41-B technical notes
- *TIA/EIA TSB51*—Authentication and signaling message encryption
- *TIA/EIA TSB64*—Support for CDMA intersystem handoff
- *TIA/EIA TSB65*—Resolutions to border cell problems

These TSBs provide complex functionality that could almost be published as interim standards in their own right.

TSB41—IS-41-B Technical Notes

TSB41 resolved several ambiguities and errors in IS-41 that resulted in incompatible implementations. The TSB documented 38 problems found in IS-41-B that were listed as *decision points*. It also provided replacement text for IS-41-B. Most of the decision points and the replacement text address very subtle portions of IS-41-B; yet they were necessary to avoid incompatibilities that could translate into poor quality of service for subscribers.

TSB51—Authentication and Signaling Message Encryption

TSB51 provided completely new functionality to support authentication, signaling message encryption, and voice privacy. The TSB specified intersystem operations to support authentication-capable MSs. The functions supported were global challenge, unique challenge, and base station challenge incorporating the CAVE algorithms, as well as updating of shared secret data (SSD) and the call history count.

TSB55—IS-41-A/B Forward Compatibility Rules

TSB55 provided modifications and additions to IS-41 Revision A network implementations to support forward compatibility to IS-41 Revisions B and C. The TSB also applied to Revision B forward compatibility to Revision C implementations, although the title of the document implies the rules apply to Revision A only. An example of these rules is that Revision A systems should not reject protocol messages from Revision B or C systems simply because additional unrecognized parameters are present in the message.

TSB64—Support for CDMA Intersystem Handoff

TSB64 provided intersystem handoff support for CDMA mobile stations. New parameters and intersystem operations were added, as well as provisions for mobile-assisted handoff (MAHO).

TSB65—Resolutions to Border Cell Problems

TSB65 provided solutions to *border cell problems,* which occur when an MS is at or near the border between two cells served by different MSCs. When this situation arises, it is possible that incoming calls to the MS can be delivered to the wrong MSC, and the MS does not receive pages.

IS-41 Revision C

IS-41-C represented the next major leap forward in North American wireless network technology. Published in February 1996, IS-41-C was the culmination of hundreds of person-years of effort. It was a major rework of the entire standard, comprising approximately 1,500 pages of text and diagrams, which is about six times bigger than IS-41-B! The motivation to develop IS-41-C came from many factors:

- The desire to provide many more features and services to wireless subscribers.
- The desire to resolve compatibility problems found in earlier revisions of IS-41.
- The desire to integrate the TSBs into a single standard specification.
- The desire to incorporate state-of-the-art technologies such as intelligent networking principles and philosophies.

IS-41-C added a new part, *IS-41.6*, that provided detailed signaling procedures to assist with the implementation of the intersystem operations. It also added support for 19 additional features (beyond the five supported by IS-41-A) and short-message services (SMS). It was backward compatible with IS-41-A and IS-41-B, and provisions were added for forward compatibility with subsequent versions.

The TSBs were enhanced and rolled into the standard in the following ways:

- TSB41 technical notes were added to the appropriate text.
- TSB51 authentication was added and enhanced.
- TSB64 intersystem handoff for CDMA was added, with modifications to CDMA-specific parameters and procedures for handoff and automatic roaming.
- TSB65 border-cell resolutions were added with additional operations supporting precall handoff between serving and border systems.

Note that the additions to IS-41-C, which include modifications to the TSBs, supersede IS-41-B systems that incorporate those TSBs.

Other capabilities incorporated into IS-41-C included the following:

- Support for *ANSI/TIA/EIA-124 (DMH),* which provides near-real-time transfer of call detail records supporting billing and fraud management.
- Enhanced network addressing.
- Intelligent network trigger points.
- More service management controls at the HLR.

IS-41-C evolved into a very sophisticated specification that supported intelligent networking concepts, 24 wireless features, short-message services, AMPS, TDMA, and CDMA radio systems, and many other capabilities. This functionality was meant to support both cellular and PCS systems as well as providing the basis for next-generation wireless systems with a great deal of flexibility.

ANSI-41 Revision D

In December 1997, IS-41 Revision C was elevated to a full national standard known as ANSI/TIA/EIA-41 Revision D, or simply ANSI-41-D. The primary difference in a national ANSI standard and an interim TIA standard is the scope of the standards bodies that vote on the document to create the standard. Because the TIA is an accredited standards body of ANSI that has limited scope, it can only create and approve interim standards (ISs). For a standard to be approved as a national ANSI standard, it must be approved by the larger voting membership of ANSI. This was accomplished with only minor clarifications to IS-41-C, and essentially the two standards are the same. Note that a full national ANSI standard carries quite a bit more influence than an interim standard, and an ANSI standard does not need to be reaffirmed every five years, as an interim standard must be.

ANSI-41 Revision E

The latest version of the ANSI-41 standard will be ANSI/TIA/EIA-41 Revision E, or ANSI-41-E. Like IS-41-C, ANSI-41-E is a major step forward in the features and capabilities provided. Also like IS-41-C, ANSI-41-E integrates many existing and evolving specifications and standards into its scope. The motivation to develop ANSI-41-E came from many factors:

- The desire to integrate many closely related TSBs and interim standards into a single standard specification.
- The desire to address multiple frequency band technology issues (i.e., PCS and cellular) that use ANSI-41 as their network technology.
- The desire to provide more features to wireless subscribers.
- The desire to incorporate a variety of new technologies into the standard (e.g., IMSI, over-the-air service provisioning, number portability, emergency services, etc.).

- The desire to enhance existing functions to improve network capabilities (e.g., authentication, internationalization, etc.).

The following standards and specifications are now included within the scope of ANSI-41-E:

- TSB-76 specification for PCS multiband technology supporting interband handoff
- IS-725 standard for over-the-air service provisioning (OTASP)
- IS-730 standard for use of high-capacity digital control channels for TDMA
- IS-735 standard for enhanced CDMA features (e.g., use of TMSI)
- IS-737 standard for circuit-switched data calls while roaming
- IS-751 standard for use of IMSI as a subscriber identifier
- IS-756 standard for wireless number portability
- IS-764 standard for calling name presentation and calling name restriction
- IS-771 standard for the wireless intelligent network (WIN) technology
- IS-778 standard for enhancements to authentication
- IS-807 standard for the internationalization of ANSI-41 (e.g., defining new global title address types for the international transmission of messages)
- IS-812 standard for ANSI-41 message segmentation
- J-STD-034 standard for emergency location services and enhanced 911 (J-STD denotes a joint standard developed between the TIA and the ANSI telecommunications standards committee T1)

ANSI-41-E is a very sophisticated specification that continues to evolve. The next-generation standard, ANSI-41-F is currently in the planning stage. This standard will support about 10 new technologies, along with enhancements to all the technologies included in ANSI-41-E.

CHAPTER 7

Structure of the ANSI-41 Standard

The extensive collection of text, tables, and figures that comprise the ANSI-41 standard can be imposing at first. Without a clear understanding of its organization, it is difficult to make effective use of the standard. Our goal in this chapter is to facilitate understanding of the ANSI-41 specification structure, including how it differs from past revisions of its predecessor IS-41.

Background

As the complexity of telecommunications systems has increased, so, too, have the demands on the organizations of people creating standards for these systems. Tools—including analysis techniques, specification processes, and even highly specialized languages—have been created to assist in the task. While some of these tools have proved their value over time, most are in a constant state of refinement as standards committee members try to find an optimum mix of clarity, brevity, and sufficiency.

Indeed, the act of writing a good standard involves a reasonable measure of art in addition to technical expertise. Since they are produced in standards committees, most standards are also laden with compromises, some for the better, others for the worse.

ANSI-41 has grown up in this environment and has managed to remain, for the most part, structurally consistent over the course of its revisions. Its structure has been largely influenced by two factors: the three-stage specification process and the three major functional areas addressed in the standard: intersystem handoff; automatic roaming; and operations, administration, and maintenance (OA&M). We discuss these influences as an introduction to how the ANSI-41 revisions are organized.

Influence of the Three-Stage Specification Process on the ANSI-41 Structure

As we discussed in Chapter 4, the three-stage specification process is popular within the telecommunications standards community. Originally developed for use in ISDN service standardization efforts, the basic technique is described in *ITU-T Recommendation I.130*. The three stages of the process are:

1. Describe the service from the user's perspective.
2. Derive a functional model for the service in terms of (a) the information flows between the functional entities involved in the service and (b) the actions required of these functional entities in order to provide the service to the user.
3. Design protocols and procedures that realize the functional information flows and actions derived in stage 2.

Application of the Three-stage Process to ANSI-41

The I.130 three-stage process is an example of a standards development tool in a constant state of refinement. The application of this tool to ANSI-41 is no exception: Some parts of the I.130 process have been kept as is, while others have been modified to better suit ANSI-41. The following sections describe some of the key differences and similarities between the I.130 and ANSI-41 processes.

Stage 1—Service Description

While stages 2 and 3 are embodied in ANSI-41, the stage 1 cellular service descriptions associated with ANSI-41 are specified in a separate standard, known as *ANSI/TIA/EIA-664* (ANSI-664). Separating the user perspective in stage 1 from the network aspects in stages 2 and 3 is a step in the right direction. This separation is one component of the three-stage process almost universally adopted by telecommunications standards committees throughout the world. The effect is to divorce, insofar as possible, service definition from implementation constraints. User needs are given priority over the limitations of the network. The result is a more market-driven development process for new service standards than would be the case if the service definition were constrained by current network limitations.

Creating a separate stage 1 standard also enables the parallel development of the network and air-interface standards. There remains the need for coordination between the various standards groups, but each should be working to fulfill the common requirements defined in stage 1.

The ANSI-664 stage 1 service definitions are structured similarly to those of the ITU-T approach specified in Annex A of *ITU-T Recommendation I.210*. Each service is defined by using the following basic format:

- Normal procedures with successful outcome
- Exception procedures or unsuccessful outcome
- Alternate procedures
- Interactions with other cellular services

Note that, as in I.210, ANSI-664 provides specifications of service interactions (e.g., between unconditional call forwarding and do-not-disturb services). This is a critical set of requirements, particularly in a multivendor environment. It is at stage 1 that the rules are established, not only for how a service should operate on its own, but also for how it should interact with other services.

A key difference between the ANSI-664 and I.210 stage 1 methods is that ANSI-664 does not employ the Specification and Description Language (SDL) diagram technique recommended in I.130 and described in Annex D of I.210. Apparently, SDL diagrams were deemed useful but not absolutely necessary for ANSI-664 purposes.

As of December 2000, the current version of ANSI-664 is Revision A (ANSI-664-A), published in July 2000. ANSI-664-A is organized as a series of individual documents, each with a part number (e.g., ANSI/TIA/EIA/664-505-A for the Call Forwarding Unconditional feature description). This is a change from ANSI-664, which was published as one large document. One advantage of the new structure is that individual parts can be added and modified without necessarily requiring all other parts to be revised. Table 7.1 lists the various parts of the ANSI-664-A series of documents. Note that parts specified in the original version of ANSI-664 have been set to Revision A (e.g., TIA/EIA/664-505-A), whereas new service descriptions are set to Revision 0 (e.g., TIA/EIA/664-529-0).

TABLE 7.1

Listing of the Parts of ANSI-664-A

Document Name	Part Number
Wireless Features Description	TIA/EIA/664-000-A
Background and Assumptions	TIA/EIA/664-100-A
Call Delivery (CD)	TIA/EIA/664-501-A
Call Forwarding—Busy (CFB)	TIA/EIA/664-502-A

continued on next page

TABLE 7.1

Listing of the Parts
of ANSI-664-A
(continued)

Document Name	Part Number
Call Forwarding—Default (CFD)	TIA/EIA/664-503-A
Call Forwarding—No Answer (CFNA)	TIA/EIA/664-504-A
Call Forwarding—Unconditional (CFU)	TIA/EIA/664-505-A
Call Transfer (CT)	TIA/EIA/664-506-A
Call Waiting (CW)	TIA/EIA/664-507-A
Calling Number Identification Presentation (CNIP)	TIA/EIA/664-508-A
Calling Number Identification Restriction (CNIR)	TIA/EIA/664-509-A
Conference Calling (CC)	TIA/EIA/664-510-A
Do Not Disturb (DND)	TIA/EIA/664-511-A
Flexible Alerting (FA)	TIA/EIA/664-512-A
Message Waiting Notification (MWN)	TIA/EIA/664-513-A
Mobile Access Hunting (MAH)	TIA/EIA/664-514-A
Password Call Acceptance (PCA)	TIA/EIA/664-515-A
Preferred Language (PL)	TIA/EIA/664-516-A
Priority Access and Channel Assignment (PACA)	TIA/EIA/664-517-A
Remote Feature Control (RFC)	TIA/EIA/664-518-A
Selective Call Acceptance (SCA)	TIA/EIA/664-519-A
Subscriber PIN Access (SPINA)	TIA/EIA/664-520-A
Subscriber PIN Intercept (SPINI)	TIA/EIA/664-521-A
Three-Way Calling (3WC)	TIA/EIA/664-522-A
Voice Message Retrieval (VMR)	TIA/EIA/664-523-A
Voice Privacy (VP)	TIA/EIA/664-524-A
Asynchronous Data Service (ADS)	TIA/EIA/664-525-0
Calling Name Presentation (CNAP)	TIA/EIA/664-526-0
Calling Name Restriction (CNAR)	TIA/EIA/664-527-0
Data Privacy (DP)	TIA/EIA/664-528-0
Emergency Services (9-1-1)	TIA/EIA/664-529-0

continued on next page

TABLE 7.1

*Listing of the Parts
of ANSI-664-A
(continued)*

Document Name	Part Number
Group 3 Facsimile Service (G3 Fax)	TIA/EIA/664-530-0
Network Directed System Selection (NDSS)	TIA/EIA/664-531-0
Non-Public Service Mode (NP)	TIA/EIA/664-532-0
Over-the-Air Service Provisioning (OTASP)	TIA/EIA/664-533-0
Service Negotiation (SN)	TIA/EIA/664-534-0
User Group (UG)	TIA/EIA/664-535-0
Short-Message Delivery	TIA/EIA/664-601-A
Wireless Messaging Teleservice	TIA/EIA/664-602-A
Wireless Paging Teleservice	TIA/EIA/664-603-A
Mobile Station Functionality	TIA/EIA/664-701-A
System Functionality	TIA/EIA/664-801-A
Subscriber Confidentiality	TIA/EIA/664-802-A
Network Services	TIA/EIA/664-803-A

Stage 2—Functional Model and Information Flows

The application of ITU-T stage 2 techniques to the ANSI-41 environment has been more limited. The ITU-T approach for ISDN, as introduced in I.130 and defined in detail in *ITU-T Recommendation Q.65*, involves a high level of abstraction and formalism, encompassing five steps. We illustrate these formal steps in a very informal manner; we use them to model a low-technology service—trash collection.

ITU-T step 1—Derive a functional model for the service in terms of abstract functional entities (FEs) and relationships between functional entities. In general, each service model would involve a new set of FEs and relationships. Our trash collection service involves four FEs: the generator of the trash, temporary storage for the trash, the collector of the trash, and the trash treatment center (see Figure 7.1).

ITU-T step 2—Define the abstract information flows between FEs required for proper service operation. Create information flow diagrams—often referred to as *ping-pong diagrams*—to illustrate these information flows for typical cases of operation of the service (see Figure 7.2).

ITU-T step 3—Create diagrams to completely describe the actions of each FE associated with each information flow. These graphic descriptions are called *specification and description language diagrams*. An SDL diagram for the temporary storage FE is provided in Figure 7.3.

ITU-T step 4—Identify all the actions required of each FE. For example, some actions associated with the collector functional entity in our example would be:

- Schedule periodic trash collection.
- Transfer trash from multiple temporary storage locations to the collector.
- Transfer collected trash to the treatment center.
- Notify the generator if a problem with the temporary storage is detected.

ITU-T step 5—Allocate the abstract FEs to physical equipment, with each set of allocations referred to as a *scenario*. Table 7.2 illustrates a couple of scenarios for our trash collection example.

Figure 7.1

In step 1 of stage 2, the functional entities and relationships are defined.

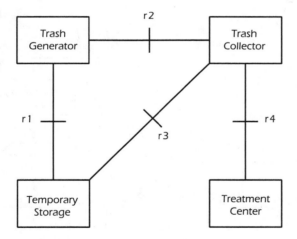

Relationships:
   r1 - Between trash generator and temporary storage.
   r2 - Between trash generator and trash collector.
   r3 - Between temporary storage and trash collector.
   r4 - Between trash collector and treatment center.

Figure 7.2
In step 2, information flow diagrams are defined, illustrating several cases of operation of the service.

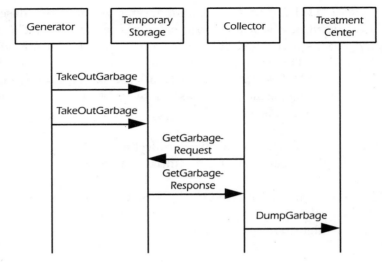

Figure 7.3
A portion of an SDL diagram for the temporary storage functional entity. Another part of the diagram would describe the actions taken when the storage is full and a message is received.

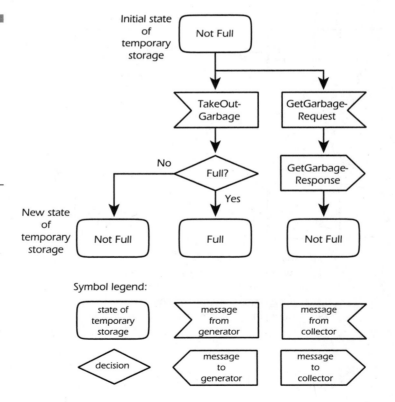

TABLE 7.2

Scenarios for the
Trash Collection
Service

Scenario	Generator	Temporary Storage	Collector	Treatment Center
1	Household	Trash Can	Sanitation Engineer	Dump
2	Business	Trash Bin	Sanitation Engineer	Incinerator

ANSI-41 takes a much more simplified approach. Although it is not formally described in any TIA document, we can consider the ANSI-41 stage 2 process to involve only two steps:

ANSI-41 step 1—Use the ANSI-41 network reference model (described in Chapter 4) as the functional basis for providing the service. Define new FEs only in cases where (a) the existing ANSI-41 FEs do not already encompass the required service-providing functions or (b) it is not desirable to add the required new functions to the existing ANSI-41 FEs. When a new FE is required, the ANSI-41 network reference model must be updated. The addition of authentication functions to ANSI-41 is an example of this step. In this case, a new FE, called the *authentication center* (AC), was added to the ANSI-41 network reference model rather than add the required authentication functions to the HLR FE's responsibilities; conversely, authentication functions were added to the existing VLR FE rather than create yet another new FE.

ANSI-41 step 2—Describe the required information flows between FEs in terms of existing or proposed ANSI-41 messages and parameters, as opposed to the Q.65 approach whereby an intermediate, abstract representation of the required information flows is defined and then mapped to actual messages and parameters in stage 3. Furthermore, ANSI-41 stage 2 information flow descriptions are in the form of ping-pong diagrams, with accompanying text to describe what is happening in each step (see Figure 7.4). The more formal SDL diagram technique is not employed.

As we can see, the ANSI-41 approach to stage 2 is similar to that of Q.65 in some aspects but different in others—not necessarily better or worse, but definitely different. Given its track record, we must assume that the approach is good enough for the job.

Stage 3—Protocols and Procedures

The I.130 goal of stage 3—the design of the messages, parameters, and procedures required for the service in question—is common to the ANSI-41 approach. In fact, I.130 does not go into any detail regarding message

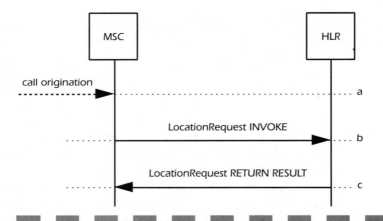

Figure 7.4   *An example of the ANSI-41 approach to stage 2. The ping-pong diagrams describe signaling scenarios in terms of actual ANSI-41 messages, like LocationRequest, rather than in abstract terms. A text description of each step accompanies the diagram. For example:*

a. A call origination is received by the originating MSC.

b. The originating MSC sends a LocationRequest INVOKE message to the HLR associated with the called mobile station.

c. The HLR determines the required call routing and returns this information to the originating MSC in the LocationRequest RETURN RESULT message.

formats and other presentation issues. Thus, ANSI-41 can be considered to be an application of the I.130 stage 3 process.

Influence of the Three Major Functional Areas on the ANSI-41 Structure

The ANSI-41 Revision D structure is virtually the same as its predecessor, IS-41 Revision C. The IS-41-C structure, in turn, is based on the organization of its predecessors, IS-41 Revisions 0, A, and B, but with some significant deviations. Therefore, it makes sense to discuss first the impact of functions on the structure of IS-41.

IS-41 encompasses three main categories of functions: intersystem handoff; automatic roaming; and intersystem operations, administration, and maintenance.

These are three distinct functional areas, and the distinctions are reflected in the structure of IS-41:

1. Intersystem handoff functions are defined in part 2 of IS-41 (IS-41.2).
2. Automatic roaming functions are defined in part 3 of IS-41 (IS-41.3).
3. Intersystem OA&M functions are defined in part 4 of IS-41 (IS-41.4).

However, all these functions share a common need: an intersystem signaling system over which IS-41 functional messages can flow. Therefore, it made sense to place the description of the IS-41 signaling systems in a separate part of the standard. This fifth part, called IS-41.5, also contains the stage 3 message and parameter definitions associated with the IS-41 functions. Again, this was a logical choice: all the implementation pieces are in a single place. However, the procedural aspects of stage 3—what to do with the messages and parameters—remained divided between IS-41.2, IS-41.3, and IS-41.4 until IS-41-C. A sixth part, IS-41.6-C, was then introduced, containing the detailed procedures for intersystem handoff and automatic roaming; OA&M was considered unique and warranted leaving its procedures in IS-41.4. Finally, IS-41.1 provides a very limited introduction to the standard. It also serves as a convenient place to present the network reference model (see Chapter 4), the framework on which the subsequent parts of the standard are based.

ANSI-41 Revision D reuses the six-part IS-41-C structure.[1] Table 7.3 provides a summary of the structure of ANSI-41 Revision D.

TABLE 7.3

The ANSI-41 Revision D Document Structure

Section of ANSI-41 Revision D	Notes
Part 1: ANSI-41.1-D Functional Overview	
Introduction	
References	
Definitions and Documentation Conventions	
Symbols and Abbreviations	
Network Reference Model	
Cellular Intersystem Services	
General Background and Assumptions	
Restrictions	

continued on next page

---

[1] We expect that the upcoming ANSI-41 Revision E will add at least one new part, ANSI-41.7-E. It should contain the definition of the distributed functional plane for the Wireless Intelligent Network (see Chapter 19), first introduced in IS-771.

TABLE 7.3

The ANSI-41 Revision D Document Structure (continued)

Section of ANSI-41 Revision D	Notes
Part 2: ANSI-41.2-D Handoff Information Flows	
Introduction	
References	
Terminology	
Intersystem Handoff Operations	This section defines the information flows for individual ANSI-41 intersystem handoff operations.
Basic Intersystem Handoff Scenarios	This section defines the information flows for common ANSI-41 handoff scenarios involving multiple ANSI-41 intersystem handoff operations.
Annex A	ANSI-41.2-D also includes an Annex that describes the handoff functionality that is in support of IS-124. See Appendix A for a description of IS-124.
Part 3: ANSI-41.3-D Automatic Roaming Information Flows	
Introduction	
References	
Terminology	
Automatic Roaming Operations	This section defines the information flows for individual ANSI-41 automatic roaming operations.
Basic Automatic Roaming Scenarios	This section defines the information flows for common ANSI-41 automatic roaming scenarios involving multiple ANSI-41 automatic roaming operations.
Voice Feature Scenarios	This section defines the information flows for common ANSI-41 automatic roaming scenarios involving voice features (e.g., call forwarding).
Short-Message Service Scenarios	This section defines the information flows for common ANSI-41 automatic roaming scenarios involving short message services.
Annex A	ANSI-41.3-D also includes an Annex that describes the assumptions associated with authentication, voice privacy, and signaling message encryption.

continued on next page

TABLE 7.3

The ANSI-41
Revision D
Document
Structure
(continued)

Section of ANSI-41 Revision D	Notes
Part 4: ANSI-41.4-D OA&M Information Flows and Procedures	
Introduction	
References	
Terminology	
Intersystem OA&M Operations	
OA&M Message Procedures for Handoff	
Inter-MSC Trunk Testing	
Issues for Further Study	
Part 5: ANSI-41.5-D Signaling Messages and Parameters	
Introduction	
References	
Terminology	
MAP Protocol Architecture	
Data Transfer Services	Both X.25-based and SS7-based data transfer (or carriage) services are defined in this section.
Application Services	
MAP Compatibility Guidelines and Rules	
Part 6: ANSI-41.6 Signaling Procedures	ANSI-41.6 is a new part, first added in IS-41 Revision C, which includes the stage 3 procedures related to inter-system handoff and automatic roaming. Most of the procedures for OA&M remain in ANSI-41.4-C; the exception is the OA&M procedures for automatic roaming, which are included in ANSI-41.6.
Introduction	
Technology and Concepts	
Basic Call Processing	
Intersystem Procedures	
Voice Feature Procedures	
Common Voice Feature Procedures	
Operation Timer Values	
Annexes	

CHAPTER 8

The ANSI-41 Protocol Architecture

For the typical mobile telecommunications network equipment vendor, ANSI-41 represents a collection of protocols for data communication—protocols the vendor must implement to support the services associated with ANSI-41. In this chapter we identify the individual protocols included in ANSI-41, the functions of each, and how they collectively fit together. This structure is called the *ANSI-41 protocol architecture*.

Data Transfer versus Information Transfer

The fundamental partition within the ANSI-41 protocol architecture lies between *application services* and *data transfer services*.

Application services make possible the transfer of information between application processes in the mobile telecommunications network. The nature of this application information—its semantics and syntax, together with its movement between application processes—defines the network's functions. For example, the basic call delivery function—allowing a subscriber to receive calls when outside of the serving area of her or his home system—would not be possible if the subscriber's identity and current location (i.e., application information) were not transferred from the visited system to the subscriber's home system.

On the other hand, from the perspective of the protocol entities providing the data transfer services, application information is nothing more than a collection of octets that must be moved from one node in the network to another. The octets could just as well correspond to a grocery list as a subscriber's location information; the "blue collar" job of data transfer would still have to be done.

Figure 8.1 illustrates this high-level view of the ANSI-41 protocol architecture as it applies to two ANSI-41 network nodes (in the figure, an MSC and an HLR) communicating via a signaling network.

Relationship with the OSI Reference Model

Since this is a book about data communications protocols, we would be remiss if we did not try to relate the ANSI-41 protocol structure to that

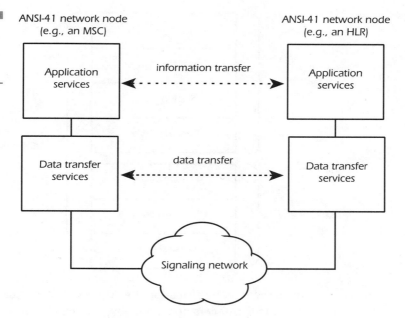

Figure 8.1
A high-level view of the ANSI-41 protocol architecture.

of the Open Systems Interconnection seven-layer reference model for data communications (OSI model), defined in the ITU X.200 series of recommendations. We do this at various points throughout the discussion in this chapter; at this stage it is enough to explain how the ANSI-41 division into application services and data transfer services fits into the OSI scheme of things.

According to ANSI-41, application services encompass the application, presentation, session, and transport layers, while data transfer services cover the network, data link, and physical layers of the OSI model (see Figure 8.2).

At this stage, OSI experts may be thinking, "They are mistaken here. The transport layer should be categorized under data transfer services—that is what the transport layer provides: end-to-end data transfer services." This is a valid point, requiring some response. The ANSI-41 protocol model is based on the Signaling System No. 7 (SS7) protocol structure, rather than the OSI model. In particular, the ANSI-41 model relies on the functions of SS7 that support transaction-oriented services. These functions are called SS7 *transaction capabilities* (TCs) and are used, for example, to provide the query-and-response class of procedure associated with retrieving information from a remote database (see Figure 8.3). The rationale for using the SS7 model was described in IS-41 Revision 0 (IS-41-0):

Figure 8.2
Relationship of the ANSI-41 protocol architecture to the layering of the OSI Reference Model.

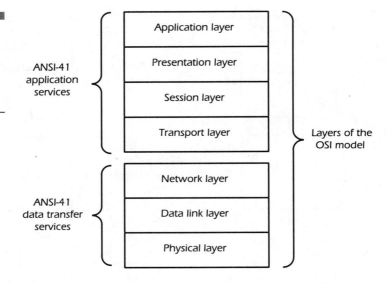

Transaction Capabilities for the Intersystem protocol are functions that control information transfer between two or more signaling nodes via a signaling network. Since the cellular radio intersystem protocol is transaction oriented, the decision has been made to adopt TC procedures similar to those defined for CCITT Signaling System No. 7 (SS7).[1]

Figure 8.3
A typical query-and-response style transaction.

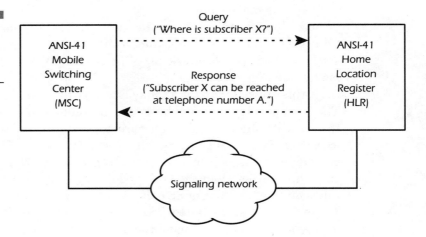

---

[1] IS-41 Revision 0 referenced the CCITT version of SS7 while subsequent versions of IS-41 and ANSI-41 are based on the ANSI version of SS7. However, both versions of SS7 specify the same basic protocol model.

SS7 transaction capabilities are defined in the 1988 version of ITU Recommendation Q.771—the version that was current when IS-41-0 was being developed—to encompass application-layer protocols and services, called Transaction Capabilities Application Part (TCAP), plus the supporting presentation, session, and transport layers, collectively called the Intermediate Service Part (ISP).[2] ANSI-41 application services (see Figure 8.4) consist of SS7 TC plus another set of application layer protocols and services called the ANSI-41 mobile application part, or MAP.[3] Thus, due to its inclusion in SS7 TC, the transport layer is considered to be part of ANSI-41 application services.

Figure 8.4
ANSI-41 application services comprise the Mobile Application Part (MAP) plus SS7 Transaction Capabilities (TC).

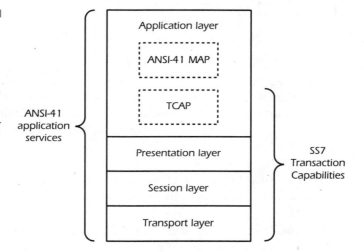

Of course, the real value to ANSI-41 of using the SS7 protocol structure is not that it places the transport layer in one category versus another; rather, it is that SS7 provides an appropriate platform on which to model (and build) the functions required for mobile network signaling. In fact, in all the revisions of ANSI-41, the transport layer is specified as the *null layer*—no transport layer protocol has ever been defined for ANSI-41.

---

[2] Q.771 refers to the Intermediate Service Part (ISP); however, Application Service Part (ASP) is the term used in the analogous ANSI standard, T1.114 (1988), on which IS-41 (post Revision 0) and ANSI-41 are based.

[3] See *ANSI-41 Application Services* for a more detailed description of ANSI-41 application services.

ANSI-41 Data Transfer Services

Data transfer services, referred to as *carriage services* prior to IS-41-C, consist of the network, data link, and physical layer services defined by the OSI model (see Figure 8.5).

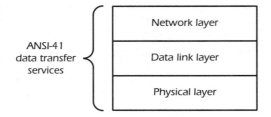

Figure 8.5
ANSI-41 data transfer service layers.

Since ANSI-41 application services are supported by SS7 protocols, the choice of data transfer protocols would seem obvious—use SS7! Of course, system design decisions are rarely that straightforward. Other factors had to be considered by the developers of the original IS-41 standard, IS-41-0:

- The conflicting goals of maximum flexibility (i.e., for feature and capacity growth) and minimum complexity[4] in the implementation of IS-41.
- The desire to deploy IS-41 in the field as quickly as possible.
- The practical need to reflect existing implementation practices in the standard.

The result of the deliberations on these issues was a "multifooted" protocol structure, whereby application services might be supported by a number of data transfer schemes (see Figure 8.6).

Two alternative protocol sets were recommended in IS-41-0 and remain part of the standard to this day:

1. *An X.25-based protocol set*. This protocol set provides a low-complexity, readily available solution. It was particularly well suited for the

---

[4] TR45.2, the subcommittee with responsibility for ANSI-41, is a technical standards group; discussion of financial cost issues associated with ANSI-41 implementation is considered taboo. In reality, of course, complexity usually implies cost, so the word *complexity* has come to be used as a suitable euphemism for cost.

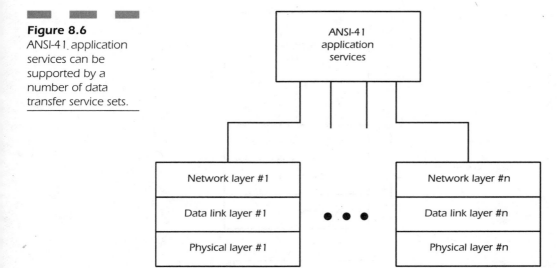

initial application of IS-41 to direct dedicated facilities between pairs of communicating systems (see Figure 8.7).

2. *An SS7-based protocol set.* This protocol set anticipates the growth in the number of subscriber calling features and processing capacity requirements. It provides an ideal protocol platform for the task of connecting multiple systems via a backbone signaling network (see Figure 8.8).

We provide an overview of these protocol sets in the following two subsections on pages 139 and 141.[5]

X.25-based Data Transfer Services

Although SS7-based services were defined in IS-41-0 and IS-41-A, both standards specified X.25-based data transfer services as "the minimum acceptable configuration" for IS-41. As a result, X.25 was the primary method of data transfer between nodes in the early stages of IS-41

---

[5] As of the end of 2000, the TR45.2 subcommittee was working on an Internet Protocol (IP) based data transfer service for ANSI-41. This capability should appear in ANSI-41 Revision F, if not earlier.

Figure 8.7
X.25 provides one set
of data transfer
services to ANSI-41
application services.

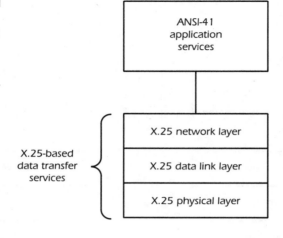

Figure 8.8
SS7 provides another
set of data transfer
services for use by
ANSI-41 application
services, where SCCP
is the SS7 Signaling
Connection Control
Part and MTP is the
SS7 Message Transfer
Part.

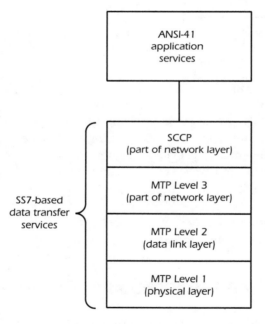

deployment. X.25 finds significant use even today, although SS7-based data transfer is the clear direction for ANSI-41's future.

X.25 specifies protocols for each of the three ANSI-41 data transfer service layers—the network, data link, and physical layers. However, X.25 was designed as a data terminal equipment to data circuit-terminating equipment (DTE–DCE) interface. This is *access signaling*, as discussed in Chapter 3; i.e., the interface is asymmetric, with one side (the

DCE) acting as the access point into a packet-switched network, through which the DTE can transfer data to another DTE.

However, for ANSI-41's purposes, X.25 is used over dedicated signaling facilities between pairs of ANSI-41 nodes. In X.25 terminology, each ANSI-41 node is considered a DTE; therefore, the ANSI-41 interface is DTE–DTE. Figure 8.9 illustrates the typical DTE–DCE use of X.25 versus ANSI-41 for DTE–DTE communications.

Figure 8.9
X.25 is generally considered an access protocol to a packet-switched network, operating between DTE and DCE. However, ANSI-41 requires a DTE–DTE interface.

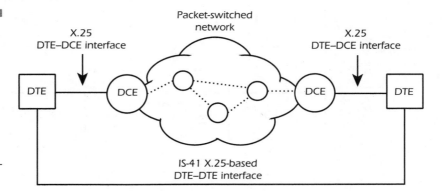

To address this requirement, ANSI-41 references an X.25-compatible version of the network layer protocol, specified in ISO 8208. ISO 8208 includes the requirements that X.25 places on DTEs, but also specifies how DTEs should operate in the DTE–DTE configuration.

Similarly, the ANSI-41 X.25-based data-link layer is aligned with ISO 7776, which is the DTE counterpart of the DCE procedures specified in X.25.

Finally, the ANSI-41 X.25-based physical layer is as specified in X.25; that is, the protocol is in accordance with X.21-bis, which defines the physical layer interface between a DTE and an ITU V-series modem.

Figure 8.10 illustrates the resulting protocol structure for an implementation supporting X.25-based data transfer services.

SS7-based Data Transfer Services

In ANSI-41, both SS7 and X.25 are identified as "technologically capable of supporting the requirements" of ANSI-41 data transfer; however, X.25 is no longer referred to as the minimum acceptable configuration. In fact, SS7 has become the ANSI-41 data transfer service of choice, particularly for large carriers.

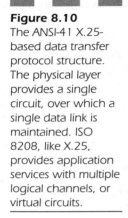

Figure 8.10
The ANSI-41 X.25-based data transfer protocol structure. The physical layer provides a single circuit, over which a single data link is maintained. ISO 8208, like X.25, provides application services with multiple logical channels, or virtual circuits.

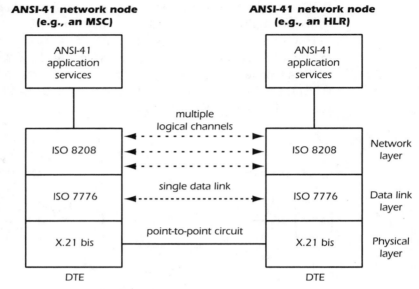

ANSI-41 SS7-based data transfer services comprise the SS7 Message Transfer Part (MTP) and Signaling Connection Control Part (SCCP), specified in ANSI T1.111 and ANSI T1.112, respectively.[6] According to the SS7 protocol structure, the MTP provides physical, data link, and a portion of the network layer service; the SCCP provides the balance of the network layer service (see Figure 8.11).[7]

For international applications, ANSI-41 also defines ITU-T SS7-based data transfer services consisting of the MTP and SCCP specified in ITU-T recommendations Q.701-710 (MTP) and Q.711-714 (SCCP), respectively. International SS7 data transfer services support was first specified in ANSI-41-D and was then expanded on in TIA/EIA/IS-807.

The SS7 MTP specifications in ANSI-41 are simply references to ANSI T1.111 and ITU-T recommendations Q.701-710—no restrictions or limitations on the use of the MTP standards is indicated.

However, the ANSI and ITU-T SCCP specifications in ANSI-41 reference *subsets* of the respective SCCP standards. Each ANSI-41 revision

---

[6] IS-41 Revisions A, B, and C and ANSI-41 Revision D reference the Issue 1 (1988) versions of the ANSI MTP and SCCP standards. IS-807 and IS-812 (see Appendix A) reference the 1996 version of T1.112 (SCCP).

[7] Although we do not go into detail regarding the functions of the MTP and SCCP layers in this book, Chapter 18 discusses some ANSI-41 implementation issues. For a more thorough treatment, see *Signaling System #7*, by Travis Russell.

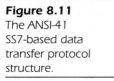

Figure 8.11
The ANSI-41
SS7-based data
transfer protocol
structure.

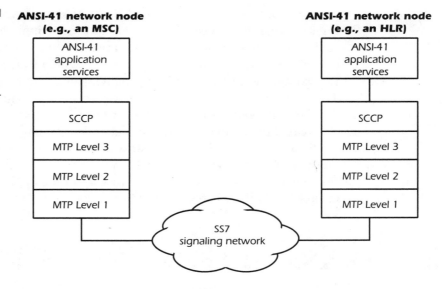

identifies the particular limitations on the use of SCCP that apply. However, one characteristic is common across all revisions: only SCCP connectionless service is used. In other words, each SCCP message is a "datagram" containing source and destination address information. The connection-oriented services defined in both the ANSI and ITU-T SCCP standards—involving connection establishment and connection termination phases—are not used in the ANSI-41 protocol architecture.

SS7 Segmentation and Reassembly

SS7 message segmentation is required when ANSI-41 messages exceed the maximum limit of the SS7 user-data parameter. The limit is variable since it depends on the information present in the SCCP called and calling party addresses and on the presence (or absence) of SCCP optional parameters. In practice, it is generally between 200 and 220 octets. Reassembly allows the segmented message to be recreated at the receiver. Segmentation and reassembly (S&R) has become an issue, primarily due to the increasing number of ANSI-41 subscriber profile parameters and the need to support the transfer of short-message service (SMS) messages via ANSI-41 over SS7.

The recent versions of ANSI (1996) and ITU-T (1993) SCCP provide support for S&R via the *XUDT message*. However, as previously pointed out, ANSI-41-D did not reference these standards and therefore made no provision for S&R. The *TIA/EIA/IS-812* standard addresses this limitation. It specifies the changes to ANSI-41-D that are required to support the lower layer S&R of ANSI-41 MAP messages. Besides changing the SCCP references, IS-812 defines a new single-bit field within the ANSI-41 *TransactionCapability* parameter, which is conveyed in a number of ANSI-41 messages. By setting this bit in an outgoing message, the sending entity (e.g., a VLR) signals that it is capable of S&R. The TransactionCapability parameter is present in enough (though not all) ANSI-41 messages to make this an effective means of communicating S&R capabilities among ANSI-41 network entities.

Of course, if the SCCP segments of the ANSI-41 messages are being analyzed by STPs within the SS7 network (e.g., for global title translation–based routing, as described in Chapter 18), these STPs must be capable of S&R. This is assumed in IS-812.

ANSI-41 Application Services

ANSI-41 specifies a single set of application services—and supporting protocols—comprising the SS7 transaction capabilities (TCs) defined in ANSI T1.114, along with the ANSI-41–specific application services and protocols, called the mobile application part (MAP), as shown in Figure 8.12. Note that, while both ANSI and ITU SS7 data transfer services are supported in ANSI-41, the application services make use of ANSI TCAP—the use of ITU TCAP for application services is not defined in ANSI-41.

Figure 8.12
The ANSI-41 application services include the MAP and SS7 Transaction Capabilities (TC). TC, in turn, comprises TCAP along with the supporting presentation, session, and transport layers.

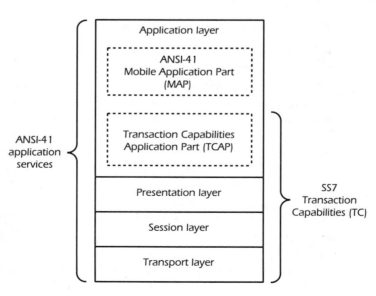

The TC transport, session, and presentation layers are null layers in ANSI-41, just as they are described in ANSI T1.114. They are included in the ANSI-41 protocol model primarily to align with the ANSI definition of transaction capabilities. In fact, a typical ANSI-41 implementation would have the TCAP entity directly use the ANSI-41 data transfer services (see Figure 8.13). Thus, it is appropriate to concentrate our discussion of ANSI-41 application services on the MAP and TCAP entities and their constituent elements (see Figure 8.14).

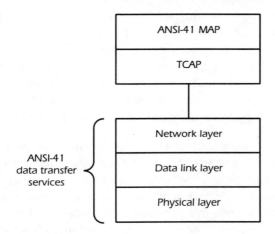

Figure 8.13
In a typical ANSI-41 implementation, TCAP would access data transfer services directly, rather than going through presentation, session, and transport layer interfaces.

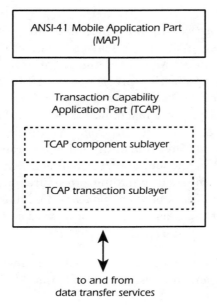

Figure 8.14
The structure of the ANSI-41 MAP and TCAP entities.

ANSI-41 Use of ANSI TCAP

As illustrated in Figure 8.14, the ANSI-41 mobile application part is supported by the ANSI Transaction Capabilities Application Part. TCAP, in turn, is composed of two sublayers:

- The *component sublayer* provides the communications tools MAP uses to realize ANSI-41 operations.
- The *transaction sublayer* provides the communications tools MAP uses to associate multiple operations as parts of a single, logical transaction between two functional entities.

The concept of an *operation* is key to ANSI-41, as evidenced by its formal title, "Cellular Radio Telecommunications Intersystem Operations." We consider an operation as encompassing a collection of application-layer communications mechanisms (see Figure 8.15):

- The operation procedures that specify the rules governing the information content and exchange between operation users; a set of related operation procedures is sometimes referred to as a *protocol machine*.
- The operation service primitives that an application process uses to access the operation's communications capabilities
- The operation protocol messages exchanged between peer protocol machines.

An application process in one functional entity uses operations to access another peer entity's application processes. For example, an application process in an MSC uses the LocationRequest operation to access a particular application process in an HLR; this would involve sending a LocationRequest Invoke message from the MSC protocol machine to the HLR protocol machine. Likewise, if the HLR application process is so defined, the HLR may send the results of the application process execution to the invoking entity using its protocol machine.

The ANSI TCAP component sublayer defines each of these operation elements: TCAP's component-handling procedures map the component-handling *service primitives* onto *components* (i.e., messages). TCAP provides six component types as the basis for operation definition, only four of which are used in ANSI-41:

1. *Invoke (Last)*. This component is used to request initiation of the remote application process. Its format and ANSI-41 encoding are shown in Table 8.1.

Figure 8.15
The elements of an operation and the relationship between operations and application processes.

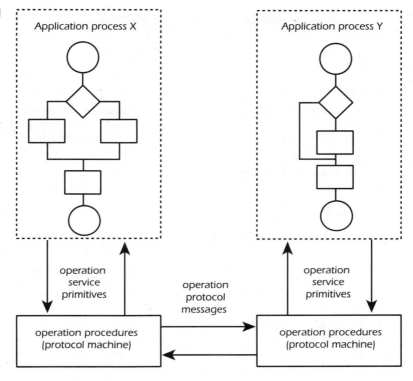

2. *Return Result (Last)*. This component is used to return the results of executing the application process. Its format is shown in Table 8.2.
3. *Return Error*. This component reports the unsuccessful completion of an invoked application process. Its format is shown in Table 8.3.
4. *Reject*. This component reports the receipt and rejection of an incorrect component or transaction. Its format is shown in Table 8.4.
5. *Invoke (Not Last)*. This component is not used in ANSI-41.
6. *Return Result (Not Last)*. This component is not used in ANSI-41.

TABLE 8.1

Format of the TCAP Invoke (Last) Component

Component Element Name	Description
Component Type Identifier	This field identifies the component as an Invoke.
Component Length	This is the length (in octets) of the Invoke component from the next element to the end of the component.
Component ID Identifier	This indicates that the Component ID follows.
Component ID Length	This is the length (in octets) of the Component ID that follows.

continued on next page

TABLE 8.1

Format of the TCAP
Invoke (Last)
Component
(continued)

Component Element Name	Description
Component ID	In the Invoke component, this is the Invoke ID—the identifier assigned by the TCAP user that allows correlation of invokes and responses).* The Invoke ID is one octet in length.
Operation Code Identifier	This indicates that the Operation Code follows. It is coded in ANSI-41 so that ANSI-41 Operation Codes can be defined outside the ANSI TCAP standard, T1.114; i.e., as private TCAP.
Operation Code Length	This is the length (in octets) of the Operation Code that follows.
Operation Code	The Operation Code is split into an Operation Family followed by an Operation Specifier; each is one octet in length. A different Operation Specifier is assigned to each ANSI-41 operation (e.g., LocationRequest).
Parameter Set Identifier	This indicates that a Parameter Set follows.
Parameter Set Length	This is the length (in octets) of the Parameter Set that follows.
Parameter Set	This is the set of parameters associated with the Invoke component of the ANSI-41 operation (e.g., MIN, ESN, etc.).

* In general, TCAP allows the Component ID to include an Invoke ID and a Linked ID; however, the Linked ID is not used in IS-41 or ANSI-41.

TABLE 8.2

Format of the TCAP
Return Result (Last)
Component

Component Element Name	Description
Component Type Identifier	This field identifies the component as a Return Result.
Component Length	This is the length (in octets) of the Return Result component from the next element to the end of the component.
Component ID Identifier	This indicates that the Component ID follows.
Component ID Length	This is the length (in octets) of the Component ID that follows.
Component ID	In the Return Result component, this is the Correlation ID (i.e., the received operation Invoke ID to which this Return Result component applies). The Correlation ID is one octet in length.

continued on next page

TABLE 8.2

Format of the TCAP
Return Result (Last)
Component
(continued)

Component Element Name	Description
Parameter Set Identifier	This indicates that a Parameter Set follows.
Parameter Set Length	This is the length (in octets) of the Parameter Set that follows.
Parameter Set	This is the set of parameters associated with the Return Result component of the ANSI-41 operation.

TABLE 8.3

Format of the TCAP
Return Error
Component

Component Element Name	Description
Component Type Identifier	This field identifies the component as a Return Error.
Component Length	This is the length (in octets) of the Return Error component from the next element to the end of the component.
Component ID Identifier	This indicates that the Component ID follows.
Component ID Length	This is the length (in octets) of the Component ID that follows.
Component ID	In the Return Error component, this is the Correlation ID (i.e., the received operation Invoke ID to which this Return Error component applies). The Correlation ID is one octet in length.
Error Code Identifier	This indicates that an Error Code follows. It is coded as in ANSI-41 so that ANSI-41 Error Codes can be defined outside of the ANSI TCAP standard, T1.114.
Error Code Length	This is the length (in octets) of the Error Code that follows.
Error Code	This indicates the reason why the associated operation did not complete successfully. A number of Error Codes are defined in ANSI-41.
Parameter Set Identifier	This indicates that a Parameter Set follows.
Parameter Set Length	This is the length (in octets) of the Parameter Set that follows.
Parameter Set	In ANSI-41, a single parameter may be contained in the Parameter Set; it is named FaultyParameter.

TABLE 8.4

Format of the TCAP
Reject Component

Component Element Name	Description
Component Type Identifier	This field identifies the component as a Reject.
Component Length	This is the length (in octets) of the Reject component from the next element to the end of the component.
Component ID Identifier	This indicates that the Component ID follows.
Component ID Length	This is the length (in octets) of the Component ID that follows.
Component ID	In the Reject component, this is the Correlation ID (i.e., the received operation Invoke ID or Correlation ID to which this Reject component applies). The Correlation ID is one octet in length.
Problem Code Identifier	This indicates that a Problem Code follows. The Problem Code encoding specified in the ANSI TCAP standard, T1.114, is used in ANSI-41.
Problem Code Length	This is the length (in octets) of the Problem Code that follows.
Problem Code	This indicates the reason why the Component or Transaction portion was rejected. The Problem Code is split into a Problem Type followed by a Problem Specifier; each is one octet in length.
Parameter Set Identifier	This indicates that a Parameter Set follows.
Parameter Set Length	This is the length (in octets) of the Parameter Set that follows.
Parameter Set	In ANSI-41, a single parameter may be contained in the Parameter Set; it is named FaultyParameter.

Likewise, the ANSI TCAP transaction sublayer provides seven "package types" to define operation associations, only five of which are used in ANSI-41. All TCAP package types used in ANSI-41 share a common format,[8] shown in Table 8.5.

1. *Query with permission.* This package initiates a TCAP transaction and informs the destination node that it may end the transaction.
2. *Query without permission.* This package initiates a TCAP transaction and informs the destination node that it may not end the transaction.

---

[8] The Abort package type has a different format, but is not used in IS-41 or ANSI-41. For a description, see *Signaling System #7*, by Travis Russell.

3. *Conversation with permission.* This package continues a TCAP transaction and informs the destination node that it may end the transaction.

4. *Conversation without permission.* This package continues a TCAP transaction and informs the destination node that it may not end the transaction.

5. *Response.* This package ends the TCAP transaction.

6. *Unidirectional.* This package is not used in ANSI-41.

7. *Abort.* This package is not used in ANSI-41.

TABLE 8.5

Format of TCAP Packages

Package Element Name	Description
Package Type Identifier	This field identifies the package type; e.g., Query with Permission.
Total TCAP Message Length	This is the total length (in octets) of the TCAP message.
Transaction ID Identifier	This indicates that the Transaction IDs follow.
Transaction ID Length	This is the length (in octets) of the Transaction IDs that follow.
Transaction IDs	Transaction IDs (TIDs) are used to enable transaction association. One or two TIDs may be present in an ANSI-41 TCAP package; each is four octets in length with the Originating TID (if present) followed by the Responding TID (if present), although one must be present in ANSI-41. Table 8.6 shows the association between TIDs and package types for ANSI-41. Figure 8.16 shows how the TIDs are used in typical ANSI-41 transactions.
Component Sequence Identifier	This indicates that a Component Sequence follows.
Component Sequence Length	This is the length (in octets) of the Component Sequence that follows.
Component Sequence	This is a sequence of one or more TCAP components; however, ANSI-41 requires (i.e., based on the figures in ANSI-41 illustrating the TCAP formats) that there be a single component per transaction.

TABLE 8.6

Association between TCAP Packages and Transaction IDs.

Package Type Identifier	Originating TID	Responding TID
Query with Permission	Present	Not present
Query without Permission	Present	Not present
Response	Not present	Present
Conversation with Permission	Present	Present
Conversation without Permission	Present	Present

Figure 8.16

Examples of three TCAP transactions showing how the Originating Transaction IDs (OTIDs) and Responding Transaction IDs (RTIDs) are manipulated.

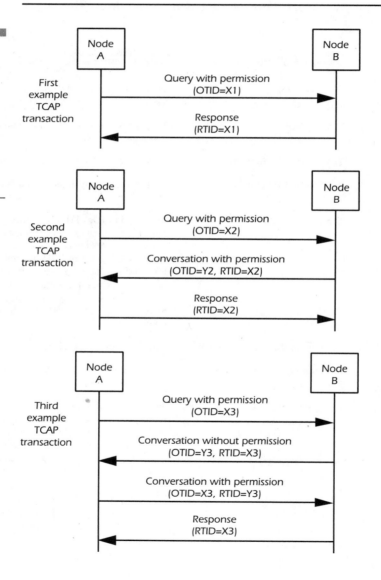

By defining standard component and package formats—and managing the procedural aspects of component and package exchange—TCAP provides consistent and versatile operation-based communication facilities to the TCAP user. In the ANSI-41 case, the TCAP user is the mobile application part.

ANSI-41 Mobile Application Part

The "mobile personality" of the ANSI-41 protocol architecture is embodied in the ANSI-41 mobile application part, or MAP. Other protocol architectures build on the SS7 transaction capabilities—along with suitable data transfer services—to support distributed telecommunications network applications (e.g., the 800-number database service). However, MAP is defined specifically for distributed mobile telecommunications network applications.

Up until now, we have shown MAP as purely an application-layer entity; this is the way it is illustrated in ANSI-41. However—once again assuming the OSI point of view—MAP can be considered to encompass two sets of functions (see Figure 8.17):

■ The communications functions—also called *application service element* (ASE) functions—within a mobile network functional entity (e.g., an HLR) that are considered to be part of the OSI application layer.
■ The data processing functions—also called *application processes*—within a mobile network functional entity that are considered to reside above the OSI application layer and that make use of the ANSI-41 ASE functions.

The MAP ASE functions are the MAP-specific operations defined using TCAP, operations such as LocationRequest, RoutingRequest, and QualificationDirective (which we discuss in later chapters). The components of these operations, e.g., LocationRequest Invoke and LocationRequest Return Result, are often referred to as ANSI-41 messages; we use this convention in this book. Together with the transaction functions offered by TCAP, the MAP ASE functions represent a broad set of communication capabilities to application processes.

The MAP application processes effectively extend the scope of the ANSI-41 standard beyond the OSI application layer, which deals exclusively with communications functions, and into the domain of application process definition. As a simple example, ANSI-41 procedures specify how

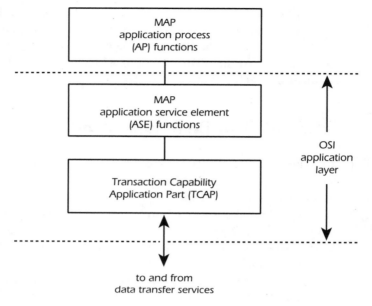

Figure 8.17
MAP functions
include application
process functions
that lie above the OSI
application layer and
application service
element functions
that lie within the OSI
application layer.

an HLR shall cancel a mobile station's service qualification in one system when the HLR is notified that the MS has registered in another system. In this and many other situations, MAP steps beyond the bounds of OSI communication. Of course, this is an absolute necessity in ANSI-41; a judicious amount of application process specification in the standard avoids unnecessary incompatibilities between implementations. In a mobile environment, consistency is particularly important. A subscriber's serving system may change from vendor A's equipment to vendor B's equipment, even during a call; however, from the point of view of the services provided to the subscriber, this change should not be detectable—the transition should be *seamless*. One of the ongoing challenges for the developers of ANSI-41 is to specify the MAP application processes to the extent that seamless service can be achieved while not constraining innovation in the individual vendor implementations.

An ANSI-41 MAP Application Service Interface

An ANSI-41 implementation designed on OSI layering concepts would likely include a clearly defined interface between the MAP application processes and application service element (ASE) functions. This interface separates the data processing functions of ANSI-41 (i.e., the processes) from the purely communications-related functions (i.e., the ASE functions), thereby off-loading communication details while providing a consistent set of communication services to a variety of application processes. However, this interface is not defined in ANSI-41 since it represents an internal interface found within an ANSI-41 implementation—clearly not an intersystem issue. Figure 8.18 illustrates a basic model for such an interface; we refer to it as the ANSI-41 MAP *application service interface* (ASI).

Figure 8.18
Model of an ANSI-41 application service interface.

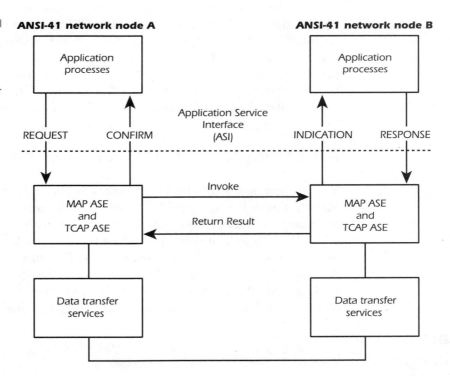

As shown in Figure 8.18, the MAP ASI consists of a set of OSI-style service primitives of the following types:

1. *REQUEST primitive.* This type of primitive allows a service user A to invoke an application process in a remote peer service user B.
2. *INDICATION primitive.* This type of primitive allows the MAP ASE to indicate to service user B that a peer service user has requested an application process invocation, thereby initiating the process in B.

3. *RESPONSE primitive.* This type of primitive allows service user B to send the results of the application process execution to service user A.
4. *CONFIRM primitive.* This type of primitive allows the MAP ASE to notify service user A of the response provided by service user B.

Each MAP ASI primitive has an associated ANSI-41 ASE function; for example, issuing an ASI primitive called MAP-LocationRequest-REQUEST results in the transmission of an *application protocol data unit* (APDU)—consisting of the ANSI-41 operation component (i.e., LocationRequest Invoke) contained in a TCAP package (i.e., query with permission)—between peer MAP ASEs. Most ANSI-41 ASE functions are confirmed and, therefore, are accessed by using all four of the ASI primitive types. The MobileOnChannel operation is the only exception in ANSI-41-D. It uses only the REQUEST and INDICATION primitives (the MobileOnChannel operation is described in Chapter 9).

An Example of MAP Operation Encoding

To illustrate how MAP and TCAP elements work together to convey application information between MAP application processes, Table 8.7 provides an example of an ANSI-41 MAP operation encoding as it would appear in a TCAP package. The example is of the Invoke component of the RegistrationCancellation operation (described in Chapter 10). As specified in ANSI-41, this component would appear in a TCAP Query with Permission package. In this example, the Invoke component has two parameters: MobileIdentificationNumber (MIN) and ElectronicSerialNumber (ESN).

TABLE 8.7

Example of MAP Registration-Cancellation Invoke Component Encoding

Package Element Name	Description	ANSI-41 Encoding (hexadecimal)
Package Type Identifier	The package type is Query with Permission.	E2
Total TCAP Message Length	This length is 32 octets in this example.	20
Transaction ID Identifier	This indicates that the Transaction ID follows.	C7
Transaction ID Length	This length is 4 octets.	04

continued on next page

TABLE 8.7

Example of MAP
Registration-
Cancellation Invoke
Component
Encoding
(continued)

Package Element Name	Description	ANSI-41 Encoding (hexadecimal)
Transaction ID	The Originating Transaction ID chosen for this example.	00 00 00 51
Component Sequence Identifier	This indicates that a Component Sequence follows.	E8
Component Sequence Length	This length is 24 octets in this example.	18
Component Type Identifier	This field identifies the component as an Invoke.	E9
Component Length	This length is 22 octets in this example.	16
Component ID Identifier	This indicates that the Component ID follows.	CF
Component ID Length	The Invoke ID length is 1 octet in this example.	01
Component ID	This is the Invoke ID chosen for this example.	03
Operation Code Identifier	This indicates that the Operation Code follows.	D1
Operation Code Length	The Operation Code is 2 octets in this example.	02
Operation Code	The Operation Code is split into the Operation Family (09) followed by the Registration Cancellation Operation Specifier (0E).	09 0E
Parameter Set Identifier	This indicates that a Parameter Set follows.	F2
Parameter Set Length	This length is 13 octets in this example.	0D
Parameter Identifier	This is the identifier for the first parameter, MIN.	88
Parameter Length	Length (in octets) of the MIN.	05
Parameter Value	The 10-digit MIN (408-296-0303) encoded per ANSI-41.	04 28 69 30 30
Parameter Identifier	This is the identifier for the second parameter, ESN.	89

continued on next page

TABLE 8.7

Example of MAP Registration-Cancellation Invoke Component Encoding (continued)

Package Element Name	Description	ANSI-41 Encoding (hexadecimal)
Parameter Length	Length (in octets) of the ESN.	04
Parameter Value	The actual 32-bit ESN (all zeros in this example).	00 00 00 00

CHAPTER 9

Basic Intersystem Handoff Functions

In this chapter, we discuss the basic ANSI-41 mobile telecommunications network functions related to intersystem handoff. These functions are divided into the following categories:[1]

- Handoff measurement
- Handoff forward
- Handoff back
- Path minimization
- Call release

We also examine the ANSI-41 application processes that support these basic network functions. While there are other handoff-related processes in ANSI-41—for authentication, voice feature, and short-message services support—we treat these separately in later chapters. In this chapter we focus on the basic functions required to hand off a call from one system to another.

ANSI-41 does not include complete specifications for each of the basic intersystem handoff processes; for example, ANSI-41 does not define the circumstances under which an MSC should attempt path minimization, leaving this as an implementation issue. This is a deliberate omission and an example of the standards approach taken in ANSI-41: "It is intended that the procedures defined address only the required intersystem transactions without infringing on the right of individual system operators and manufacturers to design their internal methods and procedures as they may deem best."[2]

In the course of describing the processes, we identify the ANSI-41 mobile application part (MAP) operations (e.g., FacilitiesDirective) used to accomplish intersystem handoff process tasks. We summarize this information at the end of the chapter. Note that we employ the ANSI-41 convention for operation component acronyms; the Invoke component acronym is in all capital letters (e.g., FACDIR), while the Return Result component acronym is in all lowercase letters (e.g., facdir). Refer to the ANSI-41 standard for the descriptions of the individual ANSI-41 operations and parameters. Also, the reader should consult the Glossary for the description of general terms (e.g., *serving system*) not explicitly defined in this chapter.

---

[1] The list of basic intersystem handoff functions reflects the authors' subjective categorization of certain ANSI-41 functions.

[2] From ANSI-41.1-D, "General Background and Assumptions."

What Is Intersystem Handoff?

Handoff encompasses a set of mobile station (MS) functions and network functions that enable an MS to move from one radio channel to another radio channel while a call is in progress. There are two categories of handoff (see Figure 9.1): intrasystem handoff and intersystem handoff.

Figure 9.1
Intrasystem handoff occurs between two cells controlled by the same MSC. Intersystem handoff occurs between two cells controlled by two different MSCs.

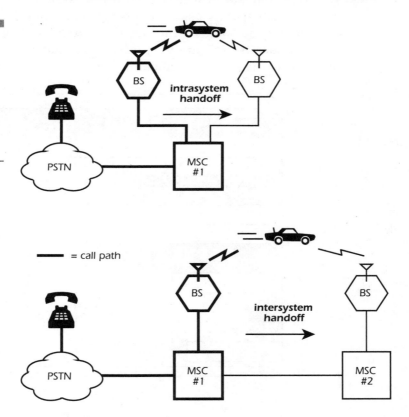

Intrasystem handoff is a handoff between two radio channels controlled by the same mobile switching center (MSC). No coordination is required between MSCs to support intrasystem handoff; therefore, intrasystem handoff is not within the scope of ANSI-41. *Intersystem handoff* is a handoff between two radio channels controlled by two different MSCs. This type of handoff requires specialized signaling between the two MSCs to coordinate the movement of the MS between the two radio channels. The ANSI-41 protocol provides this signaling.

Where Are Intersystem Handoff Functions Specified in ANSI-41?

The initial version of IS-41, IS-41-0, was primarily an intersystem hand-off standard: of the twelve intersystem operations defined in this early standard, ten were concerned with intersystem handoff and mainte-nance of the trunks used for intersystem handoff. While automatic roaming now accounts for the majority of ANSI-41 Revision D (ANSI-41-D) operations (see Figure 9.2), intersystem handoff remains an impor-tant part of the ANSI-41 standard.

Figure 9.2
Initially, most IS-41 operations supported intersystem handoff. A more balanced mix of handoff and automatic roaming operations existed in IS-41-A and IS-41-B. However, automatic roaming operations account for well over half the operations in IS-41-C and ANSI-41-D.

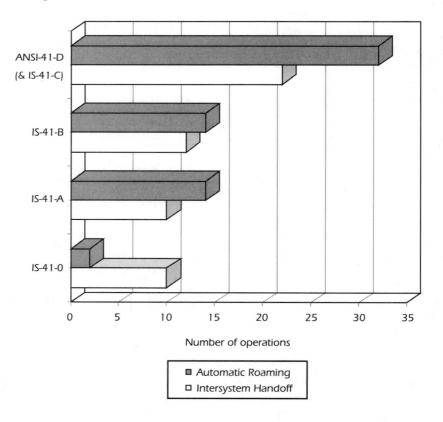

Number of operations

■ Automatic Roaming
□ Intersystem Handoff

Intersystem handoff functions are specified in three parts of ANSI-41:

1. **ANSI-41.5** provides the required formats—the bit-by-bit encoding— of all the ANSI-41 operation components, including those used for

intersystem handoff. ANSI-41.5 defines both the messages (e.g., FacilitiesDirective Invoke) and the message parameters (e.g., InterMSCCircuitID, TargetCellID).

2. **ANSI-41.6** provides algorithmic descriptions of the procedures associated with sending and receiving most ANSI-41 messages, including those used for intersystem handoff. As we pointed out in Chapter 8, ANSI-41 does not define a service interface between the MAP application processes and the MAP application service element (ASE); therefore, the procedures in ANSI-41.6 encompass both application process descriptions and ASE procedural descriptions. Furthermore, the procedures leave considerable room for implementation-dependent customization; there are numerous references in ANSI-41.6 to "local procedures" and "internal algorithms."

3. **ANSI-41.2** steps back from the protocol details contained in parts 5 and 6 and attempts to explain, by using information flow diagrams and step-by-step descriptions, how the operations are used individually and together to accomplish application process tasks. Although ANSI-41.2 is a good place to begin to tackle ANSI-41 intersystem handoff, ANSI-41.5 and ANSI-41.6 contain the definitive requirements against which an ANSI-41 implementation is evaluated.[3]

Issues Associated with Intersystem Handoff

The technical challenge of reliably executing a call handoff is significant. The intersystem issues include the following:

1. *Coordinating cell identification between neighboring MSCs.* To locate an MS for signal quality measurement and other handoff purposes, neighboring MSCs must agree on a cell identification scheme (see Figure 9.3).

2. *Coordinating inter-MSC facility identification between neighboring MSCs.* ANSI-41 handoff requires dedicated inter-MSC trunks that

---

[3] If inconsistencies are detected between parts, ANSI-41.5 and ANSI-41.6 take precedence over ANSI-41.2. Conflicting requirements between ANSI-41.5 and ANSI-41.6 are a more serious problem. In either case, the errors should be brought to the attention of the TIA TR45.2 subcommittee.

must be uniquely identifiable by the MSCs at both ends of the connection (see Figure 9.3).

3. *Supporting MS characteristics after handoff.* For example, the AMPS standard allows some variability in MS characteristics (e.g., the MS power class can be 0.6, 1.6, or 4 W). The MSCs involved in the handoff must ensure that these characteristics can be supported after the call handoff occurs.

4. *Limiting the length of the "handoff chain."* As a call is handed off from the first serving MSC (the anchor MSC) to another serving MSC and so on, a handoff chain develops—the sequence of MSCs, from the anchor to the current serving MSC actively involved in the call at a given time. The length of the handoff chain is controlled in two different ways: by counting the number of inter-MSC facilities involved in the call and limiting that number to a value set by the individual cellular service provider and by providing a handoff-back operation, which prevents "shoelace" connections from MSC-A to MSC-B and back to MSC-A.

Figure 9.3

MSC-A, MSC-B, and MSC-C must have a common understanding of cell identities (X, Y, and Z) to manage the location information that they exchange. Pairs of MSCs (e.g., MSC-A and MSC-B) also must agree on the identification of the inter-MSC trunks that connect them.

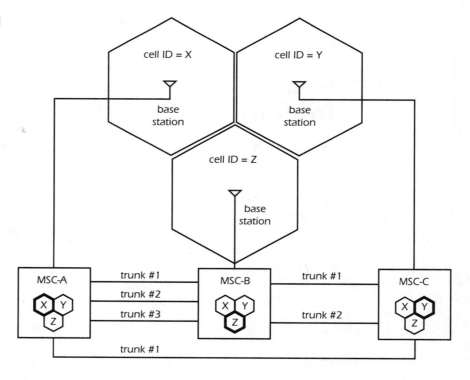

Early versions of ANSI-41 intersystem handoff (in IS-41-0 and IS-41-A) consisted of a straightforward collection of functions involving a small set of messages, parameters, and procedures required to hand off analog AMPS calls. A number of factors contributed to a rather drastic increase in the complexity of the intersystem handoff functions in ANSI-41-D compared to IS-41-A (see Figure 9.4):

- *New air interface standards*. The development of the narrowband AMPS (NAMPS) analog air-interface standard, and the time division multiple access (TDMA) and code division multiple access (CDMA) digital air-interface standards, has required protocol enhancements to manage handoffs between the different air interface types (e.g., AMPS to TDMA) in addition to handoffs between the same air interface types (e.g., CDMA to CDMA). TDMA support was first incorporated in IS-41-B; CDMA and NAMPS were added in IS-41-C.
- *Handoff path minimization techniques* (described in this chapter and initially included in IS-41-B).
- *The addition of techniques that support handoff forward prior to the call's being answered* (described in this chapter and new in IS-41-C).
- *Feature support after handoff (included in IS-41-B and extended in IS-41-C)*. MSC support after handoff of features like three-way calling and call waiting is described in later chapters.

The complexity has continued to grow beyond ANSI-41-D. For example, TIA/EIA/TSB-76 specifies the additions to IS-41-C (and ANSI-41-D) that are required to perform intersystem handoff between the cellular and PCS frequency bands, and between the various PCS bands (see Chapter 1 for a discussion of cellular and PCS frequency bands). At last count, as of the end of 2000, the FacilitiesDirective2 Invoke message was up to 49 parameters.

Handoff Measurement

Handoff measurement functions may be used during the prelude to the actual call handoff, a period that includes the following major decision-making stages:

1. Identify the need. Is a call handoff appropriate at this time?
2. Identify the candidates. Which cell(s), and associated MSC(s), should be considered for call handoff purposes?

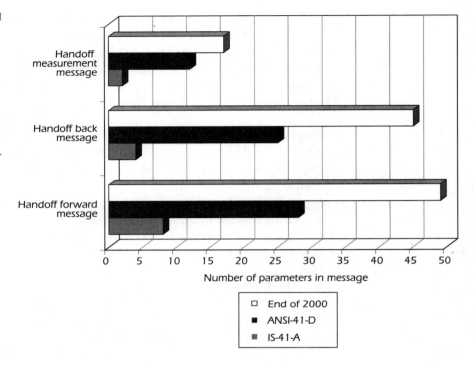

Figure 9.4
The number of
parameters in the
intersystem handoff
messages has
increased
dramatically between
IS-41-A, ANSI-41-D,
and today.

3. Evaluate the candidates. How suitable is each to handle the call?
4. Select a target. Which candidate is most suitable to handle the call?

Once these preliminaries are out of the way, the serving system can decide (1) whether to attempt call handoff and (2) what type of handoff, (forward or back) is most appropriate under the circumstances.

On the basis of received signal quality measurements, the MS, serving system, or both can determine that there is a need to perform a handoff to another channel or cell. These three strategies are known as:

- MS-controlled handoff
- Network-controlled handoff
- MS-assisted handoff (MAHO)

ANSI-41 supports network-controlled handoff and MAHO only. Analog AMPS and NAMPS MSs do not provide signal measurements to support a handoff determination; thus, they rely on network-controlled handoff techniques supported by ANSI-41. TDMA and CDMA MSs are able to provide measurements of received base station signal strength to the serving MSC without the need for the ANSI-41 handoff measurement processes.

The radio channel received-signal quality measurements taken by either the serving base station or both the serving base station and the MS (in the case of MAHO) reveal the quality of the transmission. Measurements may also be taken by adjacent base stations as the MS strays toward their coverage areas. ANSI-41's handoff measurement role is limited to requesting and conveying the received-signal quality measurements between systems to assist the serving system's evaluation of handoff candidates.

Handoff-measurement Processes

ANSI-41 handoff-measurement processes include the serving MSC handoff-measurement process and the candidate MSC handoff-measurement process, which executes in each candidate MSC.

The serving MSC handoff-measurement process may be used to evaluate a candidate MSC's suitability for the role of serving MSC, that is, to accept a call handoff. Since multiple candidates may be involved, the serving MSC is capable of requesting and collecting measurement data from one or several candidate MSCs (see Figure 9.5).

Figure 9.5

An example of the handoff measurement processes: (1) The serving system requests handoff measurements from both candidate MSC #1 and MSC #2. (2) Candidate MSC #1 returns measurement information. (3) Candidate MSC #2 measures an unacceptable signal quality level, so does not respond.

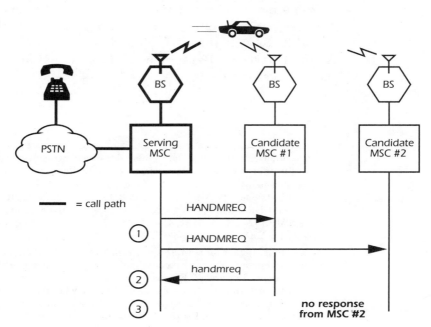

As shown in Figure 9.5, the serving MSC uses the HandoffMeasurementRequest Invoke (HANDMREQ) message[4] to request the measurement data from the candidate MSC.

The candidate MSC handoff-measurement process provides handoff suitability information to the serving MSC. This information may be either explicit or implicit. Explicit information includes measures of signal quality that the candidate MSC provides to the serving MSC. Implicit information is provided to the serving MSC simply by the candidate MSC's failure to respond to the serving MSC's request for measurement information (e.g., as candidate MSC 2 does in Figure 9.5). In the latter case, the candidate MSC may not respond because:

1. The candidate MSC cannot support the requested radio channel characteristics; for example, a TDMA digital channel may be required but is not available.
2. The signal quality that the candidate MSC measures does not meet its internal criteria for acceptability.
3. The current traffic conditions[5] on the candidate MSC render it unavailable for handoff purposes at the given time; therefore, it chooses to disregard the serving MSC's request (although this information could be explicitly provided to the serving MSC via an error response).

While not sending a handoff-measurement response to the serving MSC may be more efficient for the candidate MSC, it requires the serving MSC to wait the maximum time—by default, seven seconds in ANSI-41 Revision D—before it can be sure that all candidate responses have been received.

---

[4] Unless otherwise indicated, the HandoffMeasurementRequest2 operation may be substituted for references to the HandoffMeasurementRequest operation; e.g., either HandoffMeasurementRequest Invoke or HandoffMeasurementRequest2 Invoke messages may be used by the serving MSC. The HandoffMeasurementRequest2 operation was added to IS-41 in Revision C to support all air-interface standards, including CDMA and narrowband AMPS (NAMPS) MSs; the HandoffMeasurementRequest operation supports TDMA and AMPS MSs, but not CDMA or NAMPS.

[5] For example, radio channel resource availability influences the candidate MSC's ability to support handoff. There are many implementation-dependent strategies for allocating radio channels for handoff. For instance, some systems reserve a number of channels to support handoff from other systems, regardless of whether channels are available to new calls within that system. Other systems allocate channels for handoff only if they are available; if they are not, the call is dropped.

As shown in Figure 9.5, the candidate MSC's measurement information is conveyed to the serving MSC in the HandoffMeasurementRequest Return Result (handmreq) message.

Once the serving MSC has selected the target MSC for the call handoff, it must decide on the form of handoff that is appropriate:

- If the target MSC is already involved in the call and connected to the serving MSC via an inter-MSC circuit, a handoff back is required.
- Otherwise, the serving MSC may attempt path minimization or choose simply to perform a handoff forward.

Handoff Forward

The ANSI-41 handoff-forward functions provide the specialized form of call control signaling needed to (1) move the MS from a radio channel on the serving MSC to a compatible radio channel on the target MSC while (2) maintaining a call path between the MS and the other party to the active call by establishing a land-line circuit between the serving MSC and the target MSC.

Compatibility of serving and target radio channels has a number of aspects including:

- The ability of the target to support the desired "call mode" (e.g., AMPS, CDMA, TDMA, or NAMPS).
- The ability of the target to support the characteristics of the MS (e.g., power class, use of discontinuous transmission).
- The ability of the target to support the "confidentiality modes" desired by the service subscriber (e.g., encryption of the radio channel).

Handoff forward also has a uniquely identifiable characteristic: When successfully executed, handoff forward results in an increase in the length of the handoff chain within the limit set by the MAXHANDOFF system parameter.[6]

Finally, new provisions for call handoff prior to the call's being answered were added to IS-41 in Revision C. Two cases are covered:

---

[6] Note that the MAXHANDOFF parameter is not an ANSI-41 parameter, but rather a system parameter programmed in each MSC by the mobile telecommunications service provider.

- In the MS-originated call case, the MS is awaiting answer by the called party when handoff forward occurs. In this situation, special procedures are useful to inform the target MSC when the called party answers; for example, for call timing purposes.
- In the MS-terminated call case, the MS is being alerted when handoff forward occurs. In this situation, special procedures are useful to ensure that the MS is placed on the new target channel in the alerting state and to inform the anchor MSC when the MS answers.

ANSI-41 handoff-forward processes include the serving MSC handoff-forward process and the target MSC handoff-forward process, illustrated in Figure 9.6.

Serving MSC Handoff-forward Process

The serving MSC handoff-forward process manages handoff-forward call control in the serving MSC. The key tasks for the serving MSC are as follows:

- Select and reserve the inter-MSC circuit that will provide the landline connection between the serving MSC and the target MSC.
- Request a call handoff to the target MSC. The serving MSC uses the FacilitiesDirective Invoke (FACDIR) message[7] for this purpose.
- When notified that the target MSC accepts the call handoff request via the FacilitiesDirective Return Result (facdir) message, direct the MS to move from the current serving radio channel to the desired target radio channel.
- When notified that the MS has moved to the target channel via the MobileOnChannel Invoke (MSONCH) message, connect the other party to the MS by using the inter-MSC circuit.
- If the handoff occurs while the MS is awaiting answer (i.e., in the case of an MS-originated call), wait until the called party answers and then inform the target MSC, using the InterSystemAnswer Invoke (ISANSWER) message (see Figure 9.7).

---

[7] Unless otherwise indicated, the FacilitiesDirective2 operation may be substituted for references to the FacilitiesDirective operation; e.g., either FacilitiesDirective Invoke or FacilitiesDirective2 Invoke messages may be used by the serving MSC. The FacilitiesDirective2 operation was added to IS-41 in Revision C to support all air-interface standards, including CDMA and narrowband AMPS (NAMPS) MSs; the FacilitiesDirective operation supports TDMA and AMPS MSs, but not CDMA or NAMPS.

Figure 9.6
An example of the handoff-forward processes: (1) The serving MSC requests a handoff forward. (2) The target MSC accepts the handoff request. (3) The selected inter-MSC circuit is now ready for the handoff. (4) The serving MSC issues a handoff order to the MS. (5) The MS tunes to the new channel and is detected by the target MSC. (6) The target MSC notifies the serving MSC that the MS has been detected. (7) The serving MSC connects the call path to the inter-MSC circuit, completing the handoff.

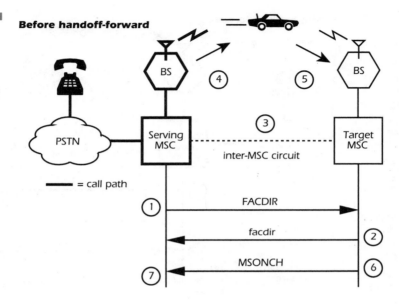

Before handoff-forward

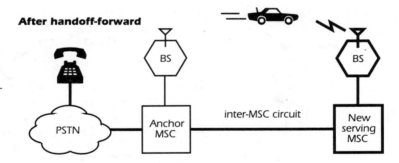

After handoff-forward

- If the handoff occurs while the MS is alerting (i.e., in the case of an MS-terminated call), wait for the ISANSWER message from the target MSC, indicating the MS has answered the call; then acknowledge the notification by sending the InterSystemAnswer Return Result (isanswer) message to the target MSC and provide answer supervision toward the calling party.

Figure 9.7
An example of the handoff-forward while the called party is alerting: (1–7) Same as Figure 9.6, except serving MSC may include the HandoffState parameter in the FACDIR message to indicate that the handoff is occurring while the called party is alerting. (8) The called party answers; the serving MSC notifies the target MSC. The target MSC acknowledges the message and the handoff is complete.

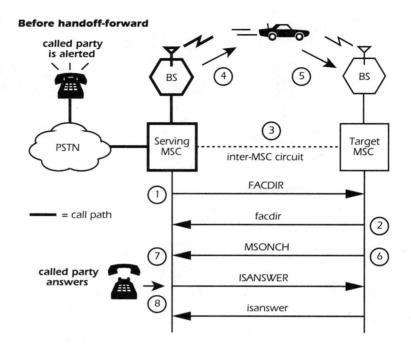

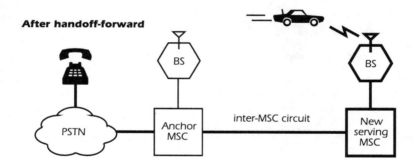

Target MSC Handoff-forward Process

The target MSC handoff-forward process manages handoff-forward call control in the target MSC. The key tasks for the target MSC are as follows:

- Respond to the serving MSC's request for a call handoff, and—in the case of handoff acceptance—inform the serving MSC of the selected target radio channel; handoff acceptance is indicated by using the facdir message.

- Wait until the MS is detected on the target radio channel; then complete the path between the radio channel and the inter-MSC circuit.
- Notify the serving MSC that the MS has successfully moved to the new radio channel; the target MSC uses the MSONCH message for this purpose.
- If the handoff occurs while the MS is alerting (i.e., in the case of an MS-terminated call), wait until the MS answers; then inform the anchor MSC, using the ISANSWER message.
- If the handoff occurs while the called party is alerting (i.e., in the case of an MS-originated call), wait for the ISANSWER message from the anchor MSC, indicating that the called party has answered the call; then acknowledge the notification by sending the isanswer message to the anchor MSC (see Figure 9.7).

Handoff Back

Use of the handoff-forward functions to hand off a call back and forth between two MSCs would result in an excessively long handoff chain, composed of circular paths—creating a *tromboning* effect—between the two MSCs (see Figure 9.8). This would deplete the inter-MSC resources available for other call handoff purposes. The ANSI-41 handoff-back functions are designed to avoid this situation.

Figure 9.8
Repeated use of handoff-forward between two MSCs results in a tromboning effect.

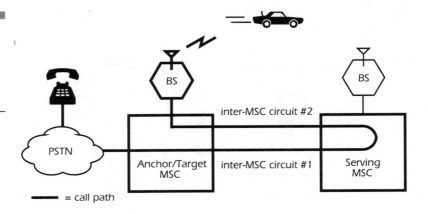

As with handoff forward, the ANSI-41 handoff-back functions are used to move the MS from a radio channel on the serving MSC to a compatible radio channel on the target MSC; unlike handoff forward, the target MSC is already connected to the serving MSC by an inter-MSC circuit. Therefore, once the MS is moved to the new channel, the unused inter-MSC circuit must be removed.

ANSI-41 handoff-back processes include the serving MSC handoff-back process and the target MSC handoff-back process, illustrated in Figure 9.9.

Figure 9.9
An example of the handoff-back processes: (1) The serving MSC requests a handoff back. (2) The target MSC accepts the handoff request. (3) The serving MSC issues a handoff order to the MS. (4) The MS tunes to the new channel and is detected by the target MSC. (5) The target MSC requests the release of the unnecessary inter-MSC circuit and the serving MSC accepts the request. (6) The inter-MSC is released and is now ready for other handoff purposes.

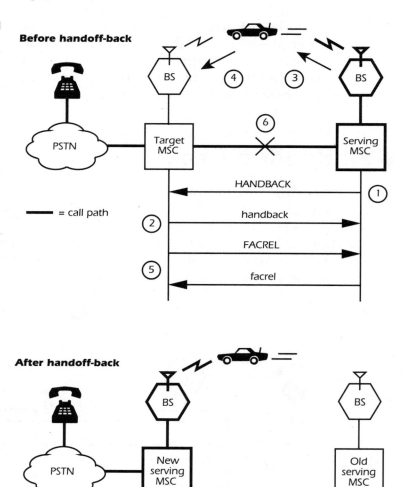

Serving MSC Handoff-back Process

The serving MSC handoff-back process manages handoff-back call control in the serving MSC. The key tasks for the serving MSC are as follows:

- Identify the inter-MSC circuit that already provides a land-line connection between the serving MSC and the target MSC; this circuit must be released once the handoff-back process is completed.
- Request a handoff-back to the target MSC; the serving MSC uses the HandoffBack Invoke (HANDBACK) message[8] for this purpose.
- When notified that the target MSC accepts the call handoff request via the HandoffBack Return Result (handback) message, direct the MS to move from the current serving radio channel to the desired target radio channel.
- When notified that the MS has moved to the target channel via the FacilitiesRelease Invoke (FACREL) message, acknowledge the message by sending a FacilitiesRelease Return Result (facrel) message to the target MSC and release the inter-MSC circuit and radio channel, along with any other resources used for the call.

Target MSC Handoff-back Process

The target MSC handoff-back process manages handoff-back call control in the target MSC. The key tasks for the target MSC are as follows:

- Respond to the serving MSC's request for a call handoff and—in the case of handoff acceptance—inform the serving MSC of the selected target radio channel; handoff acceptance is indicated by using the handback message.
- Wait until the MS is detected on the target radio channel; then complete the path between the radio channel and the other party to the call.
- Notify the serving MSC that the MS has successfully moved to the new radio channel and initiate the release of the inter-MSC circuit

---

[8] Unless otherwise indicated, the HandoffBack2 operation may be substituted for references to the HandoffBack operation; e.g., either HandoffBack Invoke or HandoffBack2 Invoke messages may be used by the serving MSC. The HandoffBack2 operation was added to IS-41 in Revision C to support all air-interface standards, including CDMA and narrowband AMPS (NAMPS) MSs; the HandoffBack operation supports TDMA and AMPS MSs, but not CDMA or NAMPS.

between the target and the old serving MSC; the target MSC uses the FACREL message for this purpose.

Path Minimization

While handoff-back functions provide a basic form of path minimization, eliminating the tromboning effect that otherwise would occur between the target and serving MSCs when a call is repeatedly handed back and forth between them, the ANSI-41 path minimization functions refer to more complex handoff optimization techniques. These techniques were introduced in IS-41 Revision B and were updated to support the CDMA and NAMPS air interfaces in IS-41-C.

In general, MSCs in the handoff chain may use path-minimization functions to eliminate unnecessary inter-MSC circuits between the anchor MSC and the target MSC. Handoff with path minimization ideally results in a single inter-MSC circuit directly between the anchor MSC and the target MSC (see Figure 9.10), or no circuits if the anchor is the target (as shown in Figure 9.11 in the next section). Alternatively, a tandem MSC may employ path minimization to eliminate unnecessary inter-MSC circuits between it and the target MSC. The latter option would be used only if the anchor MSC were unable to perform path minimization; that is, if no inter-MSC circuit between the anchor and target MSCs were available, or if the handoff chain exceeded a system-specified length related to the TANDEMDEPTH system parameter.[9] The decision to attempt path minimization, rather than the simpler handoff forward, is an internal implementation issue outside the scope of ANSI-41.

ANSI-41 path minimization processes include (1) the serving MSC path minimization process, (2) the anchor MSC path minimization process, and (3) the tandem MSC path minimization process. If the target MSC is not the anchor or tandem MSC, then the target MSC handoff-forward process is also involved, as illustrated in Figure 9.10 and described below.

---

[9] Note that the TANDEMDEPTH parameter is not an ANSI-41 parameter, but rather a system parameter programmed into each MSC by the mobile telecommunications service provider.

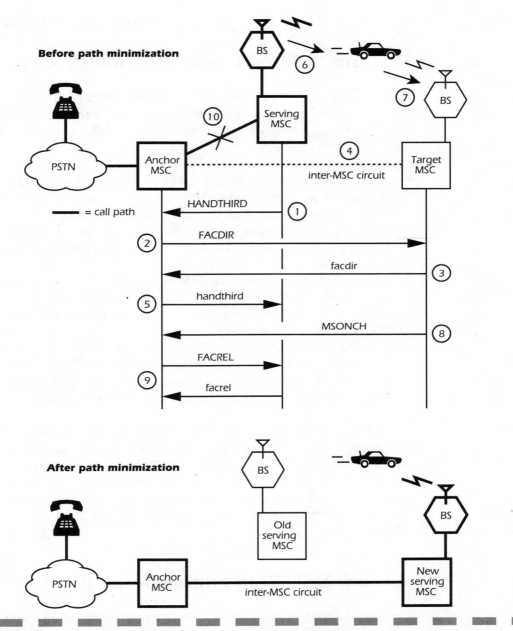

Figure 9.10 An example of the path minimization processes: (1) The serving MSC sends a path minimization request to the anchor MSC. (2) The anchor MSC attempts to determine if path minimization is possible; it sends a handoff request to the target MSC. (3) The target MSC accepts the handoff request. (4) The selected inter-MSC circuit between the anchor MSC and the target MSC is now ready for the handoff. (5) The anchor MSC accepts the path minimization request. (6) The serving MSC issues a handoff order to the MS. (7) The MS tunes to the new channel and is detected by the target MSC. (8) The target MSC notifies the anchor MSC that the MS has been detected. (9) The anchor MSC connects the call path to the inter-MSC circuit, completing the handoff. It then releases the unnecessary inter-MSC circuit to the serving MSC. (10) The inter-MSC circuit is released and is now ready for other handoff purposes.

Serving MSC Path-minimization Process

The serving MSC path-minimization process manages path-minimization call control in the serving MSC. The key tasks for the serving MSC, when there are no tandem MSCs in the handoff chain (as is the case in Figure 9.10) are as follows:

- Identify the inter-MSC circuit that already provides a land-line connection between the serving MSC and the anchor MSC; this circuit must be released once path minimization is completed.
- Request that the anchor MSC perform path minimization; the serving MSC uses the HandoffToThird Invoke (HANDTHIRD) message[10] for this purpose.
- When notified that the anchor MSC accepts the path minimization request via the HandoffToThird Return Result (handthird) message, direct the MS to move from the current serving radio channel to the desired target radio channel.
- When notified by the anchor MSC that the MS has moved to the target channel via the FACREL message, acknowledge the message by sending a facrel message to the anchor MSC and release the inter-MSC circuit along with any other resources used for the call.

If there are one or more tandem MSCs in the handoff chain, the only change in the process is that the serving MSC must send the path minimization request toward the anchor MSC by way of the tandem MSC(s). Either the anchor MSC or possibly a tandem MSC may then provide the path minimization acceptance notification (i.e., the handthird message) to the serving MSC—the serving MSC is not explicitly informed of the source of the acceptance.

Anchor MSC Path-minimization Process

The anchor MSC path-minimization process manages path-minimization call control in the anchor MSC. This process is initiated by the

---

[10] Unless otherwise indicated, the HandoffToThird2 operation may be substituted for references to the HandoffToThird operation; e.g., either HandoffToThird2 Invoke or HandoffToThird2 Invoke messages may be used by the serving MSC. The HandoffToThird2 operation was added to IS-41 in Revision C to support all air-interface standards, including CDMA and narrowband AMPS (NAMPS) MSs; the HandoffToThird operation supports TDMA and AMPS MSs, but not CDMA or NAMPS.

receipt of the HANDTHIRD message from the serving MSC (possibly via a tandem MSC). The key tasks of the process are, first, to determine if the anchor MSC is the target MSC for the handoff. If the anchor MSC is not the target (as in Figure 9.10), then the following steps are performed:

- Select and reserve the inter-MSC circuit that will provide the land-line connection between the anchor MSC and the target MSC.
- Request a call handoff to the target MSC using information (e.g., target cell identification) provided to the anchor MSC by the serving MSC in the HANDTHIRD message. The anchor MSC uses the FACDIR message to request the call handoff to the target MSC.
- When notified that the target MSC accepts the call handoff request via the facdir message, notify the serving MSC that its path minimization request is accepted; the anchor MSC uses the handthird message for this purpose.
- When notified that the MS has moved to the target channel via the MSONCH message, connect the other party to the MS, using the inter-MSC circuit established between the anchor and target MSCs.
- Initiate release of the inter-MSC circuit (or circuits, if there are one or more tandem MSCs) between the anchor MSC and the old serving MSC, using the FACREL message.

On the other hand, if the anchor MSC *is* the target MSC for the handoff, then it executes the following steps (see Figure 9.11):

- Determine whether resources, like a radio channel, are available for the call handoff; if not, reject the serving MSC's path minimization request by sending a HandoffToThird Return Error message to the serving MSC and end the process.
- Otherwise, notify the serving MSC that its path minimization request is accepted, using the handthird message.
- Wait until the MS is detected on the target radio channel; then complete the path between the radio channel and the other party to the call.
- Notify the old serving MSC that the MS has successfully moved to the new radio channel, and initiate release of the inter-MSC circuit between the anchor and the old serving MSC; the anchor MSC uses the FACREL message for this purpose.

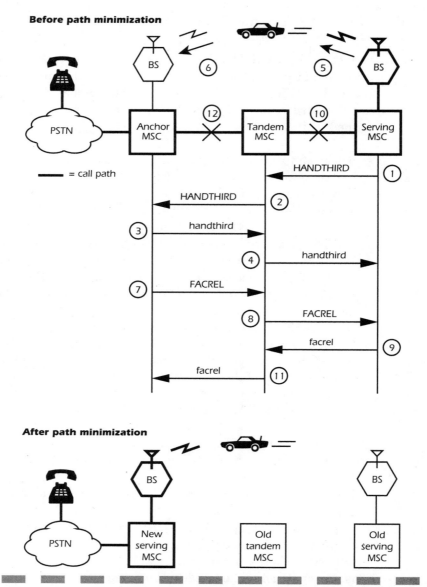

Figure 9.11 An example of path minimization when the anchor MSC is the target MSC: (1) The serving MSC sends a path minimization request toward the anchor MSC. (2) The tandem MSC relays the request to the anchor MSC. (3) The anchor MSC determines that it is the target for the handoff; therefore, it accepts the path minimization request. (4) The tandem MSC relays the response to the serving MSC. (5) The serving MSC issues a handoff order to the MS. (6) The MS tunes to the new channel and is detected by the anchor MSC. (7) The anchor MSC connects the call path to the MS, completing the handoff. It then releases the unnecessary inter-MSC circuit towards the serving MSC. (8) The tandem MSC requests release of the unnecessary inter-MSC circuit to the serving MSC. (9) The serving MSC accepts the release request. (10) The inter-MSC circuit between the tandem MSC and the serving MSC is released and is now ready for other handoff purposes. (11) The tandem MSC accepts the release request. (12) The inter-MSC circuit between the tandem MSC and the anchor MSC is released and is now ready for other handoff purposes.

Tandem MSC Path-minimization Process

The tandem MSC path-minimization process manages path-minimization call control in a tandem MSC. Like the anchor MSC path-minimization process just described, this process is initiated when the tandem MSC receives the HANDTHIRD message from the serving MSC. The key task of the process is, first, to determine if the tandem MSC is the target MSC for the handoff. This would be the case only if there were multiple tandem MSCs in the handoff chain: The serving MSC uses handoff back if the tandem MSC—to which it is directly connected—is the target MSC. If the tandem MSC is not the target, the process must determine whether it has reached the path-minimization limit, based on the TANDEMDEPTH system parameter. For this, the tandem MSC performs a calculation using the following data (see Figure 9.12):

1. The InterSwitchCount value (value X) received in the HANDTHIRD message sent by the serving MSC; this value represents the total number of inter-MSC circuits in the handoff chain (i.e., the number of MSCs in the chain less 1).
2. The number of inter-MSC circuits in the segment of the handoff chain between the anchor MSC and the tandem MSC (value Y); the tandem MSC will have stored this value previously, before it performed a handoff forward.

Figure 9.12
If the system parameter TANDEMDEPTH = 1, then tandem MSC #1 will not forward the HANDTHIRD message to the anchor MSC, since $X - Y = 2$, which is greater than TANDEMDEPTH.

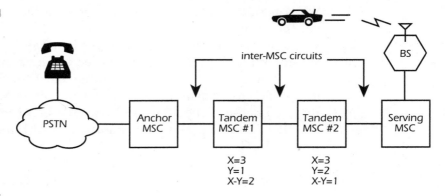

The difference between these two numbers, $X–Y$, indicates the number of inter-MSC circuits between the serving MSC and the tandem MSC; when this number is greater than the system parameter TANDEMDEPTH, the tandem MSC is not permitted to relay the path-minimization request any further down the handoff chain toward the anchor MSC. Instead, the tandem MSC must decide whether it—rather than the anchor MSC—should attempt path minimization or simply reject the path-minimization request by sending a HandoffToThird Return Error message to the serving MSC, thus ending the process.

However, if the tandem MSC is not the target and the path-minimization limit has not been reached, then the tandem MSC performs the following steps:

- Relay the HANDTHIRD message back toward the anchor MSC.
- If a handthird message is received in response, relay the message toward the serving MSC and end the process.
- However, if a HandoffToThird Return Error message is received, the tandem MSC may either relay the message toward the serving MSC, ending the process, or attempt path minimization.

If the tandem MSC chooses to attempt path minimization—based on either of the circumstances described above—it performs the following steps:

- Select and reserve the inter-MSC circuit that will provide the land line connection between the tandem MSC and the target MSC.
- Request a call handoff to the target MSC, using information (e.g., target cell identification) provided to the tandem MSC by the serving MSC in the HANDTHIRD message. The tandem MSC uses the FACDIR message to request the call handoff to the target MSC.
- When notified that the target MSC accepts the call handoff request via the facdir message, notify the serving MSC that its path minimization request is accepted, using the handthird message.
- When notified that the MS has moved to the target channel via the MSONCH message, connect the other party to the MS, using the inter-MSC circuit between the tandem and target MSCs.
- Initiate release of the inter-MSC circuit(s) between the tandem MSC and the old serving MSC, sending the FACREL message toward the serving MSC.

Finally, if the tandem MSC is the target MSC for the handoff, then it executes the following steps:

- Determine whether resources, like a radio channel, are available for the call handoff; if not, reject the serving MSC's path minimization request by sending a HandoffToThird Return Error message to the serving MSC, and end the process.

- Otherwise, notify the serving MSC that its path minimization request is accepted, using the handthird message.
- Wait until the MS is detected on the target radio channel. Then complete the path between the radio channel and the other party to the call.
- Notify the old serving MSC that the MS has successfully moved to the new radio channel, and initiate release of the inter-MSC circuit between the tandem MSC and the old serving MSC; the tandem MSC uses the FACREL message for this purpose.

Target MSC Handoff-forward Process and Path Minimization

As shown in Figure 9.10, when the target is not the anchor MSC or a tandem MSC, then the target MSC simply executes the target MSC handoff-forward process described in the section entitled Handoff Forward—the target MSC is unaware of the path minimization taking place.

Call Release

The ANSI-41 call release functions control the release at the end of a call, of the inter-MSC circuits and internal MSC resources that were used during the handoff process. Without these functions, the parties to the call could "hang up," but the handoff chain from the anchor MSC to the serving MSC would continue to exist. The call release can be initiated either by the anchor MSC; for example, based on a call release indication received from the PSTN, or by the serving MSC, based on information it receives (or does not receive, in the case of lost radio contact) from the mobile station.

Call-release Processes

ANSI-41 call-release processes include the initiating call-release process and the receiving call-release process, illustrated in Figure 9.13.

Figure 9.13
An example of the call release processes: (1) One of the parties to the call hangs up. The anchor MSC releases the unnecessary inter-MSC circuit towards the serving MSC. (2) The tandem MSC requests release of the unnecessary inter-MSC circuit to the serving MSC. (3) The serving MSC accepts the release request. The facrel message may contain billing information for use by the anchor MSC. (4) The inter-MSC circuit between the tandem MSC and the serving MSC is released and is now ready for other handoff purposes. (5) The tandem MSC accepts the release request. It forwards any billing information received in step 3. (6) The inter-MSC circuit between the tandem MSC and the anchor MSC is released and is now ready for other handoff purposes.

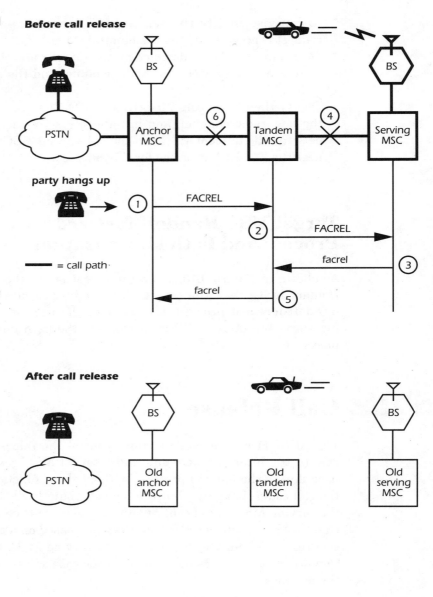

The initiating call release process sends a FACREL message to notify the receiving MSC of call release. When the initiator receives the response, it clears its internal resources used for the call (e.g., software processes and memory) and marks the inter-MSC circuit idle.

If the receiver is either the anchor MSC or the serving MSC, then the receiving call release process is completed by returning a facrel message to the initiating call release process and then marking the inter-MSC circuit idle. The receiver may initiate other processes, e.g., for features (see Chapter 12), to release the radio channel to the MS or the land-line circuit to the PSTN. However, these processes are generally beyond the scope of ANSI-41.

In the case of a tandem MSC, both types of call-release process are invoked during call release; on one side, the receiving process executes, while the initiating process executes on the other side. Also note that the initiating call-release process is nested within the receiving call-release process at the tandem MSC (see Figure 9.13). This allows information (e.g., the total number of inter-MSC segments involved in the call, which is useful for recording purposes) received by the tandem MSC to be relayed toward the anchor MSC. This ANSI-41 function was initially introduced in IS-41-C; prior to Revision C, the operations executed in an *asynchronous* fashion, as shown in Figure 9.14.

Figure 9.14
The call release processes prior to IS-41-C. Note that steps 4 and 5 are not nested within steps 1 and 2 as in IS-41-C (and later).

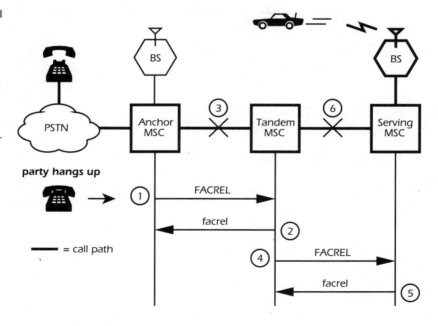

Summary of ANSI-41 Operations Used for Basic Intersystem Handoff

Table 9.1 summarizes the ANSI-41 operations used by the intersystem handoff functions described in this chapter.

TABLE 9.1

Use of ANSI-41 Operations for Intersystem Handoff

Function	ANSI-41 Operations Used for the Function
Handoff measurement	HandoffMeasurementRequest, HandoffMeasurementRequest2
Handoff-forward	FacilitiesDirective, FacilitiesDirective2, InterSystemAnswer, MobileOnChannel
Handoff-back	HandoffBack, HandoffBack2, FacilitiesRelease
Path minimization	HandoffToThird, HandoffToThird2, FacilitiesDirective, FacilitiesDirective2, FacilitiesRelease, MobileOnChannel
Call release	FacilitiesRelease

CHAPTER 10

Basic Automatic Roaming Functions

In this chapter, we discuss the basic ANSI-41 mobile telecommunications network functions related to automatic roaming. These functions are divided into the following categories:[1]

- Mobile station (MS) service qualification
- MS location management
- MS state management
- HLR and VLR fault recovery

We also examine the ANSI-41 application processes that support these basic network functions. While there are other automatic roaming-related processes in ANSI-41, as for authentication, voice feature support, and short-message services, we treat these separately in later chapters. This chapter focuses on the basic automatic roaming processes that allow an MS to obtain service in a visited system.

In the course of describing the processes, we identify the ANSI-41 mobile application part (MAP) operations (e.g., RegistrationNotification) used to accomplish basic automatic roaming process tasks. We summarize this information at the end of the chapter. Note that we employ the ANSI-41 convention for operation component acronyms; the Invoke component acronym is in all capital letters (e.g., REGNOT), while the Return Result component acronym is in all lowercase letters (e.g., regnot). Refer to the ANSI-41 standard for the descriptions of the individual ANSI-41 operations and parameters. Also, the reader should consult the Glossary for the description of general terms (e.g., serving system) that are not explicitly defined in this chapter.

Throughout this chapter we consider the serving system as a single entity encompassing the mobile switching center (MSC) and visitor location register (VLR) functional entities. This simplifies the descriptions of ANSI-41 application processes and also is representative of a large percentage of the ANSI-41 implementations currently in service. However, ANSI-41 assigns specific processing duties to the VLR that make the concept of sharing a VLR among multiple MSCs a viable alternative to the combined implementation. Therefore, while we describe automatic roaming processes in terms of a single "MSC + VLR" entity, the reader should keep in mind that the potential for separation exists and is fully defined in ANSI-41.

---

[1] The list of basic automatic roaming functions reflects the authors' subjective categorization of certain ANSI-41 functions.

What Is Automatic Roaming?

Automatic roaming encompasses a set of network functions that allow a subscriber to obtain mobile telecommunications service outside the home service-provider area. These functions are automatic in the sense that they are invoked without requiring special subscriber actions. For example, the subscriber need not dial a code or make a call to inform the home system that the subscriber has roamed into another system—this function is provided automatically when the subscriber is detected in the new system (unlike the first generation of cellular roaming services that required manual activation or a special service call on at least a daily basis).

Registration

Detection of a subscriber in a new serving system is an example of a *registration event*. Registration events, or simply registrations, are the triggers for a number of the automatic roaming functions described in this chapter (see Figure 10.1). The types of registrations supported in a network are dependent on the protocol used for the air interface between the mobile station and the base station and on the internal algorithms implemented in the serving systems. The air-interface standards for AMPS, TDMA, and CDMA support different types of registrations. The following events are common among AMPS, TDMA, and CDMA and occur most frequently:

- Mobile station power-on
- Timer-based (i.e., autonomous) registration
- Transition to a new system
- Call origination

Figure 10.1
Registration can initiate other network functions. These functions must be performed before services are provided to a subscriber.

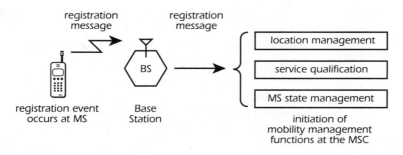

registration message

registration message

location management

service qualification

MS state management

registration event occurs at MS

Base Station

BS

initiation of mobility management functions at the MSC

Registration always occurs when the MS is initially turned on. This type of registration informs the network that the MS is now active and available to receive calls and other network services. Timer-based registration is more commonly referred to as *autonomous registration*. This type of registration occurs at periodic intervals while the MS is powered on. The interval between registrations typically ranges from about ten minutes to an hour. The periodic time value for autonomous registration is transmitted to the MS from the serving system, allowing each network individual control of the frequency of autonomous registrations. Autonomous registration helps keep the network from "losing" the current location of the subscriber while the MS is powered on and no calls are made. Registrations also occur when the MS makes a transition between systems while it is active. Registration can also occur during the process of call origination, so that the location of the MS can be updated in the network for billing of the call or the MS can again be qualified for service, if necessary.

Other events can cause registration to occur. For example, CDMA systems support registration when the mobile station is turned off (i.e., essentially a power-off deregistration) to inform the network that the MS is unavailable for service.

It is interesting to note that for systems not supporting a power-off deregistration procedure, a subscriber who turns off the MS can remain registered in a system for several minutes or even hours. The subscriber's registration is eventually canceled in that system when inactivity is detected after a time, when the subscriber registers in another system, or by a periodic maintenance procedure.

Where Are Automatic Roaming Functions Specified in ANSI-41?

Automatic roaming functions are specified in three parts of ANSI-41:

1. **ANSI-41.5** provides the required formats—the bit-by-bit encoding—of all the ANSI-41 operation components, including those used for automatic roaming. ANSI-41.5 defines both the messages (e.g., RegistrationNotification Invoke) and the message parameters (e.g., QualificationInformationCode).
2. **ANSI-41.6** provides algorithmic descriptions of the procedures associated with sending and receiving most ANSI-41 messages, includ-

ing those used for automatic roaming. As we pointed out in Chapter 8, ANSI-41 does not define a service interface between the MAP application processes and the MAP application service element (ASE); therefore, the procedures in ANSI-41.6 encompass both application process descriptions and ASE procedural descriptions. Furthermore, the procedures leave considerable room for implementation-dependent customization; there are numerous references in ANSI-41.6 to "local procedures" and "internal algorithms."

3. **ANSI-41.3** steps back from the protocol details contained in parts 5 and 6 and attempts to explain, using information flow diagrams and step-by-step descriptions, how the operations are used individually and together to accomplish automatic roaming application process tasks.

Issues Associated with Automatic Roaming

Automatic roaming accounts for the majority of ANSI-41 operations. This can be attributed to two primary factors:

- The complexity of the standard is being driven by new features, such as authentication and the dozens of ANSI-41 calling features, while the impact of new features on intersystem handoff is limited by the definition of the anchor MSC function. The anchor MSC performs most of the feature processing, with only basic information transferred across the handoff chain.
- Roaming is a higher-profile activity with service providers since their subscribers are usually much more likely to experience automatic roaming (e.g., to originate or receive a call in a visited system) than intersystem handoff (i.e., to move between systems during a call). Thus, there is more incentive to add value to the mobile communications experience in the area of automatic roaming.

When the subscriber roams outside the home system, a number of intersystem issues arise. From the visited system's perspective, automatic roaming issues include:

1. *Identification of the roaming MS's home system.* The typical mechanism used to relate roaming MSs to their home systems is the roam-

ing agreement established between the visited and home system service providers.

2. *Verification of the roaming MS's identity.* At the very least, this involves obtaining validation information from the home system. Preferably, the MS is also authenticated by the home or visited system.

3. *Identification of the roaming MS's service capabilities.* This is an aspect of service qualification that allows the visited system to be aware of the MS's subscribed services.

From the home system's perspective, automatic roaming issues include:

1. *Identification of the roaming MS's current location.* This is a key outcome of the location management processes.

2. *Identification of the roaming MS's current state.* Determine if the MS is available for call delivery and, if not, the method that the home system uses to process calls intended for the subscriber (e.g., deny the incoming call, forward the call to voice mail).

MS Service Qualification

The ANSI-41 MS service qualification function encompasses the processes that establish a roaming MS's financial accountability and service capabilities in a serving system. Service qualification information includes validation information and/or service profile information.

Validation information indicates the system—normally the home system—that is assuming financial responsibility for the roaming MS and the period for which this responsibility is assumed (e.g., one call, one hour, one day). This information assures the serving system that someone will pay for the calls that the roaming MS makes; alternatively, the information may inform the serving system that validation is denied for the MS, possibly for a specified period to prevent the serving system from repeatedly re-requesting validation information from the HLR.

Service profile information indicates the specific set of features and other service capabilities, including restrictions on these capabilities, which are associated with the roaming MS; for example, features X, Y, and Z are active for the MS, and the MS is authorized to originate local calls only. The serving system uses this information to tailor the telecommunications services it provides to the MS. An MS's HLR is the network repository for the MS's service qualification information as well as other detailed information related to the MS.

MS Service-qualification Processes

The ANSI-41 MS service-qualification processes include the serving system service qualification process and the HLR service qualification process.

The serving system service qualification process is responsible for obtaining service qualification information for each visiting MS. The serving system may obtain service qualification information either by initiating a request to the HLR or by receiving a directive from the HLR.

The serving system issues service qualification requests to the HLR under various circumstances, including:

1. When the visiting MS is initially detected
2. When the previously obtained service qualification information for a visiting MS indicates a limited service area (e.g., a single location area) and the MS moves outside the service area
3. When the previously obtained service qualification information for a visiting MS indicates limited service duration and the authorized period expires
4. If for any other reason the serving system determines that it needs to retrieve service qualification information (e.g., the MS's service qualification period has expired and a call delivery request is received for the MS).

In cases 1 and 2, the serving system packages the service qualification request along with the location update information in a RegistrationNotification Invoke (REGNOT) message (see Figure 10.2) to the HLR. In cases 3 and 4, the request should be via a QualificationRequest (QUALREQ) Invoke message[2] (see Figure 10.3).

---

[2] IS-41-C also includes the definition and procedures for the ServiceProfileRequest operation—originally included in IS-41 Revision A—with the following caveat: "The initiating of ServiceProfileRequest procedures is not recommended due to functional overlap with QualificationRequest. This transaction is not supported and may be eliminated in the future. The QualificationRequest should be used instead." In fact, the ServiceProfileRequest operation is not included in ANSI-41-D and subsequent revisions.

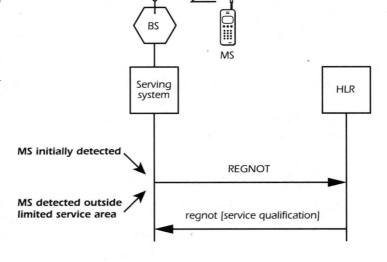

Figure 10.2
Service qualification
using the
RegistrationNotification
operation.

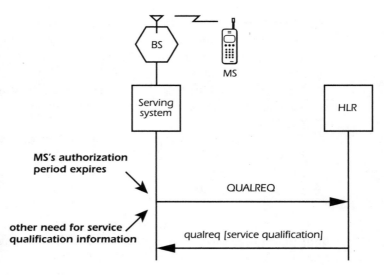

Figure 10.3
Service qualification
using the
QualificationRequest
operation.

The HLR responds to the serving system request with an appropriate message, either RegistrationNotification Return Result (regnot) or QualificationRequest Return Result (qualreq). When service qualification is successful, key ANSI-41 parameters in response to the serving system include:

- AuthorizationPeriod

- OriginationIndicator
- TerminationRestrictionCode
- CallingFeaturesIndicator

The presence of the AuthorizationPeriod parameter in the response implies the successful validation of the MS. The AuthorizationPeriod parameter also specifies the period after which the serving system must requalify the MS; this may be specified, for example, in terms of a certain number of hours. The OriginationIndicator parameter specifies the types of calls the MS is allowed to originate, such as no calls, all calls, or locals calls only. The TerminationRestrictionCode parameter indicates whether the MS is allowed to terminate calls. The CallingFeaturesIndicator parameter specifies the authorization and activity states of the MS's features.

Unsuccessful service qualification is indicated by the presence of the AuthorizationDenied parameter in the response from the HLR to the serving system. ANSI-41 does not dictate how the serving system shall handle an MS that has failed service qualification; service should obviously be refused, but particular tones or announcements to the MS are left as system implementation issues. The HLR can specify the service qualification retry interval by including the DeniedAuthorizationPeriod parameter in the response along with the AuthorizationDenied parameter. This limits subsequent service qualification requests for the refused MS.

In addition to responding to serving system–service qualification requests, the HLR issues service qualification directives in two of circumstances:

- When the roaming MS's service qualification information changes as a result of HLR administrative action (e.g., authorization has been revoked).
- When the roaming MS's service qualification information changes as a result of subscriber action (e.g., activation of a feature).

In both cases, the service qualification information is conveyed to the serving system in the QualificationDirective Invoke (QUALDIR) mes-

---

[3] IS-41-C also includes the definition and procedures for the ServiceProfileDirective operation—originally included in IS-41 Revision A—with the following caveat: "The initiating of ServiceProfileDirective procedures is not recommended due to functional overlap with QualificationDirective. This transaction is not supported and may be eliminated in the future. The QualificationDirective should be used instead." In fact, the ServiceProfileDirective operation is not included in ANSI-41-D and subsequent revisions.

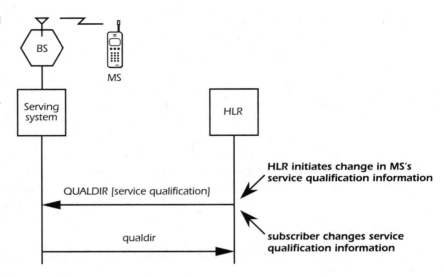

Figure 10.4
Service qualification using the QualificationDirective operation.

sage[3] (see Figure 10.4). The QualificationDirective Return Result (qualdir) message sent from the serving system to the HLR acknowledges message receipt.

MS Location Management

The ANSI-41 MS location management function has two components:

- *MS location-update processes* have the effect of creating or modifying an MS's temporary record in a visited system and updating the location information in an MS's record in the HLR.
- *MS location-cancellation processes* have the effect of deleting an MS's temporary record in a visited system and updating the location information in an MS's record in the HLR. Location cancellation is also known as *registration cancellation* or *deregistration*.

The location update and cancellation processes often interact, as when managing the movement of an MS from one serving system to another; however, this need not be the case, as we describe in this chapter.

The MS location information stored in the MS's record in the HLR is characterized by its *resolution*. ANSI-41 supports several levels of location resolution, but, at a minimum, the MS's location is resolved to the serving VLR; the HLR will always be provided a location update if the serving VLR changes. In this case, the location of the subscriber is recorded in the HLR as the MSCID of the VLR currently serving the subscriber, usually along with the VLR's network address (e.g., the VLR's SS7 point code).

MS Location-update Processes

The ANSI-41 MS location-update processes include the serving system location update process and the HLR location update process.

The basic function of the serving system location update process is to notify a visiting MS's HLR of the MS's presence in the serving system. The serving system notifies the HLR under various circumstances, including:

- When a new visiting MS (i.e., an MS for which a temporary record does not exist in the serving system) is detected in the serving system's service area.
- When the previously obtained service qualification information for a visiting MS indicates a limited service area (e.g., a single location area) and the MS moves outside the service area.

Note that the location update process is not normally triggered each time a visiting MS accesses the serving system; this would place an unnecessary signaling burden on the interface, particularly when periodic, autonomous registration is employed. While there may be other reasons for notifying the HLR on a per-system-access basis, as for authentication or service qualification purposes, the HLR normally requires location update notification only when the MS changes service areas.

If the cause of the location update is the detection of a new visiting MS, the serving system creates a new temporary record; otherwise, the serving system merely modifies the preexisting record for the MS with the new location information. In both cases, the serving system packages the location update notification along with the service qualification request in a REGNOT message[4] to the HLR (see Figure 10.2).

---

[4] Note that the RegistrationNotification operation may invoke many functions (e.g., location update, service qualification, authentication reporting, SMS functions, etc.) and sometimes these are all combined under the category of "registration." However, we have chosen to describe these diverse functions separately in this book.

When the HLR receives the REGNOT message, it checks the MS's record for location information. If the serving system recorded in the MS record is different from the requesting system, the HLR location update process triggers the location cancellation process to cancel the MS's location in the previous serving system (see Figure 10.5). If the MS is not authorized for service in the new location, the HLR denies the location update and includes the AuthorizationDenied parameter in the regnot message; otherwise, the HLR updates the MS's record with the new location information and acknowledges the location update by sending a regnot message to the serving system.

If the serving system receives a negative acknowledgment, it deletes the temporary record it created for the MS.

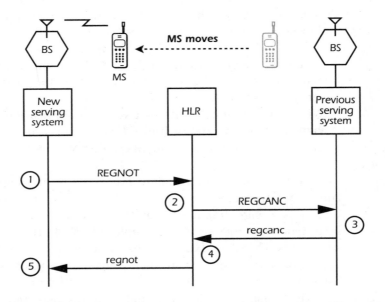

Figure 10.5   Combination of the location update and location cancellation processes: (1) A new MS moves into a service area; therefore, the serving system requests location update in the HLR. (2) The HLR has stored location information for the MS; therefore, it requests location cancellation from the previous serving system. (3) The previous serving system deletes the temporary MS record and sends an acknowledgment to the HLR. (4) The HLR acknowledges the location update. The HLR normally also provides service qualification in the regnot message. (5) The serving system provides service to the MS.

Location Update with Subscriber Confidentiality

TIA / EIA / IS-735 (see Appendix A) introduced an additional element into the serving system location update process to support the *Subscriber Confidentiality* service for CDMA mobile stations.[5] Subscriber confidentiality offers added security by allowing the serving system to assign the subscriber a *temporary mobile station identity*, or TMSI (see Chapter 5). Once assigned, the TMSI is used in the subsequent communications over the air interface with the MS, rather than the MIN or IMSI.

Subscriber confidentiality has an impact on the location update process when the MS uses a previously assigned TMSI to register in a new serving system (note that this is a CDMA feature only; TDMA MSs will not normally register in a new serving system using a TMSI). Since the previous serving system assigned the TMSI, the new serving system needs a mechanism to fetch the subscriber identification (e.g., MIN or IMSI and ESN) from the previous serving system. For this purpose, the new serving system uses the ParameterRequest Invoke (PARMREQ) message (see Figure 10.6).

MS Location-cancellation Processes

Location-cancellation processes include the serving system location cancellation process and the HLR location cancellation process.

The serving system location-cancellation process is responsible for deleting the temporary record for one or more visiting MSs in the serving system. The serving system location-cancellation process may be initiated by the HLR, in the form of a directive sent from the HLR to the serving system, or locally by the serving system itself. The latter case may occur under various circumstances, all of which are serving system options not dictated by ANSI-41:

- When the MS sends a power-down signal to the serving system (although not all MSs provide such an indication)
- When the MS misses one or more autonomous registration events
- When the serving system determines, based on some other serving system–dependent algorithm, that a single roaming MS record shall be cleared

---

[5] IS-735 also defines the *Network Directed System Selection (NDSS)* feature. NDSS is a network capability that allows a service provider, based on various customer and service provider–specified criteria, to automatically direct a subscriber to a desired serving system. If interested, please see IS-735 and the excellent analysis of the pros and cons of the service in the December 1997 issue of *Cellular Networking Perspectives* (see Bibliography).

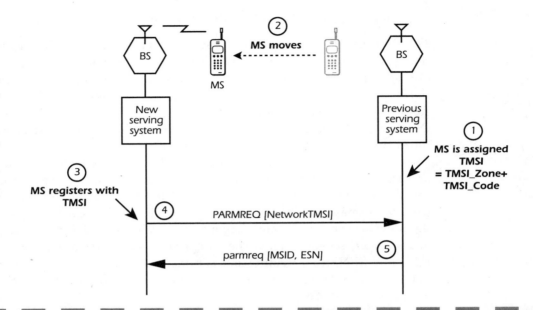

Figure 10.6 Location update with subscriber confidentiality: (1) The MS has been assigned a TMSI by its serving system, consisting of a TMSI_ZONE plus TMSI_CODE. The MS uses only TMSI_CODE as long as the TMSI_ZONE that is broadcast by the BS equals the TMSI_ZONE assigned to the MS. (2) The MS moves to a new serving system. It detects that the new BS is broadcasting a different TMSI_ZONE. (3) Therefore, the MS registers with its full TMSI value (i.e., TMSI_ZONE+TMSI_CODE). (4) Since the TMSI value is unknown to the new serving system, it sends an ANSI-41 PARMREQ message to the previous serving system, which it identifies based on the TMSI_ZONE value provided by the MS. (5) The previous serving system returns the MS's MSID (i.e., MIN or IMSI) and ESN values. The new serving system now proceeds with registration of the MS. It will normally also assign a new TMSI to the MS via an air-interface transaction.

- When the serving system determines—again, based on some other serving system–dependent algorithm—that all roaming MS records should be cleared (e.g., for serving system restoration purposes).

ANSI-41 provides two mechanisms the serving system may use to notify the HLR of the location cancellation of one or more roaming MSs:

- The ANSI-41 MSInactive Invoke (MSINACT) message may be sent from the serving system to notify the HLR of the location cancellation of a single roaming MS (see Figure 10.7). However, the message must include the DeregistrationType parameter, or else the HLR will interpret the message as merely reporting the inactive state of the MS.

■ The ANSI-41 BulkDeregistration Invoke (BULKDEREG) message may be sent from the serving system to notify the HLR of the location cancellation of all the HLR's roaming MSs currently associated with the serving system (see Figure 10.8).

Figure 10.7
Location cancellation by the serving system using the MSInactive operation.

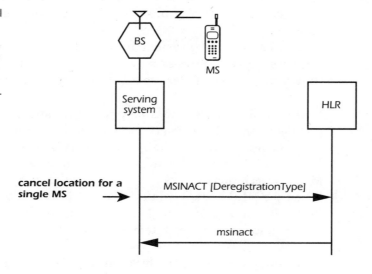

Figure 10.8
Location cancellation by the serving system using the BulkDeregistration operation.

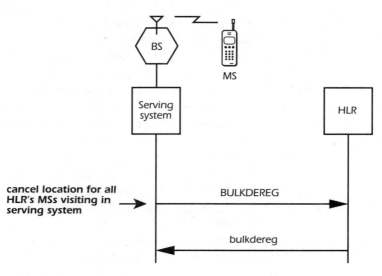

The HLR location-cancellation process responds to a serving system location-cancellation notification with an acknowledgment message: MSInactive Return Result (msinact) or BulkDeregistration Return

Result (bulkdereg). The effect of these procedures is to delete one or more temporary MS records in the visited system and to delete the transient location information in one or more MS records in the HLR. The HLR still maintains service information for each of the MSs affected (e.g., subscribed features); however, this information does not include current location, which is unknown until the next successful location update for the MS.

In addition to serving system–initiated location cancellation, the HLR location-cancellation process may issue a location-cancellation directive to the serving system under the following circumstances:

- When an MS registers in a new serving system, the location update process in the HLR triggers the HLR location cancellation process to clear the MS's record in the system previously serving the MS.
- When an administrative action is executed (e.g., withdrawal of the MS's subscription).
- When the HLR suffers a data failure, it initiates location cancellation for all its roaming MSs in all potential serving systems (assuming the HLR can identify these systems).

In the first two cases, the location-cancellation directive is conveyed to the serving system in the RegistrationCancellation Invoke (REGCANC) message (see Figure 10.5). This directive may be refused by the serving system in border cell situations (see the next section). Otherwise, the effect of this procedure is to delete the temporary MS record in the previous serving system.

For the third case, the HLR uses the UnreliableRoamerDataDirective Invoke (UNRELDIR) message (see Figure 10.9). The effect of the UnreliableRoamerDataDirective procedure is to delete all temporary MS records related to the HLR in the visited system and to delete the location information in one or more MS records in the HLR. As in the case of BulkDeregistration, the HLR still maintains service information for each of the MSs affected; however this information does not include current location, which is unknown until the next successful location update for the MS. In fact, this is the rationale behind the HLR's use of UnreliableRoamerDataDirective—to direct the serving system to assist the HLR in replenishing its database with valid roaming MS location information in as expeditious a manner as possible. Without this procedure, the serving system would not necessarily inform the HLR of a visiting MS's location after an HLR data failure, limiting the HLR's ability to deliver calls to the MS.

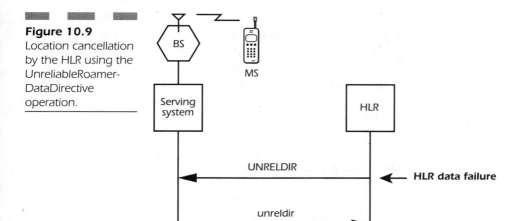

Figure 10.9
Location cancellation
by the HLR using the
UnreliableRoamer-
DataDirective
operation.

MS Location Management in Border Cell Situations

Our discussion of location-management processes has, up to this point, assumed a somewhat idealized cellular radio environment. Each time the MS wishes to access the system, it scans for and selects a single, strong control channel and then uses that channel to transmit a system access message, like a registration or call-origination message. If the receiving system already has a valid temporary record for the MS (i.e., is the current serving system), a location update to the MS's HLR is not normally necessary; the serving system simply continues to provide service to the MS.

However, if one or more other systems are also listening for system access messages on the same control channel, we have the makings of what is known as a *multiple-access problem*. Why would multiple systems be using the same control channel? In light of the frequency reuse concept described in Chapter 1, adjacent cells would not use the same channels. However, dense urban areas or topological features of the service area, such as the presence of lakes, rivers, or flat, open land, may result in unintended propagation of the MS's signal well beyond the engineered boundaries of the serving cell, into the service area of a "border system."[6] The received system-access message in the border system would be seen

---

[6] A border system does not necessarily serve cells immediately adjacent to the system in question, merely in close enough proximity that signals intended for one system are overheard by the other.

as an initial system access, causing a location update to the MS's HLR and a location cancellation from the HLR to the actual serving system (see Figure 10.10). This situation is particularly problematic when the system access is a call origination, since the effect is, at best, to place the MS on a radio channel with poorer signal quality than necessary and—more likely—to drop the call altogether when the MS cannot detect the radio channel assigned by the remote system for the call.

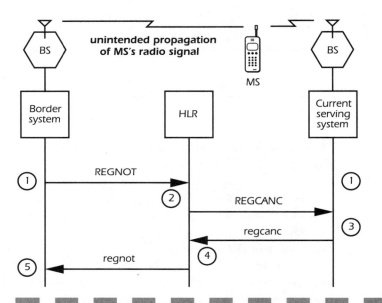

Figure 10.10 Unintended location update and location cancellation due to border cell conditions: (1) The MS is already registered on the serving system; however, a system access is overheard by a border system. The border system requests location update in the HLR. (2) The HLR has stored location information for the MS; therefore, it requests location cancellation from the current serving system. (3) The current serving system deletes the temporary MS record and sends an acknowledgment to the HLR. (4) The HLR acknowledges the location update. The HLR normally also provides service qualification in the regnot message. (5) The border system attempts to provide service to the MS.

To address this problem,[7] ANSI-41 location-management processes incorporate special procedures that allow the HLR to determine the best serving system for an MS when border cell conditions result in (1) multi-

---

[7] Note that this is only one of several border cell problems addressed by ANSI-41. Other border cell issues are discussed in later chapters.

ple location updates for a single MS system access event or (2) one or more location updates from systems with poorer received signal quality than the current serving system. These procedures require the border systems to provide additional information, including an indication that the system access occurred in a border cell and a measure of the received signal quality, to the HLR along with the location information. This information is conveyed in the ANSI-41 REGNOT message (see Figure 10.11).

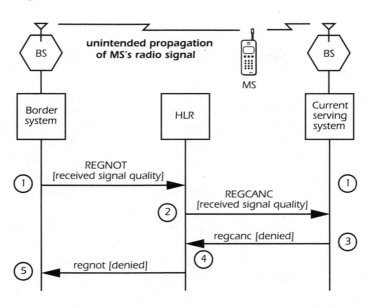

Figure 10.11

Part of ANSI-41's location management solution for border cell conditions. The steps are as described in Figure 10.10, except that the current serving system denies the location cancellation request based on its comparison of the received signal quality provided by the border system and its own recorded measurements.

The HLR location-update process makes a decision on which of several location updates is best. The HLR location-cancellation process then attempts to cancel registration in the current serving system by using the ANSI-41 REGCANC message, providing that system with the signal quality information from the best location update selected by the HLR. The serving system location-cancellation process, in turn, evaluates the location-cancellation directive from the HLR, using information it has stored regarding the last system access by the MS in question. If the serving system determines that it is the best system for the MS, it can deny the location-cancellation directive, sending the denial indication to the HLR in the ANSI-41 regcanc message. The HLR honors the location-cancellation denial from the serving system and refuses the location update from the border system; refusal is conveyed in the ANSI-41 regnot message.

Note that Figure 10.11 shows the RegistrationCancellation operation embedded within the RegistrationNotification operation in a *synchronous* fashion—steps 2 and 3 occur before step 4. Prior to IS-41-C (or, more correctly, TSB-65), these operations were *asynchronous*, so that step 4 occured before steps 2 and 3. This change was made to support the location-cancellation denial capability.

Examples of methods that the HLR can use to determine which of several location updates is best are provided in ANSI-41.6-D, Annex F.

MS State Management

The ANSI-41 MS state-management function encompasses the processes by which the call-delivery availability state of an MS is coordinated between the serving system and the HLR. The state may be either *active* or *inactive*. In the active state, the MS is available for call delivery. When in the inactive state, the MS is not available for call delivery; however, an inactive MS may be able to accept short-message service (SMS) deliveries. An MS may be deemed inactive because:

- *The MS is not registered.* When the HLR does not have a valid location for the MS, the MS is considered inactive.
- *The MS is out of radio contact.* The MS may have missed autonomous registration events due to a loss of radio contact and may have been designated inactive by the serving system.
- *The MS is intentionally inaccessible.* The MS may have gone into a mode (e.g., sleep mode) whereby it is intentionally inaccessible.

Keeping with its open nature, ANSI-41 also allows serving systems to designate an MS inactive based on internal algorithms.

MS State-management Processes

MS state-management processes include the serving system MS state-management process and the HLR MS state management process.

Generally, the MS state is set to active by the serving system and the HLR after a successful location update for the MS. Likewise, after a successful location cancellation, the HLR sets the MS state to inactive; the state may be reset to active if the location cancellation is nested within a location-update process. In this sense, the serving system MS state-management process and HLR MS state-management process make use of the same operations as the location-update and location-cancellation processes, such as RegistrationNotification, RegistrationCancellation, MSInactive, BulkDeregistration, and UnreliableRoamerDataDirective. Additionally, the MS state may be explicitly set to inactive in the serving system and the HLR in the following ways:

- When the serving system sends a valid ANSI-41 REGNOT message including a valid AvailabilityType parameter to the HLR, the serving system sets the MS's state to inactive; when the HLR successfully receives the message, it also sets the MS's state to inactive (see Figure 10.12).
- When the serving system sends a valid ANSI-41 MSINACT message to the HLR, the serving system sets the MS's state to inactive; when the HLR successfully receives the message, it also sets the MS's state to inactive (see Figure 10.13). Note that this message may also result in location cancellation for the MS if the DeregistrationType parameter is included.
- When the serving system sends a valid ANSI-41 RoutingRequest Return Result (routreq) message to the HLR, including the AccessDeniedReason (ACCDEN) parameter set to inactive, the serving system sets the MS's state to inactive (if not already inactive); when the HLR successfully receives the message, it also sets the MS's state to inactive (see Figure 10.14).

Once set, the MS's state remains inactive until a serving system sends to the HLR a valid REGNOT message that does not include the AvailabilityType parameter.

HLR and VLR Fault Recovery

The ANSI-41 HLR and VLR fault-recovery function encompasses the processes that attempt to ameliorate the effects of HLR and VLR failures in the mobile telecommunications network.

Figure 10.12
Setting the MS's state
to inactive using the
RegistrationNotification
operation.

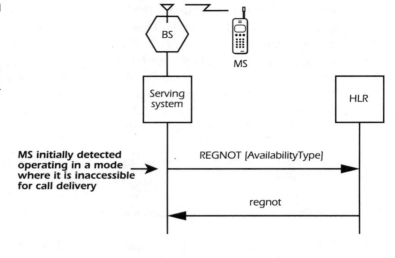

MS initially detected operating in a mode where it is inaccessible for call delivery

REGNOT [AvailabilityType]

regnot

Figure 10.13
Setting the MS's state
to inactive using the
MSInactive operation.

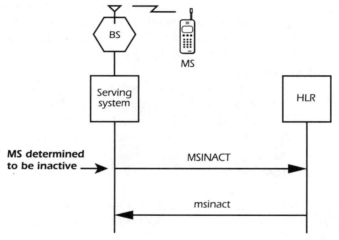

MS determined to be inactive

MSINACT

msinact

HLR Fault-recovery Processes

The HLR fault-recovery processes handle the case where the HLR experiences a failure that renders unreliable the data it uses to keep track of its roaming MSs. The UnreliableRoamerDataDirective operation is specifically designed for this purpose; the HLR sends the UNRELDIR message to each system (i.e., VLR) that it believes may be serving one or more of the HLR's MSs. On receipt, the serving system clears all temporary roaming records for MSs associated with the HLR and returns the unreldir message to the HLR as an acknowledgment (see Figure 10.15).

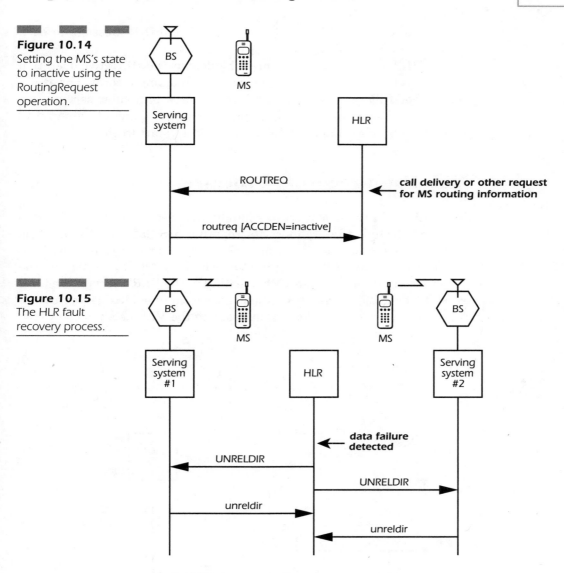

Figure 10.14
Setting the MS's state to inactive using the RoutingRequest operation.

Figure 10.15
The HLR fault recovery process.

ANSI-41 does not specify the mechanism by which the HLR determines the visited systems affected. At least two strategies are possible:

- The HLR maintains a list of all possible serving systems in a nonvolatile storage area.
- The HLR maintains a list of active serving systems (i.e., systems actually serving one or more of the HLR's MSs at any particular time) in a nonvolatile storage area.

In case of failure, the HLR sends an UNRELDIR message to each serving system on this list.

The HLR may also use the UnreliableRoamerDataDirective operation to selectively clear a single serving system's roaming records. For example, if the HLR received repeated errors from a particular serving system, it could send an UNRELDIR message to the system to direct it to clear its data related to the HLR and, in effect, "start fresh."

VLR Fault-recovery Processes

The VLR fault-recovery processes handle the case where the VLR experiences a failure that renders its roaming MS data unreliable. The BulkDeregistration operation may be used for this purpose; the VLR sends the BULKDEREG message to each HLR it believes may be the home system for one or more MSs visiting in the VLR's service area. On receipt, the HLR clears all location information for MSs served by the VLR and returns the bulkdereg message to the VLR as an acknowledgment (see Figure 10.16).

Figure 10.16
The VLR fault
recovery process.

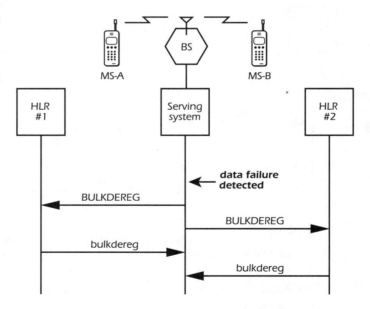

ANSI-41 does not specify the mechanism by which the VLR selects the HLRs affected. As with HLR fault recovery, at least two strategies are possible:

- The VLR maintains a list of all possible HLRs in a nonvolatile storage area.
- The VLR maintains a list of HLRs associated with active visiting MSs in a nonvolatile storage area.

In case of failure, the VLR sends a BULKDEREG message to each HLR on this list.

Summary of ANSI-41 Operations Used for Basic Automatic Roaming

Table 10.1 summarizes the ANSI-41 operations used by the basic automatic roaming functions described in this chapter.

TABLE 10.1

Use of ANSI-41 Operations for Basic Automatic Roaming

Function	ANSI-41 Operations Used for the Function
MS service qualification	RegistrationNotification, QualificationRequest, QualificationDirective
MS location management	RegistrationNotification, MSInactive, ParameterRequest, BulkDeregistration, RegistrationCancellation, UnreliableRoamerDataDirective
MS state management	RegistrationNotification, MSInactive, BulkDeregistration, RegistrationCancellation, UnreliableRoamerDataDirective, RoutingRequest
HLR and VLR fault recovery	BulkDeregistration, UnreliableRoamerDataDirective

CHAPTER 11

Authentication Functions

In this chapter, we discuss the ANSI-41 mobile telecommunications network functions related to authentication. These functions are divided into the following categories:[1]

- Shared secret data sharing
- Global challenge
- Unique challenge
- Shared secret data update
- Call-history count update
- Authentication reporting

We also examine the ANSI-41 application processes that support these network functions. In the course of describing the processes, we identify the ANSI-41 mobile application part (MAP) operations (e.g., AuthenticationRequest) used to accomplish authentication process tasks. We summarize this information at the end of the chapter. Note that we employ the ANSI-41 convention for operation component acronyms—the Invoke component acronym is in all-capital letters (e.g., AUTHREQ), while the Return Result component acronym is in all-lowercase letters (e.g., authreq). Refer to the ANSI-41 standard for the descriptions of the individual ANSI-41 operations and parameters. Also, consult the Glossary for a description of general terms (e.g., serving system) that are not explicitly defined in this chapter.

Throughout this chapter we generally consider the serving system as a single entity encompassing the mobile switching center (MSC) and visitor location register (VLR) functional entities. This simplifies the descriptions of ANSI-41 application processes and also is representative of a large percentage of the ANSI-41 implementations currently in service. However, ANSI-41 authentication assigns specific processing duties to the VLR that make the concept of sharing a VLR among multiple MSCs a viable alternative to the combined implementation. Therefore, while we describe authentication processes in terms of a single "MSC plus VLR" entity, keep in mind that the potential for separation exists and is fully defined in ANSI-41.

Note that the ANSI-41 authentication procedures are intended to support mixed-version networks; that is, the authentication center (AC) uses ANSI-41 procedures while the serving system uses TSB51 proce-

---

[1] The list of authentication functions reflects the authors' subjective categorization of certain ANSI-41 functions. Likewise, the process descriptions are the authors' subjective interpretation of the ANSI-41 procedures.

dures designed for use with IS-41-B. We do not discuss this aspect of ANSI-41; the reader is referred to the ANSI-41 standard for the descriptions of these backward-compatibility mechanisms.

Throughout this chapter we describe the authentication procedures defined in ANSI-41 Revision D (ANSI-41-D). However, we conclude with a description of the major authentication enhancements that have subsequently been standardized (i.e., in TIA/EIA/IS-778 and in TIA/EIA/IS-751) and will appear in ANSI-41 Revision E.

What Is Authentication?

Authentication is a set of functions used to prevent fraudulent access to cellular networks by phones illegally programmed with counterfeit mobile station identification (MSID) and electronic serial number (ESN) information. The functions require no subscriber intervention and provide a much more robust method of validating the true identity of a subscriber than does the basic service qualification function described in the previous chapter.

The ANSI-41 authentication functions are independent of the air-interface protocol used to access the network, and subscribers are never involved in the process. A successful outcome of authentication (i.e., an *authenticated MS*) occurs when it can be demonstrated that the mobile station (MS) and network possess identical results of an independent calculation performed in both the MS and the network. The AC is the primary functional entity in the network responsible for performing this calculation, although the serving system (i.e., the VLR) may also be allocated certain responsibilities, which we shall describe. The authentication calculations are based on a set of algorithms, collectively known as the *CAVE* (cellular authentication and voice encryption) *algorithm*.

The authentication process can be invoked by many events; however, it is performed most often when the following situations arise:

1. Registration
2. Call origination
3. Call termination

Note that the authentication process is not necessary for every instance of the events listed above. For example, during autonomous registration, authentication may be performed only if the subscriber has moved to a new location (i.e., the location management process has updated the subscriber's location at the HLR) provided that is part of the *authentication policy* in effect between the serving and home systems.

The authentication process and algorithm are based on the following two secret numbers:

- Authentication key (A-key)
- Shared secret data (SSD)

These numbers are the fundamental basis for ANSI-41 authentication and are described in the following sections.

Authentication Key (A-key)

The A-key is a 64-bit secret number that is the permanent *key* used by the authentication calculations in both the MS and the AC. The A-key is permanently installed into the MS and is securely stored (i.e., provisioned) at the AC in the network when a new subscription is obtained. In general, the method for installing the A-key in the MS is implementation dependent. However, the TIA has recommended a method for manually programming the A-key into the MS (refer to *TIA/EIA TSB50*). The TIA has also standardized a method for over-the-air A-key programming (refer to TIA/EIA/IS-725 and see Chapter 15).

Once the A-key is installed in the MS, it should not be displayable or retrievable. The MS and the AC are the only functional entities ever aware of the A-key; it is never transmitted over the air or passed between systems. The primary function of the A-key is as a parameter used in the calculation to create the SSD.

Shared Secret Data

The SSD is a 128-bit secret number that is essentially a temporary key used by the authentication calculations in both the MS and the AC. The SSD may also be shared with the serving system (specifically, the VLR functional entity within the serving system) via a number of ANSI-41 messages. Unlike the A-key, the SSD is a semipermanent value. It can be modified by the network at any time, and the network can command the MS to generate a new value.

The SSD is the result of a calculation using the A-key, the ESN, and a random number shared between the MS and the network. The SSD calculation results in two separate 64-bit values, SSD_A and SSD_B. SSD_A is the value used specifically for the authentication functions. SSD_B is the value used specifically for the encryption algorithms used for voice and data privacy and to encrypt and decrypt selected signaling messages on the radio traffic channel. Figure 11.1 illustrates the SSD generation procedure.

Figure 11.1
Deriving the SSD from the A-key. The HLR and AC functional entities are shown as a single unit due to their close functional relationship.

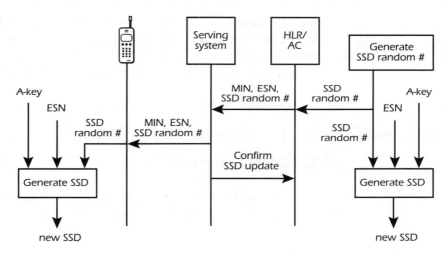

Where Are Authentication Functions Specified in ANSI-41?

Authentication functions are specified in three parts of ANSI-41:

- **ANSI-41.5** provides the required formats—the bit-by-bit encoding—of all the ANSI-41 operation components, including those used for authentication. ANSI-41.5 defines both the messages (e.g., AuthenticationRequest Invoke) and the message parameters (e.g., AuthenticationResult, RandomVariable).
- **ANSI-41.6** provides algorithmic descriptions of the procedures associated with sending and receiving most ANSI-41 messages, including those used for authentication. As we pointed out in Chapter 8, ANSI-41 does not define a service interface between the MAP application

processes and the MAP application service element (ASE); therefore, the procedures in ANSI-41.6 encompass both application-process descriptions and ASE procedural descriptions. Furthermore, the procedures leave considerable room for implementation-dependent customization; for example, there are numerous references in ANSI-41.6 to *local procedures* and *internal algorithms*.

■ **ANSI-41.3** steps back from the protocol details contained in parts 5 and 6 and attempts to explain—using information flow diagrams and step-by-step descriptions—how the operations are used individually and together to accomplish authentication application process tasks. Annex A in ANSI-41.3 includes a description of the assumptions made in the design of ANSI-41 authentication processes; these assumptions provide valuable implementation guidance.

Issues Associated with Authentication

The authentication functions are designed to validate a subscriber that is being served by either the home system or a visited system. Many issues arise in the network to support authentication, including the following:

■ A need to determine which functional entity is the *authentication controller*. The ANSI-41 authentication processes support authentication control from either the AC or the serving system (i.e., the VLR). The AC, as controller, can perform all authentication functions. The VLR, when the SSD key is shared, can perform only some authentication functions, as we shall describe.

■ Many systems still do not support authentication. Although the situation has improved since the first edition of this book, there are still systems that do not support the capability. Presumably, in these cases, the benefits of authentication do not outweigh the implementation costs involved.

■ Some mobile stations do not support authentication. The ANSI-41 authentication processes are dependent on the MS's having the capability to perform the algorithms and support the authentication key. While all mobile stations produced in the last several years are authentication capable, there still exist some older units without this capability.

■ Management of the A-key is an issue. A primary consideration for supporting authentication is A-key management, or how the key is kept secure, how it is provided to the MS, and how it is maintained. Since the entire authentication process is dependent on the A-key's remaining secret, adequate management of the A-key is of paramount importance. The basic requirements of an A-key management protocol are the following:

   – The A-key must be handled by a minimum number of people—none is best!
   – The AC must be designed as a very secure system.
   – The A-key must not be readable by any party.
   – Changes to the A-key in the MS and AC must be performed in a secure manner.

The methods for fulfilling these requirements have not been completely addressed by ANSI-41, although the over-the-air service provisioning (OTASP) feature (see Chapter 15) provides some useful A-key management functions. Beyond OTASP, A-key management is generally left to the network operator and its vendors to resolve.

Shared Secret Data Sharing

The ANSI-41 SSD sharing function encompasses the processes by which the authentication center and the serving system (i.e., the VLR) manage the sharing of authentication responsibilities for a visiting MS. Sharing authentication responsibilities is only possible if the SSD, in addition to being shared between the AC and the MS, is shared between the AC and the serving system; this is referred to as *SSD sharing* or *shared SSD* in ANSI-41. Additionally, ANSI-41 specifies that when SSD is shared with the serving system, the call history count (COUNT) information is shared as well.[2]

SSD sharing gives the serving system significant local control over the authentication of a visiting MS. Specifically, it can control the following network functions:

■ Global challenge, for all but the initial system access when SSD sharing is not yet established

---

[2] Assuming the call history count is part of the authentication "tool set" used by the home system.

- Unique challenge, again for all but the initial system access
- The base station challenge portion of an AC-initiated SSD update (see "SSD Update")
- COUNT update

Serving-system control of these network functions reduces the authentication-related signaling traffic between the serving and home systems and the associated call processing delays.

SSD sharing processes include the AC SSD sharing process, the serving system SSD sharing process, and the HLR SSD sharing process. These processes support the following tasks:

1. Notifying the AC of the serving system's capabilities for SSD sharing.
2. "Turning on" and "turning off" SSD sharing.
3. AC retrieval of shared COUNT information.

Notifying the AC of the Serving System's SSD Sharing Capabilities

The serving system has the responsibility to notify the AC of its capabilities for SSD sharing so that the AC can determine whether sharing is possible. This is normally done when the serving system first communicates with the home system, for example, on initial system access by a visiting MS. The serving system's SSD sharing capabilities are conveyed in the ANSI-41 SystemCapabilities (SYSCAP) parameter. Two signaling scenarios are possible:

- *The serving system determines that the MS is authentication capable.* This can be determined by the serving system on the basis of the air-interface protocol used to access the system (for example, TDMA or CDMA, which inherently support authentication), on the fact that a global challenge was in effect and the MS returned authentication parameters, or on specific provisioning of the serving system by the network operator (e.g., all mobile stations are assumed to be authentication capable). In this case, the first message sent from the serving system to the HLR is normally the ANSI-41 AuthenticationRequest Invoke (AUTHREQ) message, including the SYSCAP parameter. The HLR relays this message to the AC[3] (see Figure 11.2), thereby notify-

---

[3] All communication between the serving system and the AC is via the HLR.

ing the AC of the serving system's SSD sharing capabilities. An exception would be if a global challenge were in effect but the MS did not return authentication parameters. This should be treated as an authentication failure. The possible methods of reporting the failure are described in "Authentication Reporting."

- *The serving system determines that the MS is not authentication-capable.* In this case, the first ANSI-41-D message sent from the serving system to the HLR is the RegistrationNotification Invoke (REGNOT) message, containing the SYSCAP parameter. However, ANSI-41 does not define a standard way for the HLR to convey this information to the AC; therefore, when the AC is separated from the HLR (via the ANSI-41 standard "H-interface"), the AC is not informed of the serving system's SSD sharing capabilities on initial system access by a nonauthentication–capable MS. Of course, this is not a problem if the MS is truly not capable of authentication. However, if information in the HLR database indicates that the MS is in fact authentication ready, the HLR must use the information in the REGNOT message to determine whether a fraudulent MS is attempting to access the system. This scenario is not explicitly specified in ANSI-41-D (see the discussion of clarified handling of the initial system access in "Authentication Enhancements in IS-778").

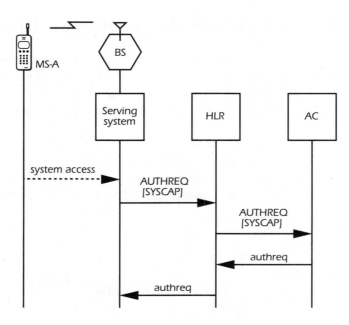

Figure 11.2
Conveying the serving system's SSD sharing capabilities to the AC.

ANSI-41 also does not explicitly define the procedures to follow if the serving system's ability to handle shared authentication responsibilities changes during the period when an MS is visiting in the serving system's area, such as, due to a failure in the subsystem that performs the authentication calculations. We can speculate on at least two approaches to this scenario:

- If the serving system is still able to support AC-controlled authentication procedures, then the serving system can notify the AC of its modified SSD sharing capabilities at the next authentication event (e.g., call origination).
- Alternatively, the serving system can initiate the location-cancellation process for one or more visiting MSs in the system, as described in Chapter 10; that is, using the ANSI-41 MSInactive Invoke message including the DeregistrationType parameter, or using the ANSI-41 BulkDeregistration Invoke message. Use of MSInactive is the preferred approach since (a) it allows the serving system to convey the current value of COUNT to the AC (see "Call History Count Update"), and (b) it allows selective location cancellation of only those MSs for which the serving system is sharing authentication responsibilities, rather than all visiting MSs.

Turning SSD Sharing On and Off

Given that the serving system is capable of sharing authentication responsibilities, the AC initiates SSD sharing by sending the ANSI-41 SSD and COUNT parameters to the serving system. The AC may use one of four ANSI-41 messages for this purpose (see Figure 11.3):

- AuthenticationDirective Invoke (AUTHDIR)
- AuthenticationRequest Return Result (authreq)[4]
- AuthenticationStatusReport Return Result (asreport)[4]
- AuthenticationFailureReport Return Result (afreport)[4]

---

[4] Of course, the information may be conveyed in the specified Return Result message only as part of the AC's response to the corresponding Invoke message (e.g., the authreq message is in response to the AUTHREQ message).

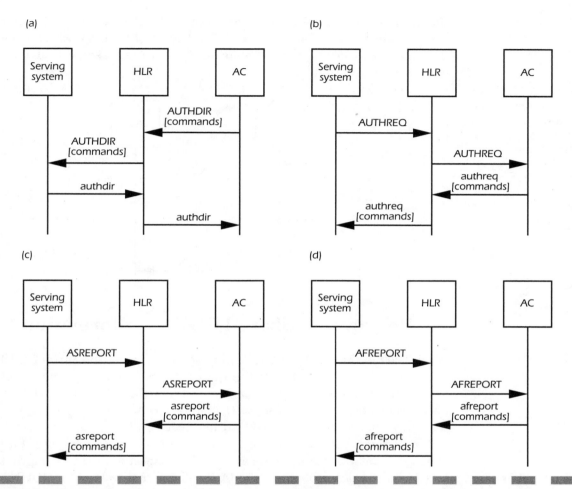

Figure 11.3 The AC may send authentication-related commands to the serving system, including the command to initiate SSD sharing, using four ANSI-41 methods: (a) the AUTHDIR message, (b) the authreq message, (c) the asreport message, and (d) the afreport message.

Once SSD sharing is turned on, the AC may turn it off by sending the SSDNotShared (NOSSD) parameter to the serving system. Although ANSI-41 specifies that the AC may use any of the four ANSI-41 messages listed above to turn off SSD sharing, the AUTHDIR message appears to be the most practical for the following reason: When turning off SSD sharing, the AC must retrieve the current COUNT from the serving system. The response to the AUTHDIR—the AuthenticationDirective Return Result (authdir) message—includes the COUNT parameter;

therefore, it provides a convenient method of COUNT retrieval under these conditions (see Figure 11.4). AC retrieval of COUNT when using the other messages to turn off sharing (i.e., authreq, asreport, or afreport) is more problematic.

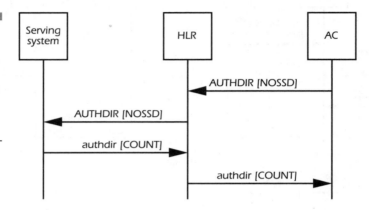

Figure 11.4
To turn off SSD sharing, the AUTHDIR message should be used since it allows the COUNT to be retrieved from the serving system.

AC Retrieval of Shared COUNT Information

When SSD is shared, the AC gives the serving system the current SSD and COUNT values for the MS in question. The serving system is not autonomously able to initiate a change in the MS's SSD—this is strictly an AC-controlled function (see "SSD Update"). However, the serving system may—based on its internal authentication algorithms—order the MS to increment its COUNT value without informing the AC (see "Call History Count Update"). Therefore, the final aspect of SSD sharing that has to be managed between the AC and serving system is COUNT retrieval—how the AC gets the (potentially revised) COUNT information (1) from the serving system when the AC turns off SSD sharing and (2) from the previous serving system when the MS moves to a new serving system. Four general scenarios are handled:

1. *COUNT retrieval when the AC turns off SSD sharing* was described in the previous section.
2. *The MS moves from one authentication-capable system to another authentication-capable system* (see Figure 11.5). In this case, the initial ANSI-41 message sent from the new serving system to the home system is normally an AUTHREQ message.[5] When the AC receives this message, it sends a CountRequest Invoke (COUNTREQ) message to the previous serving system. The previous serving system

replies with its stored value of COUNT in the CountRequest Return Result (countreq) message, and normal AuthenticationRequest processing continues.

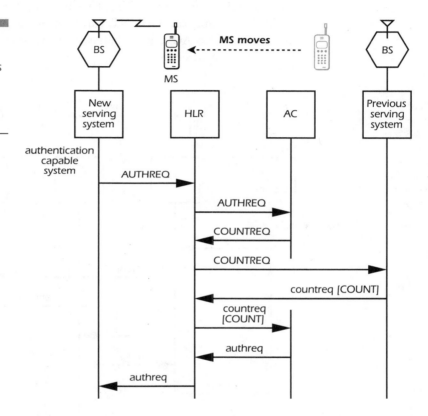

Figure 11.5
COUNT retrieval when the MS moves from one authentication-capable system to another.

3. *The MS moves from an authentication-capable system to a system that is not authentication capable* (see Figure 11.6). In this case, the initial ANSI-41 message sent from the new serving system to the home system is the REGNOT message associated with the service qualification and location update processes (see Chapter 10). The HLR does not forward this message to the AC, so the AC cannot use the CountRequest method described for case 1. Rather, the HLR initiates the location-cancellation process and sends a RegistrationCancellation Invoke

---

[5] If the first message sent from the new serving system is a RegistrationNotification Invoke message and not an AUTHREQ message, then the procedure described for case 3 is used. However, if the MS is recognized as authentication-capable, the new serving system normally will first attempt to authenticate the MS—possibly with a parallel service qualification check—before proceeding with the location update process.

(REGCANC) message to the previous serving system. The serving system inserts the current COUNT value in the RegistrationCancellation Return Result (regcanc) message sent to the HLR. On receipt of regcanc, the HLR sends an MSInactive Invoke (MSINACT) message to the AC, including the COUNT value received from the previous serving system. The AC acknowledges the message thus completing the COUNT retrieval process.

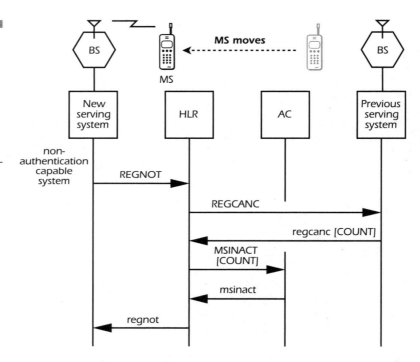

Figure 11.6
COUNT retrieval when the MS moves from an authentication-capable system to a system that is not authentication-capable.

4. *The serving system initiates location cancellation of the MS.* Chapter 10 describes two situations under which this would occur:

   (a) In the first case, the serving system uses the MSINACT message to cancel location for a single MS. The HLR forwards the MSIN-ACT message to the AC, including the COUNT value (see Figure 11.7). The AC acknowledges the message, completing the COUNT retrieval process.

   (b) In the second case, the serving system uses the BulkDeregistration Invoke (BULKDEREG) message to cancel location for all an HLR's MSs currently visiting in the serving system. ANSI-41 does not support COUNT retrieval in this case; that is, when the serving system uses the BULKDEREG message to cancel location for multiple MSs, the associated COUNT information is lost. This is one argument for using the BULKDEREG message for serving system restoration purposes only.

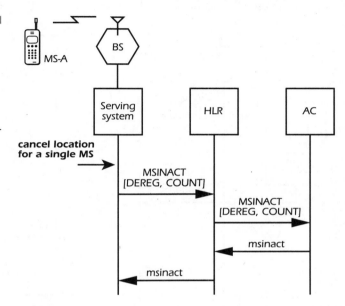

Figure 11.7
COUNT retrieval when the MS's location is canceled by the serving system using the MSInactive operation.

Another case where ANSI-41 does not support COUNT information retrieval from the serving system occurs when the HLR initiates the location cancellation of all its roaming MSs due to HLR data failure (see Chapter 10). The HLR uses the UnreliableRoamerDataDirective Invoke message for this purpose. The response to this message does not support transfer of COUNT information; therefore, this information is lost.

Global Challenge

The ANSI-41 global-challenge function encompasses the processes in which:

- The serving system presents a numeric authentication challenge to all MSs that are using a particular radio control channel.
- The ANSI-41 AC (or the serving system if SSD is shared) verifies that the numeric authentication response from an MS attempting to access the system is correct.

This is called a *global challenge* because the challenge indicator and the random number used for the challenge are broadcast on the radio control channel and so are used by *all* MSs accessing that control channel. Figure 11.8 depicts the basic global-challenge procedure for authentication when SSD is not shared.

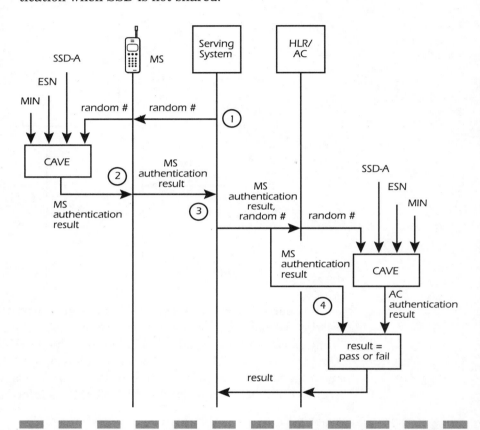

Figure 11.8 *Basic global challenge authentication process when SSD is not shared: (1) The serving system generates a random number that it sends to the MS in the control channel's overhead message train (OMT). (2) The MS calculates an authentication result and transmits that result back to the serving system when it accesses the system for registration, call origination, or page response purposes. (3) If SSD is not shared, the serving system forwards the authentication result and the random number to the AC. (4) The AC independently calculates an authentication result and compares it to the result received from the MS. If the results match, the MS has successfully responded to the challenge. If they do not match, the MS may be considered fraudulent and service may be denied. If SSD is shared, the serving system performs the calculations.*

Global-challenge processes include the serving system global-challenge process, the border system global-challenge process, and the AC global-challenge process.

Serving System Global-challenge Process

The details of the serving system global-challenge process depend on whether SSD is shared between the serving system and the AC. The steps in the process are as follows:

1. Issue the numeric global challenge to the MSs in the serving system's service area.
2. Collect the authentication information received from each MS that attempts to access the system.
3. Prevalidate (as explained below) the MS authentication information. Then, do either step 4 (if SSD is not shared) or step 5 (if SSD is shared).
4. Forward the MS's global-challenge response to the AC for evaluation, and complete the global-challenge process as directed by the AC.
5. Evaluate the MS's global-challenge response locally, and complete the global-challenge process based on the results of the local evaluation.

Issuing the Global Challenge

In the following discussion of the mechanics of the global-challenge process, we make a couple of assumptions:

- The serving system and base station are capable of and configured for executing the global challenge.
- The MS is authentication capable, and authentication is required for the MS.

The first assumption is purely a deployment issue, and is outside the scope of ANSI-41. In fact, systems can be designed to use only the unique challenge method of authentication. The second assumption raises a couple of issues related to the initial system access: (a) how the serving system knows if an MS is authentication capable, and (b) how the serving system knows if authentication is required. We are aware of three general strategies:

- Avoid the issue—when the serving system first detects a visiting MS, assume it is authentication capable and that authentication is required, and process system access attempts accordingly. The procedures are designed so that the home system can notify the serving system that the authentication of subsequent system access is not required—for example, if the MS is actually *not* authentication capable—when the subscriber's profile is transferred from the home system to the serving system as part of the service-qualification process. The ANSI-41 AuthenticationCapability parameter provides this indicator. This approach is explicitly called for in TIA/EIA/IS-778, whereas ANSI-41-D was not explicit. In fact, the serving system can use the IS-778 procedures to obtain the subscriber profile prior to—or in parallel with—authentication. These procedures are described at the end of this chapter in "Authentication Enhancements in IS-778."

- Make the decision based on information exchanged between the home system operator and serving system operator in a roaming agreement and stored in the serving system; the *roaming agreement* can include an indication of the MS's authentication capability. As with the first strategy, the home system can override the roaming agreement indication during the service qualification process and can direct the serving system to enforce or bypass authentication of the MS. While this approach is possible, it is unlikely to be used since it would burden the establishment and maintenance of roaming agreements.

- Make the decision based on information provided by the MS over the air interface; for example, if the MS attempts to access the system using the EIA/TIA-553 AMPS protocol, the serving system assumes that the MS is not authentication capable, whereas if a TDMA or CDMA protocol is used, the serving system assumes that the MS is authentication capable. Again, the home system can use the service qualification process to override the serving system determination (e.g., for the latest vintage of AMPS MSs that support authentication).

The serving system issues the numeric global challenge (see Figure 11.9) by selecting a 32-bit random number (RAND) and transmitting the RAND, along with the authentication information element (AUTH), set to 1, in the overhead message train (OMT) on the radio control channel.

Prevalidating the MS Information: Normal Case

Under normal circumstances, an MS wishing to access the system:

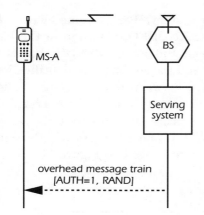

Figure 11.9
Issuing the global authentication challenge.

1. Reads the OMT to (a) determine whether a global challenge is in effect for access to the system (i.e., the AUTH bit is set to 1), and (b) acquires the RAND challenge value
2. Executes the cellular authentication and voice encryption (CAVE) algorithm for the global challenge
3. Transmits the system access message (e.g., register, call origination, SMS origination, or page response) to the serving system including
   (a) The calculated authentication result (AUTHR)
   (b) The current value of COUNT stored in the MS
   (c) The 8 most-significant bits of the 32-bit RAND, called *RANDC*.

This information is collected by the serving system and prevalidated. Prevalidation involves checking that the RANDC received from the MS corresponds to the eight most-significant bits of the RAND transmitted by the serving BS. There are at least three possible reasons for differences between RAND and RANDC:

- The MS is fraudulent and is attempting to "replay" a valid system access message that was previously recorded.
- The MS is valid but is using a RAND it received from one system to access a neighboring system (see Figure 11.10); this is another border cell problem addressed in ANSI-41 and is described below.
- The MS did not receive the RAND transmitted by the serving base station (BS) and so chose to use the default value, RAND = 0.

If the serving system determines—based on internal algorithms—that the mismatch may be due to a border cell condition, it initiates a request for the border system's RAND value by sending a RandomVariableRequest Invoke (RANDREQ) message to the border system. The RANDREQ message includes the RANDC that the serving system received from the MS and the serving cell identification (the latter being information coordinated between the two systems during provisioning). The border system checks its records of current and recently used RAND values in the area corresponding to the serving cell and responds with a matching RAND, if one exists, in the RandomVariableRequest Return Result (randreq) message (see Figure 11.10).

Figure 11.10
Resolving a potential
border cell problem
by using the
RandomVariable-
Request operation.

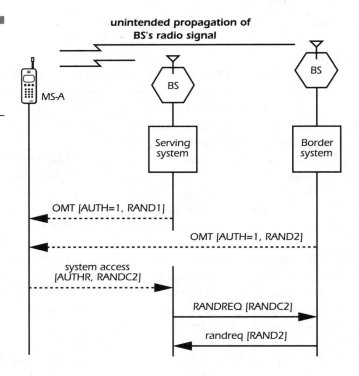

If either a valid RANDC is received from the MS or a valid RAND is received from the neighbor MSC, the serving system moves on to the global challenge evaluation stage. However, if a RAND is not returned by the neighbor MSC or the serving system determines that the RANDC failure is not due to a border cell problem, then the serving system considers the access to be fraudulent. The serving system handling of the MS access in this case is based on *local recovery procedures*. For example, the serving system may apply access-denial treatment to the MS. If the system access attempt is a call origination or a flash request during a call, denial could be in the form of an announcement or tone, followed by call disconnect. Regardless of the MS treatment, the serving system initiates the authentication reporting process (see "Authentication Reporting"), indicating that the failure is due to a RANDC mismatch.

The serving system global-challenge process also initiates authentication reporting if the MS fails to provide the AUTHR and COUNT authentication information when it is challenged to do so.

Prevalidating the MS Information: Border Cell Case

Under border cell conditions, it is possible for an MS to (1) detect a page request on a control channel from the serving system, (2) rescan and select a control channel from a border system, and (3) respond to the page on the control channel from a border system. If a global challenge is issued on the border system's control channel, the MS's challenge response is based on the RAND transmitted on that control channel. ANSI-41 provides intersystem paging procedures (see Chapter 12) that allow successful call delivery under these circumstances. These procedures also support the transfer of the MS global challenge response from the border system to the serving system, via the InterSystemPage2 Return Result (ispage2) message (see later in this section for a description of the border system global challenge process).

If the serving system receives the RAND and RANDC parameters in the ispage2 message from the border system, prevalidation of the authentication information involves checking that the RANDC corresponds to the eight most-significant bits of the RAND. If this is the case, the serving system moves on to the global-challenge evaluation stage; otherwise, the serving system considers the access to be fraudulent. The serving system initiates the authentication reporting process, indicating that the failure is due to a RANDC mismatch.

Global Challenge when SSD is not Shared

Given no failures up to this point, the subsequent handling of the global challenge information from the MS depends on whether the serving system is currently sharing SSD with the AC. SSD sharing would not be established:

- If this were the initial system access by the MS in the serving system's service area (i.e., SSD sharing would not yet have been established between the serving system and the AC).
- If the serving system were not capable of SSD sharing.

■ If the AC's authentication policy did not allow SSD sharing.

In these cases, the serving system sends the MS global-challenge information in an AUTHREQ message to the AC. The serving system[6] then handles the response from the AC, that is, the authreq message. This message can trigger the following additional processes in the serving system, depending on the message's contents:

■ SSD sharing
■ Unique challenge
■ SSD update
■ COUNT update

Additionally, the message may contain the ANSI-41 DenyAccess parameter, which directs the serving system to release the system resources it has allocated for the MS access in question; this may include the release of the call in progress, if one exists.

If the MS is successfully authenticated and no additional authentication operations are requested, the serving system terminates the global challenge process and continues with normal system access processing, which may involve call origination, location update, or other serving system processes (see Figure 11.11).

Global Challenge when SSD is Shared

If the serving system is sharing SSD, it assumes responsibility for evaluating the global-challenge information it receives from the MS. Evaluation involves five tests:

1. Check that the MSID and ESN reported by the MS are valid.[7]
2. Check that the type of air interface used by the MS to access the system is valid (e.g., is the same as the type stored in the serving system's database, if so provisioned).
3. Check that the AUTHR and COUNT information was provided by the MS.
4. Execute the CAVE algorithm, and verify that the AUTHR generated equals the AUTHR received from the MS.

---

[6] Note that the call may have been handed off when the initial serving system—now the anchor system—receives the response from the authentication controller.

[7] This check is not performed in IS-778.

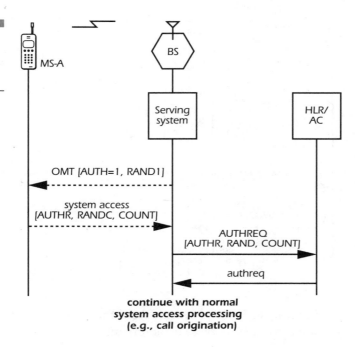

Figure 11.11
Successful global
challenge when SSD
is not shared.

OMT [AUTH=1, RAND1]

system access
[AUTHR, RANDC, COUNT]

AUTHREQ
[AUTHR, RAND, COUNT]

authreq

**continue with normal
system access processing
(e.g., call origination)**

5. Finally, check that the received COUNT value matches the expected value stored in the serving system's database record for the MS.

If any of these checks fails, the serving system global-challenge process initiates the authentication reporting process, passing an indication of the cause of the failure. Otherwise, the serving system considers the MS to have passed the global challenge.

The serving system processing beyond this point depends on the particular authentication policies in effect. If the global challenge is considered sufficient for authenticating the MS in this case, the serving system may simply terminate the global-challenge process and continue with system-access processing (see Figure 11.12). Alternatively, the serving system may pass control to either the unique challenge process or the COUNT update process to perform further authentication functions.

Border System Global-challenge Process

The border system global-challenge process deals with two distinct tasks:

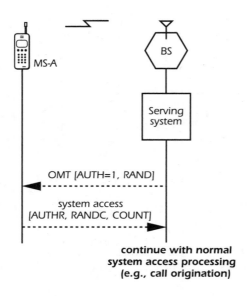

Figure 11.12
Successful global
challenge when SSD
is shared.

OMT [AUTH=1, RAND]

system access
[AUTHR, RANDC, COUNT]

**continue with normal
system access processing
(e.g., call origination)**

- Handling of RAND requests from the serving system
- Handling of MS access attempts under intersystem paging conditions

The first task is initiated when the border system receives a RAN-DREQ message from the serving system, including the RANDC that the serving system received from the MS and the serving-cell identification (the latter is information coordinated between the two systems during provisioning). The border system checks its records of current and recently used RAND values in the area corresponding to the serving cell and responds with a matching RAND, if one exists, in the randreq message.

One form of the second task is initiated when (a) the border system receives an IntersystemPage Invoke (ISPAGE) message for an MS and then (b) the border system receives a valid page response with global-challenge information (that is, COUNT, AUTHR, and RANDC) from the specified MS. The responsibility of the process in this case is to authenticate the MS by sending an AUTHREQ message to the MS's home system. If this is successful, the border system continues with the normal procedures. If it is unsuccessful, the border system initiates "local recovery procedures," which could involve sending an indication to the serving system that the MS is denied access in the IntersystemPage Return Result (ispage) message.

A very different form of the second task is initiated when (a) the border system receives an IntersystemPage2 Invoke (ISPAGE2) message for an MS and then (b) the border system receives a valid page response with global-challenge information (that is, COUNT, AUTHR, and RANDC) from the specified MS. The responsibility of the process in this case is to package the received authentication information—together with the RAND value the border system is currently using—in the IntersystemPage2 Return Result (ispage2) message and send the message to the serving system.

AC Global-challenge Process

The AC global-challenge process is initiated when the AC receives an AUTHREQ message from the serving system that includes the ANSI-41 global-challenge information parameters (RAND, AUTHR, and COUNT). The AC global-challenge process is responsible for evaluating the information in the AUTHREQ message.

The AC first checks that the MSID and ESN reported by the MS are valid. If this test fails, the AC sends an authreq message to the serving system, including the ANSI-41 DenyAccess parameter, and terminates the process. This message directs the serving system to release the system resources it has allocated for the MS access in question; this may include the release of the call in progress, if one exists.

The AC then checks that the type of air interface used by the MS to access the system (as indicated by the ANSI-41 TerminalType parameter included in the AUTHREQ message) corresponds to one of the values stored in the AC's database. If it does not, ANSI-41 specifies that the AC may choose to deny the MS access to the network (using the DenyAccess parameter in the authreq message, as previously described) and terminate the global-challenge process.

Otherwise, the AC executes the CAVE algorithm and verifies that the AUTHR generated by the AC equals the AUTHR received from the MS via the serving system. If the two values do not match, the AC may choose to deny the MS access to the network (using the DenyAccess parameter in the authreq message, as previously described) and terminate the global-challenge process.

If the AC's internal algorithms are such that authentication processing continues even in the event of an AUTHR mismatch, the AC's next task is to check that the received COUNT value matches the expected value—at least to within an AC-determined range of confidence.[8] How-

---

[8] It is possible that the network may lose COUNT synchronization with the MS. For example, when the network requests a COUNT update, the MS may increment its COUNT value and send an acknowledgment that is not received by the network due to poor radio transmission conditions. System failures (e.g., in the VLR or AC) may also result in loss of COUNT information. This must be factored into the COUNT verification procedure.

ever, if SSD is shared with a previous serving system, then the AC must retrieve the current COUNT value from that system, as described above in "Shared Secret Data Sharing." Again, if a COUNT mismatch is detected, ANSI-41 allows the AC to select access-denial treatment (i.e., the AC sends an authreq message to the serving system, including the DenyAccess parameter) or to continue authentication processing. Note that the COUNT check is only performed during the global-challenge process—this is the only time when the MS provides the COUNT information to the network. Therefore, an authentication strategy that includes COUNT would necessarily include the global challenge.

The AC processing beyond this point depends on the particular authentication policies in effect. If the MS passes the global challenge (i.e., the AUTHR and COUNT are correct), then the AC may simply terminate the global-challenge process and complete the transaction with the serving system by sending it an empty authreq message (see Figure 11.11). Alternatively, the AC global-challenge process may pass control to any one of the following additional processes in the AC (described elsewhere in this chapter):

- SSD sharing
- Unique challenge
- SSD update
- COUNT update

Unique Challenge

The ANSI-41 unique-challenge function encompasses the processes by which:

1. The ANSI-41 authentication controller directs the serving system to present a numeric authentication challenge to a single MS that either is requesting service from the network or is already engaged in a call.
2. The anchor system may choose to pass the authentication challenge request to the current serving system (i.e., in the case of a handed-off call).
3. The serving system presents the numeric authentication challenge to the MS and verifies that the numeric authentication response provided by the MS is correct (Figure 11.13 depicts the basic unique challenge procedure for authentication when SSD is not shared).

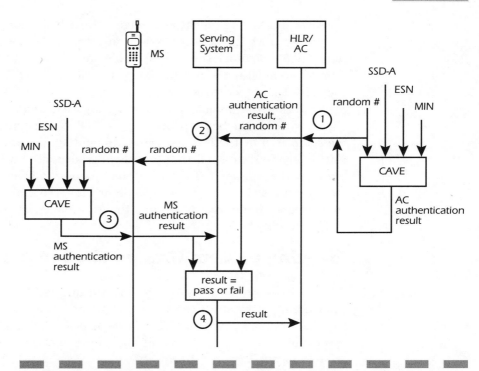

Figure 11.13 *Basic unique-challenge authentication process when SSD is not shared: (1) The AC generates a random number and uses it to calculate an authentication result. The AC sends both the random number and the authentication result to the serving system. (2) The serving system forwards the random number to the MS. (3) The MS calculates an authentication result and sends it to the serving system. (4) The serving system compares the result from the AC with the result from the MS. If the results match, the MS is considered to have successfully responded to the challenge. If they do not match, the MS may be considered fraudulent and service may be denied. Either way, the serving system reports the results to the AC. If SSD is shared, the serving system may initiate the unique challenge process and would only report a failure to the AC.*

The unique challenge is so named because the challenge indicator and the random number used for the challenge are directed to a *particular* MS (rather than to all MSs, as in the global challenge process).

In fact, the unique challenge may be used as a follow-up to the global challenge. Whether the global challenge succeeds or fails, the unique challenge serves as a double-check on the authenticity of the MS. Since the random number for the challenge is changed for each operation, the unique challenge provides a more secure authentication test than the global-challenge function.

Also, unlike the global challenge, which has a single mechanism of initiation (i.e., the serving system autonomously transmits the global-challenge indicator), the unique challenge can be initiated under a number of circumstances and by using a variety of ANSI-41 messages. We describe these techniques in this section.

Note that the unique challenge is supported under call-handoff conditions, whereas the SSD update and the COUNT update operations cannot be completed by the initial serving system (i.e., the anchor system) if the MS is involved in a call that has been handed off to another system.[9] To support the call-handoff scenario, the unique-challenge processes include the serving system unique-challenge process, the nonanchor serving system unique-challenge process, and the AC unique-challenge process.

Serving System Unique-challenge Process

The details of the serving system unique-challenge process differ depending on whether the unique challenge is controlled by the AC or by the serving system itself; the latter is possible if SSD is shared. However, note that the AC may initiate a unique challenge whether SSD is shared or not.

When Controlled by the AC

Generally, the serving system begins the AC-controlled unique-challenge process when it receives the ANSI-41 RandomVariable-UniqueChallenge (RANDU) and AuthenticationResponseUniqueChallenge (AUTHU) parameters from the AC in one of four ANSI-41 messages:

1. AuthenticationDirective Invoke (AUTHDIR)
2. AuthenticationRequest Return Result (authreq)
3. AuthenticationStatusReport Return Result (asreport)
4. AuthenticationFailureReport Return Result (afreport)

Case 1 is shown in Figure 11.14. In case 2, the authreq message may be a response to an AUTHREQ message sent from the serving system global-challenge process to the AC; alternatively, the serving system unique-challenge process may send the AUTHREQ message to the AC if

---

[9] COUNT update after handoff is supported in IS-778.

the MS requires authentication and the global challenge is not in effect. In this sense, the serving system is the initiator of the unique challenge.

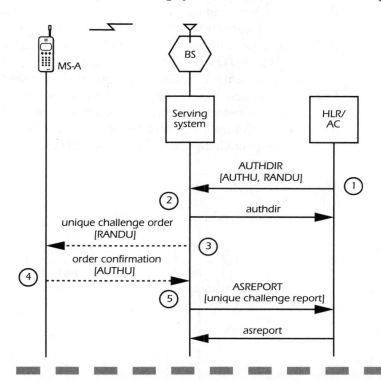

Figure 11.14 The AC-initiated unique-challenge process: (1) The AC selects RANDU, calculates AUTHU, and sends both parameters to the serving system in an AUTHDIR message. (2) The serving system acknowledges the unique challenge request by sending an authdir message to the AC. (3) The serving system sends a Unique Challenge Order, including RANDU, to the MS in a mobile station control message. (4) The MS responds with its calculated AUTHU value in an order confirmation message.[10] (5) The serving system compares the AUTHU from the MS with the AUTHU from the AC and generates a unique-challenge report indicating either success (i.e., the two AUTHUs match) or failure. The serving system sends the report to the AC in the ASREPORT message.

A complicating factor in the message exchange shown in Figure 11.15 is intersystem handoff, when the initial serving system—the anchor system—is no longer the current serving system. ANSI-41 compensates for

---

[10] The Unique Challenge Order may be sent from the BS to the MS via a control channel, an analog voice channel, or a digital traffic channel. Likewise, the response from the MS to the BS may use any of these channel types.

this condition with the AuthenticationDirectiveForward operation. When a unique challenge is requested for an MS involved in a call, and the call has been handed off, the anchor system forwards the RANDU and AUTHU parameters to the current serving system—possibly via one or more tandem systems—using the AuthenticationDirectiveForward Invoke (AUTHDIRFWD) message.[11] The response from the serving system—a unique-challenge report in the AuthenticationDirectiveForward Return Result (authdirfwd) message—is passed to the authentication reporting process, and the unique-challenge process terminates (see Figure 11.15 and our discussion of the nonanchor serving system unique-challenge process later in this chapter).

Figure 11.15
The AC-initiated unique challenge process when the MS is involved in a call that has been handed off from the anchor system to a nonanchor serving system. Message flows are similar to Figure 11.14, with the addition of the Authentication-DirectiveForward exchange between the serving system and anchor system.

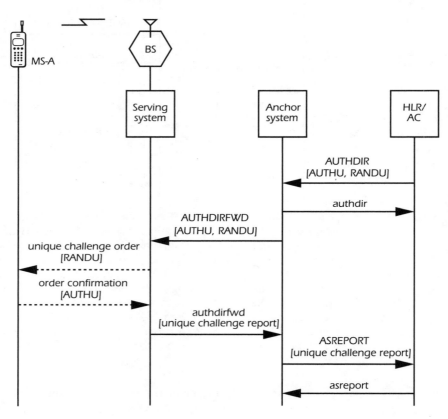

---

[11] The only exception to this procedure occurs when the unique challenge is part of an SSD update process; in this case, if the call has been handed off, the unique challenge is not attempted.

A final complicating factor exists if either the anchor system or the serving system cannot complete the unique-challenge operation. For example:

■ The BS cannot send the order to the MS or receives no response from the MS due to a loss of radio contact or because the MS is turned off by the user.
■ The call is handed off to a non–authentication-capable system.

In this case, the unique-challenge report generated by the anchor or serving system indicates either that the challenge was not attempted or that no response was received from the MS.

Note that, in addition to the unique challenge, the AUTHDIR, authreq, asreport, and afreport messages can trigger a number of other authentication processes in the serving system, including:

■ SSD update if the ANSI-41 RandomVariableSSD (RANDSSD) parameter is included.
■ COUNT update if the ANSI-41 UpdateCount (UPDCOUNT) parameter is included.
■ SSD sharing if the ANSI-41 SSD parameter is included.
■ Termination of SSD sharing if the ANSI-41 SSDNotShared (NOSSD) parameter is included.
■ Release of system resources for the current system access by the MS in question if the ANSI-41 DenyAccess (DENACC) parameter is included (this is at the option of the serving system).

When more than one of these processes is requested in the same message, ANSI-41 defines the required sequence of process execution (see Figure 11.16).

When Controlled by the Serving System

When SSD is shared, the serving system (i.e., the initial serving system in the case of a call handoff) may initiate a unique challenge at any time, based on its internal authentication algorithms. In this case, the serving system selects the RANDU value for the challenge and uses the CAVE algorithm to calculate the corresponding AUTHU value. Otherwise, the operation of the serving system–initiated unique-challenge process is the same as for the AC-initiated case. However, the authentication reporting process operates differently—reports are passed to the AC only in case of authentication failure, rather than on success or failure.

Figure 11.16
The sequence in
which the serving
system processes
authentication
commands received
in a message from
the AC.

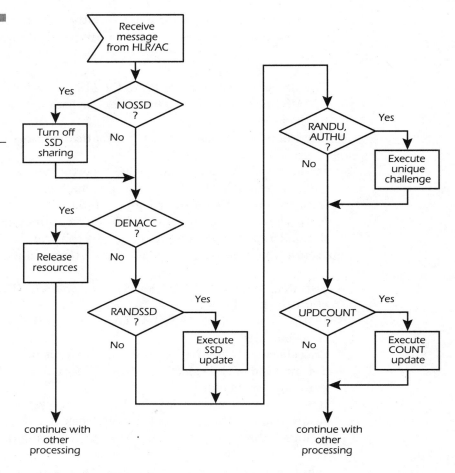

Figure 11.16
The sequence in which the serving system processes authentication commands received in a message from the AC.

Nonanchor Serving System Unique-challenge Process

In ANSI-41-D, the nonanchor serving system (referred to simply as the serving system in the following description) has only one authentication-related task—handling the unique-challenge process. In IS-778, it also handles the COUNT update process, as described later in this chapter.

The process begins when the serving system receives an AUTHDIRFWD message from the anchor system, including the ANSI-41 RANDU and AUTHU parameters, and proceeds as follows (refer back to Figure 11.15):

- The serving system directs the serving BS to send a Unique Challenge Order, including RANDU, to the MS in a mobile station control message.
- The MS responds with its calculated AUTHU value in an order confirmation message.
- The serving system compares the AUTHU from the MS with the AUTHU from the anchor system and generates a unique-challenge report indicating one of the following:
  - Unique challenge successful (i.e., the two AUTHUs match)
  - Unique challenge not successful (i.e., the two AUTHUs do not match)
- The serving system sends the execution report to the anchor system, and the process terminates.

If the serving system cannot complete the unique-challenge operation, for example, due to loss of radio contact, the unique-challenge report indicates either that the challenge was not attempted or that no response was received from the MS.

AC Unique-challenge Process

The AC initiates the unique-challenge process based on its internal authentication algorithms. Possible triggers for the process include, but are not limited to:

- A failed global challenge
- A successful global challenge in a high-fraud area
- A call origination, call termination, or in-call flash request
- A manual request via the AC's user interface

When SSD is not shared, the AC (i.e., the home service provider) must trade off the enhanced level of security provided by regular unique challenges with the added load on the signaling network between the home system and visited systems.

The AC selects the RANDU value for the challenge and uses the CAVE algorithm to calculate the corresponding AUTHU value. The AC sends these parameters to the serving system in one of four ANSI-41 messages:

- AuthenticationDirective Invoke (AUTHDIR)
- AuthenticationRequest Return Result (authreq)
- AuthenticationStatusReport Return Result (asreport)

■ AuthenticationFailureReport Return Result (afreport)

If the AC uses the AUTHDIR message, it waits for an AuthenticationDirective Return Result (authdir) message from the serving system; otherwise, the AC notifies the authentication reporting process that a unique-challenge report is expected from the serving system and terminates the unique-challenge process.

SSD Update

The ANSI-41 SSD update function encompasses the processes by which the SSD in an MS is changed to a new value under the direction of the AC. Note that only the AC may initiate this operation, even when SSD is shared with the serving system.

On the network side, ANSI-41 authentication procedures specify that a unique challenge is executed immediately after the SSD update to confirm that the target MS has successfully changed its SSD. On the MS side, the various authentication-capable air-interface standards (e.g., IS-95, IS-136) specify that the MS initiates a *base station challenge* procedure when the MS is directed by the base station to change its SSD.[12] The base station challenge allows the MS to authenticate the base station issuing the SSD update; this prevents a fraudulent base station from disrupting the normal network operation by forcing the MS's SSD out of alignment with the network's SSD.

The SSD update processes include the AC SSD update process and the serving system SSD update process, in addition to the AC, serving system, and nonanchor serving system unique-challenge processes previously described. The base station challenge functionality is included in the AC and serving system SSD update process descriptions.

AC SSD Update Process

The AC initiates the SSD update process based on its internal authentication algorithms; however, since the SSD update process requires significant network signaling, the AC does not normally change an MS's SSD without justification. For example, since the SSD and COUNT

---

[12] This is the only circumstance under which the base station challenge is used.

information may be transferred between systems using ANSI-41, there is the possibility that a valid authentication set (that is, MSID, ESN, SSD, and COUNT) could fall into the wrong hands. A criminal with knowledge of the CAVE algorithm and a valid authentication set would be able to simulate a legitimate MS—at least until the SSD was modified. Without knowledge of the A-key associated with the legitimate MS—presumably a far more secure piece of information than the SSD—the criminal would not be able to change the SSD correctly.

The first step in the AC SSD update process is for the AC to execute the CAVE algorithm using the MS's A-key, ESN, and a random number, called the RandomVariableSSD (RANDSSD). The result is the new, "pending" SSD.

The next step depends on whether the AC chooses to share the new SSD with the serving system. If it chooses to share SSD, the AC sends the RANDSSD and the new SSD to the serving system in one of four ANSI-41 messages:

- AuthenticationDirective Invoke (AUTHDIR)
- AuthenticationRequest Return Result (authreq)
- AuthenticationStatusReport Return Result (asreport)
- AuthenticationFailureReport Return Result (afreport)

The AC then terminates the SSD update process and turns over control to the authentication reporting process, which waits for a report from the serving system. With both the RANDSSD and the new SSD, the serving system can:

- Update the SSD in the MS
- Respond to the base station challenge from the MS
- Verify the update by issuing a unique challenge to the MS

If SSD is not shared with the serving system, the AC must manage the unique challenge and base station challenge that accompany the SSD update. Therefore, the AC initiates the unique-challenge process (see Figure 11.17)—it selects the RANDU value for the challenge and uses the CAVE algorithm to calculate the corresponding AUTHU value. The AC sends the RANDU, AUTHU, and RANDSSD parameters in the message to the serving system; again, any of the four messages identified above can be used (the use of the AuthenticationDirective operation is shown in Figure 11.17). The AC notifies the authentication reporting process that an SSD update report is expected from the serving system.

However, the AC also waits for a BaseStationChallenge Invoke (BSCHALL) message from the serving system. If the AC receives the BSCHALL message, it executes the CAVE algorithm. Inputs are the random number selected by the MS and included in the BSCHALL message, the MS's MSID and ESN, and the new SSD. The AC returns the result of the calculation, AUTHBS, to the serving system in the BaseStationChallenge Return Result (bschall) message and then terminates the SSD update process, leaving the handling of the report from the serving system to the authentication reporting process.

Serving System SSD Update Process

The serving system begins the SSD update process when it receives the ANSI-41 RandomVariableSSD (RANDSSD) parameter from the AC in one of four ANSI-41 messages:

- AuthenticationDirective Invoke (AUTHDIR)
- AuthenticationRequest Return Result (authreq)
- AuthenticationStatusReport Return Result (asreport)
- AuthenticationFailureReport Return Result (afreport)

The message also contains either the new SSD, if the AC chooses to share it with the serving system, or the RANDU and AUTHU parameters, if the new SSD is not shared.

The serving system unique-challenge process works as follows (see Figure 11.17):

- If the MS is involved in a call and the call has been handed off, or if the SSD update cannot otherwise be attempted (e.g., loss of radio contact), then the serving MSC generates an SSD update report (indicating that the SSD update was not attempted), passes the report to the authentication reporting process, and terminates.
- Otherwise, the serving system directs the serving BS to send an SSD Update Order, including RANDSSD, to the MS in a mobile station control message.
- The MS responds with a Base Station Challenge Order, including the random number selected by the MS, RANDBS.
- If SSD is shared, the serving system executes the CAVE algorithm with inputs of RANDBS, the MS's MSID and ESN, and the new SSD. The result of the calculation is AUTHBS. If SSD is not shared (as

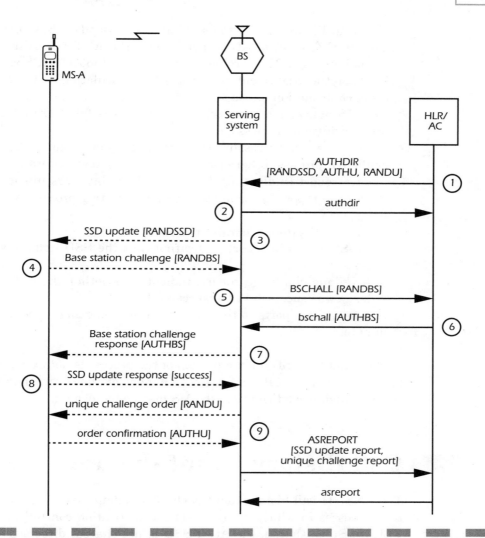

Figure 11.17 The SSD update process when the new SSD is not shared with the serving system: (1) The AC selects RANDSSD and calculates the new SSD. It then selects RANDU, calculates AUTHU, and sends the RANDSSD, RANDU, and AUTHU parameters to the serving system in an AUTHDIR message. (2) The serving system acknowledges the SSD update request by sending an authdir message to the AC. (3) The serving system sends an SSD Update Order, including RANDSSD, to the MS in a mobile station control message. (4) The MS calculates the new SSD. It then selects RANDBS and calculates AUTHBS using the new SSD. It sends a Base Station Challenge Order to the serving system, including RANDBS. (5) The serving system forwards the challenge to the AC in the BSCHALL message. (6) The AC calculates AUTHBS, using the received RANDBS and the new SSD, and sends the result to the serving system in the bschall message. (7) The serving system forwards the base station challenge response from the AC to the MS. (8) The MS then confirms the SSD update. (9) The serving system executes a unique challenge using the RANDU and AUTHU values received in step 1. It sends a report, including the SSD update and unique challenge results, to the AC in the ASREPORT message.

shown in Figure 11.17), the serving system sends a BSCHALL message to the AC. The AC responds with the AUTHBS result in the bschall message. The serving system returns the AUTHBS value (i.e., based on the local calculation or the AC's calculation) to the MS in an order confirmation message.

- The MS indicates a successful or an unsuccessful SSD update in an order confirmation message.
- If the update is successful, the serving system executes the unique-challenge process; otherwise, the serving system generates an SSD update report (indicating that the SSD update was unsuccessful), passes the report to the authentication reporting process, and terminates.
- The serving system generates two reports:
  - An SSD update report, indicating that the SSD update was successful
  - A unique challenge report, indicating whether the unique challenge was successful or unsuccessful

  It passes the reports to the authentication reporting process and terminates.

Note that the SSD update portion of the process is not executed if the MS is involved in a call that has been handed off; however, after sending the SSD update confirmation, the MS may be handed off.

Call History Count Update

The ANSI-41 call history count (COUNT) update function encompasses the processes in which the ANSI-41 authentication controller (i.e., the VLR if SSD is shared, the AC if SSD is not shared) directs the MS to update (i.e., increment by 1) its stored COUNT value.

The COUNT is a 6-bit parameter (i.e., a counter that increases from 0 to 63 and then wraps around to 0) that is intended to provide additional security in case the A-key or SSD is compromised. The current value of COUNT is maintained by both the MS and the authentication controller. The respective counts should generally be the same—they may not always match exactly due to radio transmission problems or system failures in the network. If the respective counts differ by a large enough range, or frequently do not match, the AC may assume that a fraudulent condition exists and take corrective action. Note that a COUNT mismatch detection does not conclusively indicate that the particular MS accessing the system is fraudulent—only that a "clone" may exist.

The ANSI-41-D COUNT update processes include the serving system COUNT update process and the AC COUNT update process. To support the new COUNT update after a call-handoff scenario, IS-778 requires a nonanchor serving system COUNT update process that is analogous to the process used for unique challenge.

Serving System COUNT Update Process

The details of the serving system COUNT update process differ depending on whether the update is initiated by the AC or by the serving system itself; the latter is possible if SSD is shared. However, note that the AC may initiate a COUNT update whether SSD is shared or not. Figure 11.18 depicts the COUNT update procedure initiated by the AC.

When Controlled by the AC

The serving system begins the AC-controlled COUNT update process when it receives the ANSI-41 UpdateCount (UPDCOUNT) parameter from the AC in one of four ANSI-41 messages:

- AuthenticationDirective Invoke (AUTHDIR)
- AuthenticationRequest Return Result (authreq)
- AuthenticationStatusReport Return Result (asreport)
- AuthenticationFailureReport Return Result (afreport)

The serving system COUNT update process works as follows:

- If the COUNT update cannot be attempted (e.g., loss of radio contact), then the serving MSC generates a COUNT update report, indicating that the COUNT update was not attempted, passes the report to the authentication reporting process, and terminates. This procedure also applies in the case of a ANSI-41-D serving system when the MS is involved in a call and the call has been handed off; however, IS-778 adds support for COUNT update after handoff, as described below.
- Otherwise, the serving system sends a Parameter Update Order to the MS in a mobile station control message.
- The MS increments its stored value of COUNT and responds with an order confirmation message.

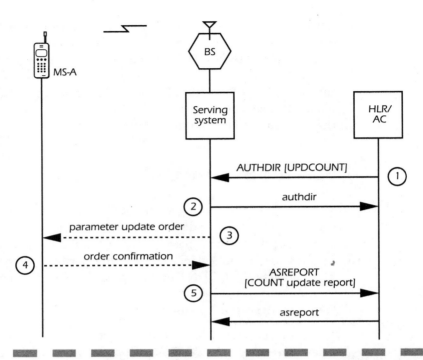

Figure 11.18 The AC-initiated COUNT update process: (1) The AC initiates the COUNT update by sending the UPDCOUNT parameter to the serving system in an AUTHDIR message. (2) The serving system acknowledges the COUNT update request by sending an authdir message to the AC. (3) The serving system sends a parameter update order to the MS in a mobile station control message. (4) The MS updates its stored COUNT value and responds with an order confirmation message. (5) The serving system sends a COUNT update report indicating either success (i.e., confirmation from the MS) or failure. The serving system sends the report to the AC in the ASREPORT message. On receipt, the AC updates its stored COUNT value for the MS.

- If the serving system receives the order confirmation message, it generates a COUNT update report indicating the COUNT update was successful; otherwise, it generates a COUNT update report indicating that there was no response to the COUNT update.
- The serving system passes the report to the authentication reporting process and terminates.

IS-778's support for COUNT update after handoff is shown in Figure 11.19. When a COUNT update is requested for an MS involved in a call, and the call has been handed off, the anchor system forwards the UPDCOUNT parameter to the current serving system—possibly via one or more tandem systems—using the AuthenticationDirectiveForward Invoke (AUTHDIRFWD) message. The response from the serving system—a COUNT update report in the AuthenticationDirectiveForward Return Result (authdirfwd) message—is passed to the authentication reporting process, and the COUNT update process terminates (see our discussion of the nonanchor serving system COUNT update process later in this chapter).

Figure 11.19
The AC-initiated COUNT update process when the MS is involved in a call that has been handed off from the anchor system to a nonanchor serving system. Message flows are similar to Figure 11.18, with the addition of the Authentication-DirectiveForward exchange between the serving system and anchor system.

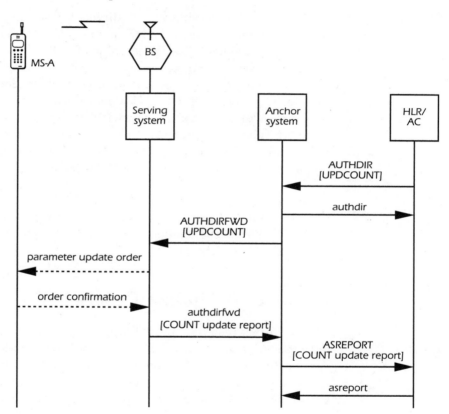

When Controlled by the Serving System

When SSD is shared, the serving system may initiate a COUNT update at any time, based on its internal authentication algorithms. Otherwise, the operation of the serving system–controlled count update process is the same as for the AC-controlled case. However, the authentication reporting process operates differently—reports are passed to the AC only in the case of authentication failure, rather than on both success and failure.

Nonanchor Serving System COUNT Update Process

The process begins when the serving system receives an AUTHDIRFWD message from the anchor system, including the ANSI-41 UPDCOUNT parameter, and proceeds as follows (see Figure 11.19):

- The nonanchor serving system directs the serving BS to send a Parameter Update Order to the MS in a mobile station control message.
- The MS increments its stored value of COUNT and responds with an order confirmation message.
- If the nonanchor serving system receives the order confirmation message, it generates a COUNT update report indicating the COUNT update was successful; otherwise, it generates a COUNT update report indicating that there was no response to the COUNT update.
- The nonanchor serving system sends the execution report to the anchor system, and the process terminates.

If the serving system cannot complete the unique-challenge operation, e.g., due to loss of radio contact, the unique-challenge report indicates either that the challenge was not attempted or that no response was received from the MS.

AC COUNT Update Process

The AC initiates the COUNT update process based on its internal authentication algorithms. The triggers for the process should be random so that the COUNT is not predictable by an illegitimate user.

To execute the COUNT update, the AC sends the ANSI-41 Update-Count parameter to the serving system in one of four ANSI-41 messages:

- AuthenticationDirective Invoke (AUTHDIR)
- AuthenticationRequest Return Result (authreq)
- AuthenticationStatusReport Return Result (asreport)
- AuthenticationFailureReport Return Result (afreport)

The AC count update process then passes control to the authentication reporting process, which waits for a COUNT update report from the serving system.

Authentication Reporting

The ANSI-41 authentication-reporting function encompasses the processes by which the success or failure of various authentication processes is conveyed from the serving system to the AC. The processes that provide reports are:

- Global challenge
- Unique challenge
- SSD update
- COUNT update

The authentication reporting processes include the serving system authentication reporting process and the AC authentication reporting process.

Serving System Authentication-reporting Process

We describe the mechanics of the serving system authentication reporting process in terms of three characteristics:

1. *Reporting triggers*. When is the process initiated?
2. *Reporting method*. How are reports conveyed to the AC?
3. *Report content*. What is in a report?

The first two characteristics are correlated: the reporting method depends on the reporting trigger.

Reporting Triggers and Methods

Each of the global challenge, unique challenge, SSD update, and COUNT update processes has associated outcomes—success or failure. While these outcomes are normally recorded, not all are reported; not every authentication process outcome is conveyed from the serving system to the AC. The five report-triggering factors are:

1. The nature of the authentication function,—global challenge, unique challenge, SSD update, or COUNT update.
2. The authentication version, either TSB51 or ANSI-41.

3. The authentication process result, either success or failure.
4. The authentication initiator, either the AC or the serving system.
5. The status of SSD sharing, either shared or not shared.

ANSI-41 is backward compatible with the authentication procedures defined in TSB51; however, the reporting methods defined in the two documents differ slightly. In particular, TSB51 specifies a single ANSI-41 operation (called SecurityStatusReport in TSB51 and renamed AuthenticationFailureReport in ANSI-41) to report the outcome of both AC-initiated (success or failure) and serving system–initiated (failure only) authentication processes. ANSI-41 specifies the Authentication-FailureReport Invoke (AFREPORT) message for reporting the failure of serving system–initiated authentication processes and the AuthenticationStatusReport Invoke (ASREPORT) message for reporting the success or failure of AC-initiated authentication processes.

The operation initiator is generally the AC when SSD is not shared; the exception is the global challenge, which is always initiated by the serving system. When SSD is shared, either the AC or the serving system can initiate the unique challenge and COUNT update operations. Only the AC may initiate the SSD update operation.

Table 11.1 lists the ANSI-41 reporting methods (i.e., reporting messages generated) under various conditions based on the five factors identified above.

TABLE 11.1

ANSI-41
Authentication
Reporting Methods

Function	Initiator and Version	Message Sent to AC on Successful Outcome	Message Sent to AC on Failed Outcome
global challenge	ANSI-41 or TSB51 serving system	AUTHREQ (no report)	AUTHREQ, REGNOT, or AFREPORT *
unique challenge	ANSI-41 AC	ASREPORT	ASREPORT
unique challenge	TSB51 AC	AFREPORT	AFREPORT
unique challenge	ANSI-41 serving system	no report	AFREPORT
unique challenge	TSB51 serving system	no report	AFREPORT
SSD update	ANSI-41 AC	ASREPORT	ASREPORT

continued on next page

TABLE 11.1

ANSI-41
Authentication
Reporting Methods
(continued)

Function	Initiator and Version	Message Sent to AC on Successful Outcome	Message Sent to AC on Failed Outcome
SSD update	TSB51 AC	AFREPORT	AFREPORT
COUNT update	ANSI-41 AC	ASREPORT	ASREPORT
COUNT update	TSB51 AC	AFREPORT	AFREPORT
COUNT update	ANSI-41 serving system	no report	AFREPORT
COUNT update	TSB51 serving system	no report	AFREPORT

* The handling of the global challenge failure case when an authentication-capable MS registers and does not provide a global-challenge authentication response is somewhat ambiguous in ANSI-41-D. For example, one can interpret ANSI-41-D to allow each of the listed messages to convey the authentication failure report. However, IS-778 is clear in specifying the use of the AUTHREQ message when this failure occurs and the AuthenticationCapability of the MS is unknown (see "Authentication Enhancements in IS-778").

Report Content

Another difference between ANSI-41 and TSB51 authentication reporting lies in the content of reports. The AFREPORT message in TSB51 reports on both the success and failure of authentication processes; therefore, the TSB51 ReportType parameter supports both types of indications. The ANSI-41 ReportType parameter is reduced in scope, supporting only failure indications.

ANSI-41 also adds new, individual report-type parameters for each of the authentication processes the AC can initiate: UniqueChallenge-Report, SSDUpdateReport, and CountUpdateReport.

AC Authentication Reporting Process

A significant aspect of the AC authentication reporting process, which is modified in ANSI-41 compared to TSB51, relates to the timing of the reports the AC receives from the serving system. In TSB51, when the AC initiates an authentication operation, the timing of the report is indeterminate; there is no specification that the report shall be generated within x seconds of the receipt of the authentication-related request (see Figure 11.20).

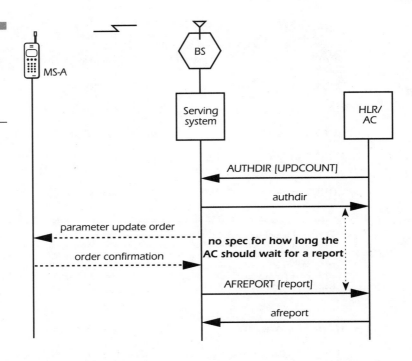

Figure 11.20
Prior to IS-41-C, the timing of the report generated by the serving system was not specified.

ANSI-41 defines the Authentication Status Report Timer (ASSRT) that specifies the number of seconds (default value is 24) the AC should wait for a report before executing recovery procedures. The ASSRT timer is started in one of two ways:

- When the authdir message is received by the AC, indicating the serving system's acknowledgment of the authentication-related request
- When the authreq, asreport, or afreport message is sent by the AC, including an authentication-related request (e.g., unique challenge).

Authentication Enhancements in IS-778

A number of enhancements to the ANSI-41-D authentication functions were standardized in TIA/EIA/IS-778. These will be rolled into ANSI-41 Revision E and include:

- Handling of authentication-capable mobile stations when the home system is not authentication-capable
- Clarified handling of the initial system access
- Handling of suspicious call originations

Of course, there are various corrections, clarifications, and enhancements of a more minor nature, some of which we have already described in this chapter (e.g., COUNT update after handoff, described in the section "Call History Count Update"). For more information on these, the reader should consult IS-778.

The need to support IMSI-based mobile stations (see Chapter 5), originally standardized in TIA/EIA/IS-751, also introduces some authentication-related issues. We discuss these issues in the sections that follow.

Handling of Non–authentication-capable Home Systems

The issue here is the handling of authentication-capable mobile stations roaming in authentication-capable serving systems, when the home system is *not* authentication capable. This situation was the natural result of the industry's transition to the distribution of authentication-capable mobile stations, in many cases in advance of network authentication readiness. The ANSI-41 enhancement is this case is a straightforward correction of an oversight in the ANSI-41-D serving system procedures related to the AuthenticationRequest operation.

When an authentication-capable mobile station initially registers in an authentication-capable serving system, the first thing the serving system does is send an AUTHREQ message to the subscriber's home system or the HLR. If the home system has not been upgraded to support authentication (and, therefore, there is no AC), the HLR would not recognize the AUTHREQ message and would normally respond with an AuthenticationRequest Return Error message containing the error code "Operation Not Supported" or a TCAP Reject message. Unfortunately, these responses are not properly handled in the MSC and VLR procedures specified in ANSI-41.6-D, Sections 4.4.1 and 4.4.2, respectively. Instead, the MSC is specified to treat these situations as authentication failures and apply access denial treatment.

The IS-778 fix for this situation is straightforward. If the serving system receives the "Operation Not Supported" error code or TCAP Reject message, it should proceed with registration of the MS and not attempt to authenticate the MS with the home system on subsequent system accesses.

Clarified Handling of Initial System Access

There are two problems associated with the initial system access by a mobile station.

The *first problem* is related to a fundamental issue: how does the serving system determine whether an MS is authentication-capable or not? As we discussed in "Shared Secret Data Sharing," the serving system will normally send one type of message (i.e., AUTHREQ) to the home system if the MS is considered to be authentication capable, and another type of message (i.e., REGNOT) if the MS is considered not authentication capable. If the serving system is always sending AUTHREQ messages on initial system access, but most MSs are not authentication capable (as was the case in the early phases of authentication deployment), the result is a lot of unnecessary signaling. On the other hand, if the serving system assumes that an MS is not authentication capable (e.g., because it does not respond to the global challenge), and sends a REGNOT message to the HLR, then the AC is not given the opportunity to deal with potentially fraudulent system accesses—this authentication burden falls on the HLR, at least when the HLR and AC are separate systems connected via the standard H-interface. Since the majority of MSs today are authentication capable, it makes sense to err on the side of security. Towards this end, IS-778 specifies that the serving system shall always send an AUTHREQ message to the home system, regardless of whether the MS is considered authentication capable or not.

The *second problem* starts when the MS originates a call, but before registration with the home system has taken place. This could be due to the mobile station characteristics or the behavior of the serving system. Either way, the fact that registration with the home system has not taken place means that the serving system does not have the subscriber's profile, and in particular the subscriber's calling privileges, which are needed to properly handle the call.

The ANSI-41-D procedures call for the serving system to execute the AuthenticationRequest procedures, followed by registration. Unfortunately, this does not allow the serving system to handle the call origination expeditiously. The result is that even honest, bill-paying customers can be denied that initial call until the serving system has completed its authentication and registration processes.

In some vendors' minds, ANSI-41-D already provided the solution to this problem, in the form of the QualificationRequest operation. When the unexpected call origination occurs, fire off the QualificationRequest Invoke (QUALREQ) message to the HLR, get the subscriber's profile in the response (including—as a bonus—the AuthenticationCapability parameter, which could indicate that authentication of the MS is not required), handle the call origination based on the subscriber's calling privileges, while also proceeding with the normal authentication and registration processes. The problem is solved.

Unfortunately, not all vendors supported this solution, and interoperability problems resulted. The TR45.2 group addressed this issue in IS-778 by modifying the procedures explicitly to allow the *early* QualificationRequest approach we just described. In fact, the procedures went one step further by explicitly allowing the QualificationRequest messaging and the AuthenticationRequest messaging to proceed in parallel, as illustrated in Figure 11.21. Note that, while the figure illustrates the global challenge case, overlapping of the QualificationRequest messaging with the unique challenge signaling is also supported.

Handling of Suspicious Call Originations

This IS-778 enhancement is primarily designed to protect against a form of the so-called *replay attack*. In a replay attack, a fraudulent MS listens for a valid system access by a legitimate MS and then replays the authentication portion of the access in an attempt to outsmart the network.

For example, this could work—albeit, in a limited sense as described below—if the network only used the global challenge portion of the authentication procedures, described earlier in this chapter. The fraudulent MS would listen for a registration containing the legitimate MS's MSID, ESN, AUTHR, and COUNT values. The AUTHR value represents the MS's calculated CAVE algorithm output with inputs MSID, ESN, SSD, and the RAND value being broadcasted by the serving system. As long as the RAND value is not changed by the serving system, the fraudulent MS could register later (e.g., after the legitimate MS has deregistered) by replaying the values of MSID, ESN, AUTHR, and COUNT, with the latter properly increased.

Figure 11.21
Overlapping of the Authentication-Request operation with the QualificationRequest operation to handle initial call origination system access prior to registration.

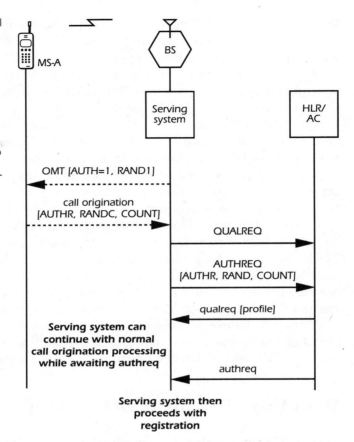

This approach is limited because even the global-challenge procedures include some additional security measures for call originations; such as the CAVE algorithm, which specifies that the last six digits of the called party number be used as input, replacing the MSID digits, when the subscriber is attempting to originate a call. Therefore, unless the last six digits of the party the criminal wants to call are the same as the last six digits of the party previously called by the legitimate subscriber, the replay attack will be fruitless.

One would think that this would be limiting enough, but one should never underestimate the criminal mind. At least one such mind came up with the following work-around: "If the legitimate subscriber previously called the number 234-5678, then all I have to do is add 345678 (i.e., the last six digits dialed) to the number that I want to call (e.g., 1-212-555-1212, to come up with 1-212-555-1212-345678) and I could be in business!" How could this possibly work? The answer lies in how the serving switch performs digit analysis and call routing. If the first digit it receives is 1, then the switch only needs to look at the next ten digits to route the call. The other digits can be disregarded. One can think of various 800-numbers (i.e., that are greater than eleven digits in length) that illustrate this mechanism.

So, in this case, the serving system would receive the 17-digit number and perform its analysis based on the first eleven digits. It would send the dialed digits and the other authentication parameters to the AC, which would verify that the CAVE calculations match up and allow the call to proceed.

While there are significant limitations associated with the mechanism—even beyond those we have mentioned—the potential for exploitation of this authentication weakness exists. What the IS-778 enhancement provides is a way for the serving MSC to flag this anomaly—i.e., that extraneous digits have been provided—to the authentication controller (i.e., the VLR or the AC, which are generally not expected to do any kind of extensive digit analysis on the called party number). The serving MSC does this by including the new *SuspiciousAccess* (SUSACC) parameter in the AUTHREQ message that it sends to the authentication controller (see Figure 11.22). According to IS-778, when the authentication controller receives this parameter, it should initiate a unique challenge. This would certainly expose the perpetrator of the fraud.

Authentication of Mobile Stations that Support IMSI

As we discussed in Chapter 5, IS-751 specifies the additions to ANSI-41-D that are needed to support the use of the IMSI for mobile identification purposes. This leads to a few authentication-related questions:

■ *Question*: If the MS has a IMSI but does not have a MIN, what does it use as input to the CAVE algorithm when MIN (or a portion thereof) is expected?

Answer: The MS uses portions of the IMSI in place of the MIN. Refer to TIA/EIA-136-005-B for the description applicable to TDMA MSs and TIA/EIA-95-B for the description applicable to CDMA MSs.

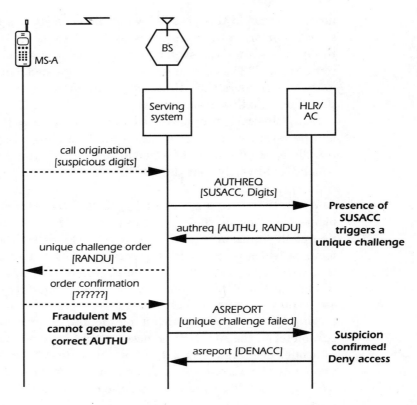

Figure 11.22
Handling of authentication of suspicious call originations. This example assumes there is no global challenge in effect.

- *Question*: If the MS has both a IMSI and a MIN, but uses the IMSI to access the network, what does it use as input to the CAVE algorithm when MIN is expected?
  Answer: Both CDMA and TDMA MSs will always use the MIN as CAVE input if a MIN is available.
- *Question*: If the MS uses the IMSI to access the network and is using the MIN as CAVE input, and SSD is shared with the serving system, how does the serving system get the MS's MIN so that it can properly execute the CAVE algorithm?
  Answer: The HLR will return the MIN to the serving system in the RegistrationNotification Return Result message.

Summary of ANSI-41 Operations Used for Authentication

Table 11.2 summarizes the ANSI-41 operations used by the authentication functions described in this chapter.

TABLE 11.2

Use of ANSI-41 Operations for Authentication

Function	ANSI-41 Operations Used for the Function
SSD sharing	AuthenticationDirective, AuthenticationFailureReport, AuthenticationRequest, AuthenticationStatusReport, CountRequest, MSInactive, RegistrationCancellation, RegistrationNotification
Global challenge	AuthenticationRequest, IntersystemPage2, RandomVariableRequest
Unique challenge	AuthenticationDirective, AuthenticationDirectiveForward, AuthenticationFailureReport, AuthenticationRequest, AuthenticationStatusReport
SSD update	AuthenticationDirective, AuthenticationFailureReport, AuthenticationRequest, AuthenticationStatusReport, BaseStationChallenge
Call history count update	AuthenticationDirective, AuthenticationFailureReport, AuthenticationRequest, AuthenticationStatusReport
Authentication reporting	AuthenticationFailureReport, AuthenticationStatusReport

CHAPTER 12

Call Processing Functions

In this chapter, we discuss the ANSI-41 mobile telecommunications network functions related to call processing. These functions are divided into the following categories:[1]

■ Call origination functions
■ Call termination functions
■ Feature control functions

We describe each of the call processing functions defined in ANSI-41 and provide examples of how the ANSI-41 mobile application part (MAP) operations (e.g., LocationRequest) are used to accomplish call processing tasks. We summarize this information at the end of the chapter. Of course, our descriptions are not intended to be a complete definition of how each of the ANSI-41 call processing functions works—for that, we defer to the standard.

Note that we use the ANSI-41 convention for operation component acronyms; that is, the Invoke component acronym is in all-capital letters (e.g., LOCREQ), while the Return Result component acronym is in all-lowercase letters (e.g., locreq). Refer to the ANSI-41 standard for the descriptions of the individual ANSI-41 operations and parameters. Also, consult the Glossary for a description of general terms (e.g., serving system) not explicitly defined in this chapter.

Throughout this chapter we consider the serving system as a single entity encompassing the mobile switching center (MSC) and visitor location register (VLR) functional entities. This simplifies the descriptions of the ANSI-41 call processing functions and also is representative of a large percentage of the ANSI-41 implementations currently in service. However, keep in mind that the potential for separation of the MSC and VLR exists and is fully defined in ANSI-41.

What Is ANSI-41 Call Processing?

Call processing is the broad category of functions that establish, maintain, and release calls to and from the mobile subscriber. It consists of the operations performed from the initial acceptance of an incoming call through the final disposition of the call. The call processing functions

---

[1] The list of call processing functions reflects the authors' subjective categorization of certain ANSI-41 functions.

establish and release terrestrial transmission paths for calls; they also invoke and manage call features that provide a specific variation on the way a call is treated.

In this chapter, our discussion of ANSI-41 call processing encompasses the following categories of functions:

- *Call origination functions*—the intersystem functions that support the mobile-originated calling capabilities, including the handling of call origination features
- *Call termination functions*—the intersystem functions that support the mobile-terminated calling capabilities, including the handling of call termination features
- *Feature control functions*—the intersystem functions that support the subscriber's ability to modify certain feature parameters (e.g., feature activation status, call-forwarding, forward-to number).

Where Are Call Processing Functions Specified in ANSI-41?

Call processing functions are specified in three parts of ANSI-41:

- **ANSI-41.5** provides the required formats—the bit-by-bit encoding—of all the ANSI-41 operation components, including those used for call processing. ANSI-41.5 defines both the messages (e.g., LocationRequest Invoke) and the message parameters (e.g., MobileIdentification-Number).
- **ANSI-41.6** provides algorithmic descriptions of the procedures associated with sending and receiving most ANSI-41 messages, including those used for call processing. This part of ANSI-41 also includes a significant level of detail on the application-process logic associated with call origination, call termination, and voice feature control.
- **ANSI-41.3** steps back from the protocol details contained in parts 5 and 6 and attempts to explain, using information flow diagrams and step-by-step descriptions, how the operations are used individually and together to accomplish call processing application tasks.

Issues Associated with Call Processing

Many call processing functions are beyond the scope of the ANSI-41 standard, and many aspects of call processing are closely coupled to the basic ANSI-41 automatic roaming and intersystem handoff functions described in Chapters 9 and 10. These aspects include:

- Delivering calls to a roaming subscriber. Location and status information needs to be obtained, involving the location management and MS state management functions described in Chapter 10.
- Controlling call origination from a roaming subscriber. The subscriber's authorized calling capabilities must be provided to the serving system, involving the service qualification functions described in Chapter 10.
- Controlling inter-MSC circuits during intersystem handoff, involving the various intersystem handoff functions described in Chapter 9.

There are also some general issues that have an impact on a number of the call processing functions including:

- All features are optional, and many are not supported everywhere. Subscribers who change to a new system via either roaming or intersystem handoff may lose features not supported by that system.
- Feature codes used to control features are not standardized. A feature code used to activate call forwarding, for instance, may not be the same in a visited system as in the home system.
- Feature interaction.

Feature Interaction

Feature interaction is a complex issue. It is a situation that arises when a subscriber is authorized for multiple features. Four aspects of feature interaction need to be analyzed:

- Definition of a feature interaction
- Definition of feature precedence
- Definition of feature treatment
- Definition of which functional entities control an interaction

A *feature interaction* means that if multiple features are simultaneously active, these features need to be analyzed in order to apply the appropriate call treatment. At this point, it does not matter which feature is analyzed first; they are all reviewed, and a logical decision is made about how to treat the call.

Feature precedence usually refers to one feature's superseding another. That is, if two features are active and one feature takes precedence over the other, the subordinate feature need not be factored into the algorithm to determine call treatment; the fact that it is active can be ignored.

Feature treatment means that certain actions are taken on a call based on calling- and called-party numbers, active features, and other events that can affect the call. For example, if call forwarding—no answer (CFNA) is active and conditions are such that CFNA is invoked (e.g., the call is not answered), then CFNA treatment means to forward the call.

Depending on the individual features involved in the interaction, the MSC, the HLR, or a combination of the MSC and HLR controls the interaction. Control of the interaction includes determining feature precedence as well as the call treatment to apply.

An example of an HLR-controlled feature interaction with precedence is the interaction between call forwarding—unconditional (CFU) and call waiting (CW). If a mobile-terminated call arrives, the HLR controls the interaction between CFU and CW. If CFU is active, CW is never even checked for activation. This means that CFU takes precedence over CW. It does not matter whether the subscriber has the feature, or if it is active. It is simply never analyzed in this case.

An example of an MSC-controlled feature interaction with precedence is the interaction between call transfer (CT) and three-way call (3WC). If a subscriber has established a three-way call and then disconnects, normal 3WC processing disconnects all the parties involved in the call. However, if CT is also active, the MSC transfers the call to the two other parties in the three-way call when the controlling subscriber disconnects. For this aspect of the interactions between the CT and 3WC features (i.e., disconnect treatment), CT takes precedence over 3WC.

Feature interaction analysis is a very important function to maintain seamlessness. It is necessary to address each call scenario involving every feature for proper call processing. Some of the feature interactions are specified in ANSI-41. Other feature interactions are implementation dependent; those that are HLR-controlled can be seamlessly provided without standardized, feature-specific serving system support. The management of feature interactions, however, is required to ensure proper and efficient logic processing at both the HLR and the serving system.

Call Origination Services

The simplicity of the process of making a mobile call—turning the phone on, entering digits, and pressing the SEND key—masks an enormous amount of call processing that takes place in the mobile telecommunications network. Prior to Revision C, IS-41 had relatively little impact on this processing—the Revision B services were mainly call termination–related (i.e., call delivery, call forwarding, call waiting). This situation changed in IS-41-C and ANSI-41-D, and will likely continue to change in subsequent revisions of ANSI-41 as "wireless intelligent network" concepts (see Chapter 19) are applied to the standard and call origination becomes a more network-controlled—rather than a serving system–controlled—process.

In general, ANSI-41 call origination services are the network functions that enable, restrict, supplement, or otherwise have an impact on an MS's ability to originate a call while roaming outside the home service area. ANSI-41 supports the following call origination services:

- Basic call origination
- Subscriber personal identification number (PIN) access (SPINA)
- Subscriber PIN intercept (SPINI)
- Calling number identification restriction (CNIR)
- Calling name restriction (CNAR)
- Message waiting notification (MWN)
- Voice message retrieval (VMR)
- Call transfer (CT)
- Three-way calling (3WC)
- Conference calling (CC)
- Priority access and channel assignment (PACA)
- Preferred language (PL)
- Voice privacy (VP), data privacy (DP), and signaling message encryption (SME)

The MWN, PL, VP, DP, and SME features are not solely call origination features—they also include call termination support capabilities.

Basic Call Origination

A portion of a simplified call origination process in a serving system is shown in Figure 12.1. The protocol and procedures in ANSI-41 affect this model in two areas: digit analysis and call origination authorization.

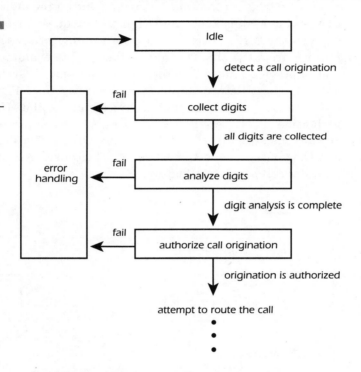

Figure 12.1
Part of a simplified call origination process in a serving system.

ANSI-41 Impact on the Digit Analysis Process

ANSI-41 defines two basic categories of *triggers* that can be applied to the digit analysis process: call origination triggers and feature request triggers. The call origination triggers are "armed" when the serving system receives the ANSI-41 OriginationTriggers parameter. This parameter is part of the MS's service profile information provided by the home system during the service qualification process (see Chapter 10). Some of the potential triggers are:

- All call originations
- Call origination to an international destination
- Call origination to the subscriber's own number, termed a *revertive call*

- Call origination to a number beginning with the * digit
- Call origination to a number beginning with the # digit
- Call origination with n digits, where n may be set from 0 to 14 digits
- Call origination with 15 or more digits

When a trigger condition is detected during the digit analysis process, the serving system sends an OriginationRequest Invoke (ORREQ) message to the MS's HLR, containing (among other things) the digits received from the MS. This allows the HLR to evaluate the dialed digits and to provide routing instructions to the serving system via the OriginationRequest Return Result (orreq) message (see Figure 12.2). Generally, the evaluation and routing procedures in the HLR are associated with some subscriber-specific features, such as subscriber PIN intercept or voice message retrieval. Thus, the OriginationTriggers parameter and OriginationRequest operation can be considered the building blocks for other call processing features.

Figure 12.2
An example of call origination trigger processing using the OriginationRequest operation.

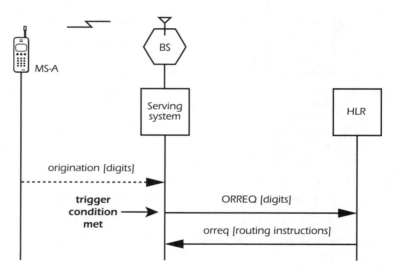

The basic "feature request trigger" does not require explicit arming; when the serving system detects a call origination to a number beginning with the * digit—this is called a *feature code string*—it sends a FeatureRequest Invoke (FEATREQ) message to the MS's HLR. However, note that the corresponding origination trigger takes precedence over the feature request trigger; if the origination trigger for the * digit is armed, then the serving system sends an ORREQ message rather than a FEATREQ message to the HLR. See "Feature Control Services" for additional discussion of the feature request processes.

ANSI-41 Impact on the Call Origination Authorization Process

The key ANSI-41 parameters that affect the basic call origination authorization process are:

- AuthorizationDenied
- AuthorizationPeriod
- OriginationIndicator

This information is obtained through the service qualification processes described in Chapter 10 (e.g., using the RegistrationNotification or QualificationDirective operation). The HLR sends either the AuthorizationDenied or the AuthorizationPeriod parameter to the serving system, but not both. If authorization is denied, then the subscriber receives access denial treatment (e.g., a special tone or announcement) from the serving system when call origination is attempted. If authorization is allowed, the AuthorizationPeriod parameter indicates the period for which the authorization applies:

- Per call
- For n calls, hours, days, or weeks, where n ranges from 0 to 255
- For a period set in an agreement between the home and serving systems
- Indefinitely (i.e., until denied or deregistered)

When the authorization period expires, for example, after 24 hours, the next call origination or other system access request by the MS triggers the service qualification process in the serving system (see Figure 12.3).

The OriginationIndicator (ORIGIND) parameter specifies the types of calls the MS is allowed to originate, for example:

- No calls (i.e., call origination not allowed)
- Local calls only
- Only calls to numbers that start with a specified string of digits
- Only local and national long-distance calls
- All calls—local, national, and international
- Only calls to a specified number

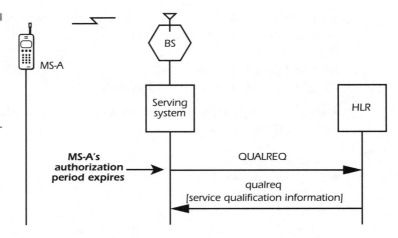

This last call type supports a *hotline* service; regardless of the number received from the MS, the serving system establishes a call to the single number provided by the HLR.[2] Also note that each serving system may define locally allowed calls (e.g., to numbers such as 911, *911) that are not affected by the value of the OriginationIndicator parameter.

Subscriber PIN Access

The subscriber PIN access (SPINA) feature supports a limited form of subscriber control over the ANSI-41 OriginationIndicator (ORIGIND) parameter value. The subscriber—by performing either a single-step or multistep feature control procedure that includes the entry of a personal identification number—can indirectly toggle the value of the ORIGIND parameter between no calls (SPINA active) and the subscriber's normally assigned origination authorization level (SPINA inactive). The HLR usually conveys the revised ORIGIND parameter to the serving system by using the QualificationDirective operation (see Figure 12.4). In effect, SPINA provides a network-based MS locking mechanism that is presumably more secure than MS-based locks.

---

[2] While this value of the OriginationIndicator has been defined since IS-41 Revision A, it was only in IS-41-C that the serving system processing for this case was explicitly defined; therefore, serving system implementations of prior versions of IS-41 may not operate the same as IS-41-C and ANSI-41 products.

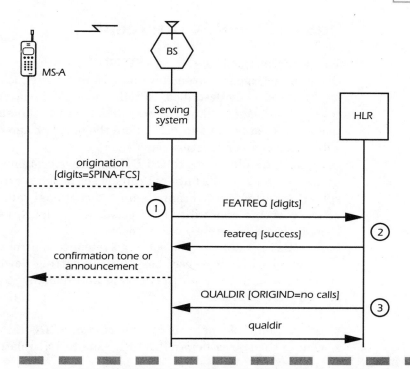

Figure 12.4 An example of SPINA feature activation using the single-step feature control procedure: (1) The subscriber enters the SPINA activation feature code string (SPINA-FCS) and presses the SEND key; the SPINA-FCS is in the form "*FC+PIN." The serving system detects the FCS and sends the received digits to the HLR in the FEATREQ message. (2) The HLR verifies that the PIN is valid and signals operation success in the featreq message. The serving system notifies the subscriber that the request was successful. (3) The HLR modifies the subscriber's service qualification to disable call origination using the QUALDIR message.

While the SPINA feature is described in ANSI-664 and ANSI-41 primarily as a subscriber-controlled on/off switch for call originations, it is also possible for the home service provider to use the feature in an *automatic lock* mode as a fraud-deterrent measure. In other words, the SPINA feature is automatically activated by the service provider and must be deactivated on a per-call basis by the subscriber (in effect, the feature becomes a derivative of the subscriber PIN intercept feature described in the next section). However, this application of SPINA diminishes in importance when authentication—a much more effective fraud-deterrent measure—is ubiquitously deployed in the network.

Subscriber PIN Intercept

The subscriber PIN intercept (SPINI) feature is an application of the OriginationTriggers parameter and OriginationRequest operation building blocks already described. SPINI allows the home service provider to set the conditions, like triggers, under which the subscriber may be requested to enter a valid PIN before the network agrees to complete the subscriber's call origination request.

Figure 12.5 illustrates the SPINI feature operation when (a) the origination trigger for international calls is set in the serving system, (b) the subscriber attempts to originate an international call, and (c) the interaction with the subscriber is directed by the HLR, using the ANSI-41 RemoteUserInteractionDirective operation.

It is also possible for the serving system to control the SPINI-related interaction with the subscriber, thereby cutting down on the network signaling traffic between the serving and home systems. For this, the serving system must:

- Inform the HLR that it is capable of local SPINI feature control.
- Obtain the subscriber's SPINI PIN and SPINI-related call origination triggers.

The serving system accomplishes this information transfer in the course of qualifying the MS for service. The serving system's local SPINI capability is conveyed in the ANSI-41 TransactionCapability parameter (TRANSCAP), and the subscriber's SPINI PIN and triggers are returned to the serving system in the SPINIPIN and SPINITriggers parameters, respectively. With this information, the serving system can control the SPINI-related call origination processing (see Figure 12.6).

Calling Number Identification Restriction

The calling number identification restriction (CNIR) feature allows a subscriber to prevent the presentation of the subscriber's calling number identification (CNI) information to the call-terminating party. The subscriber may be the party who originated the call or a party who is redirecting the call (e.g., the subscriber is busy and the call forwarding—busy feature is active).

As with many features in ANSI-41, protocol support for the CNIR feature has three components:

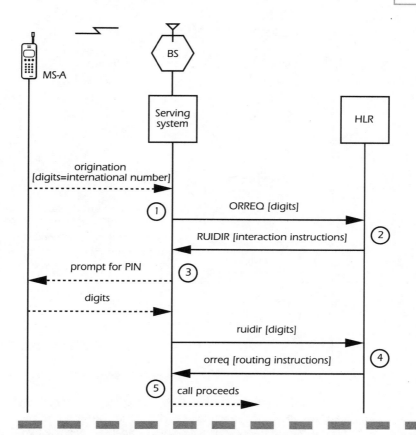

Figure 12.5 *An example of SPINI feature invocation using the* RemoteUserInteractionDirective operation: (1) The subscriber enters an international phone number and presses the SEND key. The "call origination to an international destination" trigger is set; therefore, the serving system sends an ORREQ message to the HLR including the dialed digits. (2) The HLR chooses to query the subscriber for a PIN before allowing the call to proceed; therefore, it sends interaction instructions to the serving system in the RUIDIR message. (3) The serving system connects the call to a system capable of user interaction (e.g., an interactive voice response unit), prompts the subscriber for a PIN, collects the digits entered by the subscriber, and sends them to the HLR in the ruidir message. (4) The HLR verifies that the PIN is valid and returns call routing instructions in the orreq message. (5) The serving system continues with call origination, as directed by the HLR.

1. *Conveying the subscriber's default CNIR feature activity status to the serving system.* The ANSI-41 CallingFeaturesIndicator parameter contains this information; the HLR sends this to the serving system during the service qualification process.

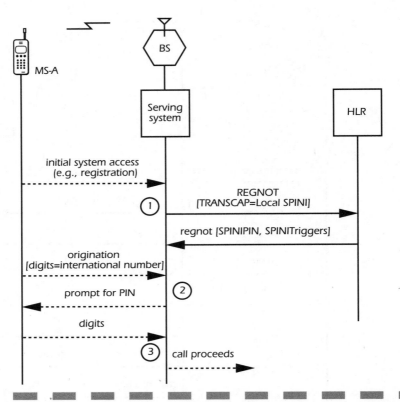

Figure 12.6 *An example of local SPINI feature handling by the serving system: (1) The subscriber registers in a local SPINI-capable serving system. The HLR provides the subscriber's SPINI PIN and call origination triggers to the serving system. (2) The subscriber enters an international phone number and presses the SEND key. The "call origination to an international destination" trigger is set; therefore, the serving system connects the call to a system capable of user interaction (e.g., an interactive voice response unit), prompts the subscriber for a PIN, collects the digits entered by the subscriber, and verifies that they match SPINI PIN digits received from the HLR. (3) The serving system allows the call origination to proceed.*

2. *Feature control.* The CNIR feature may be activated or deactivated on a per-call origination basis, using the per-call feature control procedure described in "Feature Control Services" later in this chapter.
3. *Feature invocation,* such as marking the subscriber's CNI information to indicate the status of the CNIR feature for the call.

For the purpose of the third function, each of the following ANSI-41 number identification parameters includes a CNIR indicator field (i.e., presentation allowed or presentation restricted):

- CallingPartyNumberDigits1
- CallingPartyNumberDigits2
- CallingPartyNumberString1
- CallingPartyNumberString2
- RedirectingNumberDigits
- RedirectingNumberString

When the called party is an MS served by a system connected to the subscriber's serving system using ANSI-41, the subscriber's CNI parameters can be conveyed to the serving system by using ANSI-41 messages (see Figure 12.7); otherwise, a call control protocol with number identification support (e.g., the SS7 ISDN user part, ISUP) is required between the subscriber's system and the called party's system to transport the CNI information.

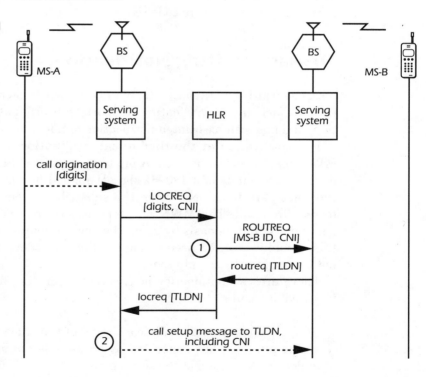

Figure 12.7
Methods of conveying calling number identification (CNI) information: (1) The CNI information can be sent from the originating system (i.e., MS-A's serving system) to the terminating system (i.e., MS-B's serving system) using ANSI-41 parameters in the normal call delivery messages, LOCREQ and ROUTREQ. (2) The CNI information can also be sent from the originating system to the terminating system using an appropriate call control protocol (e.g., ISUP).

Finally, ANSI-41 supports a nonstandard (i.e., not defined from a stage-1 perspective in ANSI-664) capability whereby a legally authorized called subscriber (e.g., an emergency services operator) can override the calling subscriber's CNIR feature and gain access to the CNI information—even when CNIR is activated for the calling subscriber. This is referred to as the CNIR override feature.

Calling Name Restriction

The calling name restriction (CNAR) feature was initially defined in TIA/EIA/IS-764 and will be incorporated in ANSI-41-E. This feature allows a subscriber to prevent the presentation of the subscriber's calling name identification (CNA) information to the call-terminating party.

CNAR is analogous to CNIR in its operation. In fact, the CNAR feature control codes may be the same as those of CNIR, allowing the subscriber to turn off both number and name presentation to the called party via a single feature code request.

Message-waiting Notification

Message-waiting notification (MWN) is not solely a call origination feature; in fact, the MWN feature supports call origination, call termination, and non–call-associated notification options.

For notification at the time of call origination or termination, the MWN *pip tone* option is used. A pip tone consists of an audible tone (i.e., four 100-ms bursts of a 480-Hz signal) applied by the serving system on the voice path to the MS when the subscriber attempts to originate or terminate a call. The MWN pip tone feature may be deactivated on a per-call origination basis by using the per-call feature control procedure; this capability can be used to ensure that a modem-driven data call is not interrupted by the pip tone.

There are two options for notification when the MS is idle (i.e., not involved in a call):

- An *alert pip tone* nominally consisting of the same four signal bursts as the pip tone, but generated by an audible annunciator in the MS under the command of the serving system.
- A *message-waiting indication* on the MS—it may be a simple message-waiting lamp or a digital display of the number of messages waiting for retrieval—that is driven based on signaling messages from the serving system.

The HLR conveys the MWN information to the serving system in the ANSI-41 MessageWaitingNotificationCount (MWNCOUNT) and MessageWaitingNotificationType (MWNTYPE) parameters as part of the service qualification process. The serving system uses this information to provide the appropriate MWN indication to the MS.

In the case of call handoff, the revised MWN information may be sent down the handoff chain from the anchor system to the serving system by using the ANSI-41 InformationForward Invoke (INFOFWD) message (see Figure 12.8). However, the anchor system will initiate this operation only if the MWN is of the message-waiting indication type described above; the pip tone and alert pip tone methods do not apply in the case of an active call that has been handed off.

Note that the method for the signaling to the HLR that MWN is required (e.g., from a voice message system) is not standardized in ANSI-41. This is left to vendor-proprietary interfaces or to standard interfaces developed by other industry groups (e.g., for wireline applications).

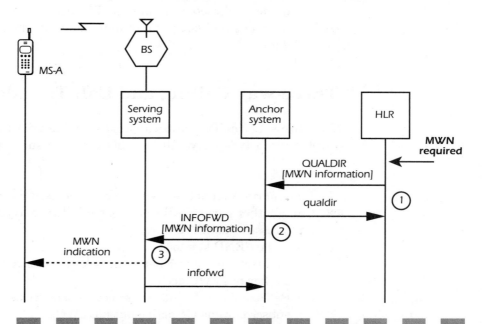

Figure 12.8 An example of message waiting notification (MWN): (1) The HLR determines that MWN is required for MS-A. The HLR sends the MWN information (i.e., MWNTYPE and MWNCOUNT parameters) to the anchor system. (2) MS-A is involved in a call that has been handed off; therefore, the anchor system sends the MWN information to the serving system in the INFOFWD message. (3) The serving system notifies MS-A via the appropriate notification method.

Voice Message Retrieval

The voice message retrieval feature allows service providers to offer subscribers easy access to their voice mailboxes for message retrieval purposes. Two forms of subscriber access are supported:

- The subscriber dials a feature code string (for example, *VM).
- The subscriber dials his/her own number (i.e., a *revertive* call).

The ANSI-41 implementation of the first access method is based on the feature control with call routing procedure described in "Feature Control Services" (see Figure 12.9). Implementation of the second access method is based on the OriginationTriggers parameter and Origination-Request operation described above (see Figure 12.10).

The VMR feature also allows the service provider to route subscriber calls to the voice message system by using a form of call delivery and temporary local directory numbers (TLDNs), rather than associating each subscriber's mailbox with an individual directory number (see Figures 12.9 and 12.10).

Three-way Calling and Call Transfer

The three-way calling and call transfer features allow a subscriber involved in a two-party call (i.e., between the subscriber and party A) to:

1. Place party A on hold—the subscriber presses the SEND key.
2. Call another party B—the subscriber enters party B's phone number and presses the SEND key.
3. Press the SEND key to
   (a) Connect the subscriber with both parties A and B (if the subscriber has 3WC active).
   (b) Release party B and reconnect the subscriber to party A (if the subscriber has CT active but not 3WC).

However, the subscriber presses the END key—rather than the SEND key—to release the subscriber and connect party A to B (if the subscriber has CT active).

ANSI-41 does not define procedures that apply solely to the three-way calling and call transfer features, stating that these features "are con-

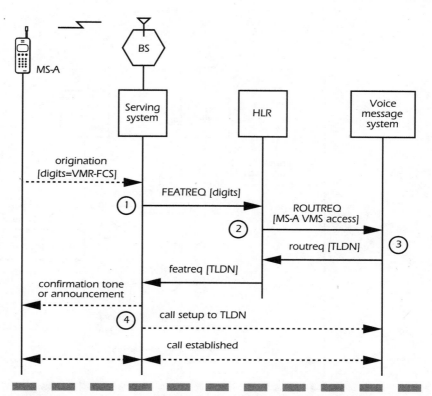

Figure 12.9 An example of VMR feature invocation using the feature code procedure: (1) The subscriber enters the VMR invocation feature code string (VMR-FCS), that may include the VMR password, and presses the SEND key. The serving system detects the FCS and sends the received digits to the HLR in the FEATREQ message. (2) The HLR sends a ROUTREQ to the subscriber's voice message system (VMS), identifying the subscriber (e.g., MS-A's voice mailbox number and VMR password) requesting VMR routing information. (3) The VMS returns a TLDN to the HLR and the HLR relays the TLDN to the serving system. (4) The serving system establishes a call to the VMS using the TLDN, and the subscriber is connected to the voice mailbox.

trolled by the call processing in the Anchor MSC in a manner consistent with IS-53."

However, the 3WC and CT features require ANSI-41 support in two generic areas:

- *To convey the subscriber's 3WC and CT feature activity status to the serving system.* The ANSI-41 CallingFeaturesIndicator parameter contains this information; the HLR sends this to the serving system during the service qualification process.

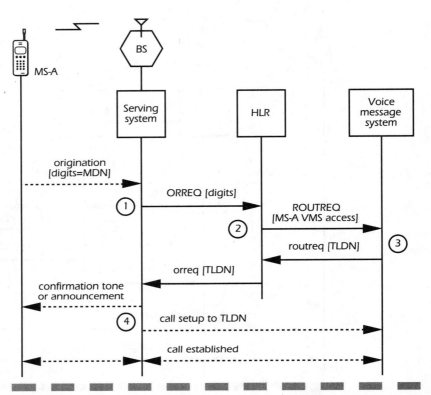

Figure 12.10 An example of VMR feature invocation using the revertive call procedure: (1) The subscriber enters his or her own mobile directory number (MDN), possibly along with the VMR password, and presses the SEND key. The serving system detects the revertive call. The "revertive call" trigger is set; therefore, the serving system sends the received digits to the HLR in the ORREQ message. (2) The HLR sends a ROUTREQ to the subscriber's voice message system (VMS), identifying the subscriber (e.g., MS-A's voice mailbox number and VMR password) and requesting VMR routing information. (3) The VMS returns a TLDN to the HLR and the HLR relays the TLDN to the serving system. (4) The serving system establishes a call to the VMS using the TLDN, and the subscriber is connected to the voice mailbox.

■ *Feature invocation under call handoff conditions.* Since these features are controlled by the anchor system (this is the ANSI-41 philosophy), the serving system must notify the anchor system when the subscriber either presses the SEND key (possibly with other dialed digits) or presses the END key.

The serving system uses the ANSI-41 FlashRequest Invoke (FLASHREQ) message to forward the notification to the anchor system when the SEND key is pressed by the subscriber (see Figure 12.11); this message also includes any other digits dialed by the subscriber. When the subscriber presses the END key, the serving system initiates the call release process; i.e., it sends a FacilitiesRelease Invoke (FACREL) message to the anchor system.

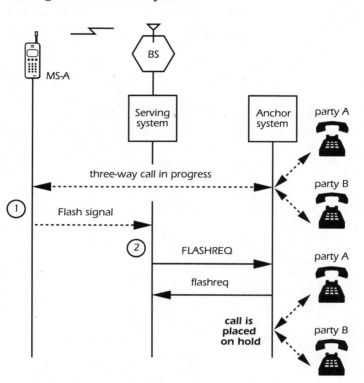

Figure 12.11
An example of the use of the FlashRequest operation for three-way calling: (1) The subscriber is involved in a three-way call that has been handed off from the anchor system to a nonanchor serving system. The subscriber presses the SEND key. (2) The serving system sends a FLASHREQ message to the anchor system. The anchor system places party A and party B on hold.

Conference Calling

Conference calling (CC) is another multiparty feature, like 3WC and CT; however, it requires specific ANSI-41 protocol elements to support two key capabilities of the feature:

- The ability of the service provider to define the maximum number of conferees allowed for any particular conference call, rather than fix the number for all calls and all subscribers. For this purpose, the conference call is initiated with a feature code string provided by the subscriber—either at the start of a call (see Figure 12.12) or during an active two-party call—and conveyed from the serving system to the HLR in the FEATREQ message, as described in "Feature Control Services." The featreq message contains the ANSI-41 ConferenceCall-ingIndicator (CCI) parameter that specifies the maximum number of conferees for the call. The presence of the CCI parameter in the featreq message informs the serving system that the subscriber has invoked the CC service and to process the call accordingly.

- The ability of the subscriber to drop the last party added to the conference call. This is useful if a busy signal, wrong number, or voice message system is reached during the attempt to add a party to a conference call. This function is also initiated with a feature code string from the subscriber—the "drop last party" feature code string—which the serving system sends to the HLR in a FEATREQ message. The HLR returns a featreq message to the serving system, containing the ANSI-41 ActionCode parameter set to the value *conference calling drop last party*. The serving system executes the drop-last-party operation, and the conference call continues.

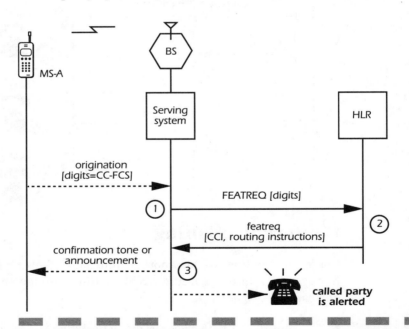

Figure 12.12 *An example of conference call initiation: (1) The subscriber enters the CC invocation feature code string (CC-FCS), which includes the phone number of the first party to add to the call, and presses the SEND key. The serving system detects the FCS and sends the received digits to the HLR in the FEATREQ message. (2) The HLR authorizes the conference call and returns the ConferenceCallingIndicator (CCI) parameter and call routing instructions in the featreq message. (3) The serving system notifies the subscriber that the request was successful, begins CC processing, and routes the call as directed by the HLR.*

Additionally, the feature is defined so that once the conference call has been authorized by the HLR, new parties can be added to the call under either HLR control or serving system control. The HLR-controlled case is shown in Figure 12.12; each new party's number is preceded by a feature code that directs the serving system to query the HLR for routing instructions. With serving system control:

1. The subscriber presses SEND to place the current conferees on hold.
2. The subscriber enters the new party's number—without a feature code—and presses SEND.
3. The serving system immediately establishes a call to the new party; no query (i.e., FEATREQ message) is sent to the HLR.
4. The subscriber presses SEND to add the held conferees back onto the call.

This approach may be supplemented with the call-origination triggers to give the HLR selected control over adding new parties. For example, if the international destination trigger is armed and the subscriber dials an international number as the next party to add to the conference call, the serving system will launch an ORREQ message to the HLR, allowing the HLR the opportunity to authorize the addition explicitly.

Finally, CC works under call-handoff conditions in the same manner as the 3WC and CT features; that is, the serving system uses the FLASHREQ message or FACREL message to forward the SEND key or END key notifications, respectively, to the anchor system.

Priority Access and Channel Assignment

The priority access and channel assignment (PACA) feature enables a mobile subscriber to have priority access to voice or traffic channels on call origination. When channels are not available, the subscriber is queued at the radio system on a first-come, first-served and a priority basis. There can be up to 16 priority levels, with 1 being the highest. The subscriber is considered to be busy while a call is queued. Note that this feature is not intended to be a pay feature; some mobile subscribers can essentially request higher-priority service over others. The feature is meant to provide service priority for emergency response personnel (e.g., police, fire, medical) or extraordinarily important individuals, such as government emergency officials. The PACA feature requires special MS support, not currently defined in any air-interface standard.

The PACA feature has two modes of activation that affect the ANSI-41 protocol support procedures:

- The feature is automatically requested on each call origination.
- The feature is manually requested by the subscriber.

The HLR conveys this subscriber profile attribute to the serving system in the ANSI-41 PACAIndicator parameter; the HLR sends this to the serving system during the service qualification process. With manual activation, the subscriber must precede the destination address with the PACA activation feature code; the serving system sends the entire string to the HLR for PACA authorization in the FEATREQ message. If this is approved, the HLR sends a single-call PACA invocation indicator to the serving system in the ANSI-41 OneTimeFeatureIndicator (OTFI) parameter, as described in "Feature Control Services." With automatic activation, the serving system treats each call by the subscriber as a potential PACA call—if PACA condition exists (an emergency situation), the call is given the treatment prescribed for the subscriber's priority level (placed into a queue for the given priority level or assigned an available channel from a set of channels reserved for the given priority level).

Preferred Language

The preferred language (PL) feature provides a standardized means of notifying the serving system of a visiting subscriber's preferred language for recorded announcements, directory assistance, and other serving system interactions.

The HLR conveys this subscriber attribute to the serving system in the ANSI-41 PreferredLanguageIndicator parameter; the HLR sends this to the serving system during the service qualification process. Normally, this attribute is fixed in the HLR's database; however, ANSI-41 supports the ability to change the value for an authorized MS—e.g., an MS used in a phone rental operation—using the single-step feature control mechanism described in "Feature Control Services."

Voice Privacy, Data Privacy, and Signaling Message Encryption

The voice privacy (VP) and signaling message encryption (SME)[3] features provide an enhanced degree of privacy by encrypting the signals transmitted over the radio channel between the MS and the serving base station. The VP feature provides encryption of the subscriber's voice conversations, while the SME feature encrypts selected parameters in the signaling messages sent on an analog voice channel or digital traffic channel, such as a signaling message that conveys a subscriber's personal identification number from the MS to the BS. The VP feature is only available in the digital modes of operation (TDMA and CDMA), while the SME feature can be applied to both digital and analog modes, provided authentication is supported in the MS and the network. The VP mask used to encrypt the voice signals is generated using the CAVE algorithm with SSD_B as input (see Chapter 11); the SME feature uses SSD_B and CAVE to generate a key that is then used in the Cellular Message Encryption Algorithm (CMEA) to encrypt the select signaling message elements. Although we discuss the VP and SME features together, the subscriber has no control over SME—it is a service provider option—and the two features may be independently provided to the subscriber.

The data privacy (DP) feature was initially defined in *TIA/EIA/IS-737*. This feature is applicable to TDMA systems only. DP is analogous to VP, except that DP was intended to provide encryption of the subscriber's *data* sessions. Like VP and SME, the DP feature makes use of SSD_B; however, DP uses a new algorithm to generate the data key and the encryption mask, called the ORYX algorithm.

To support these features, both the authentication controller[4] and the MS generate the appropriate keys and masks on a per-call origination or termination basis (see Figure 12.13). The MS and the serving BS use these parameters to encode and decode the signals transmitted on the radio channel.

ANSI-41 support for these features falls into three categories:

---

[3] The signaling message encryption acronym, SME, should not be confused with the same acronym used for short message entity in Chapter 13.

[4] As described in Chapter 11, the authentication controller is the serving system (i.e., the VLR) if SSD is shared or the authentication center (AC) if SSD is not shared.

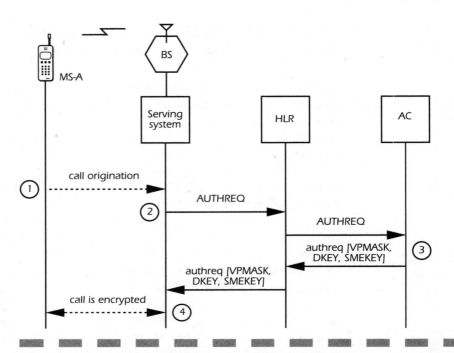

Figure 12.13 Generating the VP mask, data key, and SMEKEY when SSD is not shared: (1) The subscriber originates a call. The MS generates a VPMASK (assume this is a TDMA MS) or data key (if TDMA data call) and SMEKEY that it will use for the call. (2) The serving system initiates MS authentication, as described in Chapter 11. Since SSD is not shared, the serving system sends an AuthenticationRequest Invoke (AUTHREQ) message to the HLR, which relays the message to the AC. (3) The AC performs the authentication functions described in Chapter 11. Since the system access is a call origination, it also generates a VPMASK, DKEY, and SMEKEY for the call and sends these to the HLR. If the MS is subscribed to the VP and DP features, the HLR relays the VPMASK and DKEY to the serving system; otherwise, the HLR discards these elements. (4) The serving system uses the VPMASK (if voice) or DPKEY (if data) and the SMEKEY received from the HLR to encrypt and decrypt the information exchanged with the MS over the radio channel.

1. *Conveying the subscriber's VP and DP feature activity status to the serving system.* The ANSI-41 CallingFeaturesIndicator parameter contains this information; the HLR sends this to the serving system during the service qualification process. If the MS is subscribed to VP and DP and requests one of these features, then it is invoked—provided the serving system can support it. Since the SME feature is not subscriber controlled, it is not part of the subscriber's service profile information.

2. *Conveying the appropriate VP, DP, and SME masks to the anchor system.* The procedures in ANSI-41 state that the authentication controller generates the VP, DP, and SME masks when the MS attempts a call origination or termination and successfully passes the global authentication challenge. The masks are conveyed from the authentication controller to the anchor MSC in the AuthenticationRequest Return Result (authreq) message in the following ANSI-41 parameters:

(a) For TDMA operation, the voice privacy mask is in the VoicePrivacyMask (VPMASK) parameter.

(b) For CDMA operation, the voice privacy mask is in the CDMAPrivateLongCodeMask (CDMAPLCM) parameter.

(c) For TDMA operation, the data privacy key is in the DataKey (DKEY) parameter.

(d) For both modes of operation, the signaling message encryption key is in the SignalingMessageEncryptionKey (SMEKEY) parameter.

3. *Conveying the appropriate VP mask, DP key, and SME key to the serving system under call-handoff conditions.* To maintain support for the VP, DP, and SME features when a call handoff occurs, the parameters are passed to the new serving system in each of the following ANSI-41 messages:

(a) FacilitiesDirective Invoke

(b) FacilitiesDirective2 Invoke

(c) HandoffBack Invoke

(d) HandoffBack2 Invoke

(e) HandoffToThird Invoke

(f) HandoffToThird2 Invoke

(g) InterSystemSetup Invoke

The ANSI-41 ConfidentialityModes (CMODES) parameter is also passed in these messages to inform the serving system that zero or more of the VP, DP, and SME features is desired for the call; the response message from the serving system also includes the CMODES parameter, set to indicate whether the features can be supported by the serving system for the call.

Finally, we should point out that the CMEA and ORYX algorithms have been determined to be insufficiently strong and will be phased out in the future. The TR-45 ad hoc authentication group (AHAG) has developed Enhanced Privacy and Encryption (EPE) algorithms that will better secure communications in TDMA (ANSI-136-B and later) systems.

For details, the reader should see the July and August 2000 issues of the *Wireless Security Perspectives* newsletter (see Bibliography).

Call Termination Services

In general, ANSI-41 call termination services are those network functions that enable, restrict, supplement, or otherwise affect an MS's ability to receive a call while roaming outside the home service area. ANSI-41 supports the following call termination services:

- Call delivery (i.e., basic call termination)
- Call forwarding—unconditional (CFU)
- Call forwarding—default (CFD)
- Call forwarding—no answer (CFNA)
- Call forwarding—busy (CFB)
- Call waiting (CW)
- Do not disturb (DND)
- Calling number identification presentation (CNIP)
- Calling name presentation (CNAP)
- Selective call acceptance (SCA)
- Password call acceptance (PCA)
- Mobile access hunting (MAH)
- Flexible alerting (FA)

Call Delivery

The call delivery (CD) feature can be considered the most fundamental form of call redirection supported by ANSI-41.[5] With CD, the call is redirected from its original called number—the called MS's mobile directory number[6] (MDN)—to another telephone number temporarily assigned to

---

[5] Call delivery has been supported by IS-41 since Revision A. Prior to the availability of automatic call delivery, calls to roaming subscribers required the calling party to dial a special "roamer port" number to receive a second dial tone; then, the calling party would dial the MS's directory number.

[6] The MS's mobile directory number is the number to call to reach the MS. In current practice, this number normally corresponds to the MS's mobile identification number (MIN) that is programmed into the mobile station equipment; however, ANSI-41 allows these two numbers to be different.

the MS by the serving system, specifically for call delivery purposes. The system that performs the call-delivery redirection is called the *originating system*. The originating system is often the MS's home system. However, another system may be specifically provisioned to recognize calls to mobile subscribers and to initiate the call delivery processing on behalf of the home system; such a system is generally termed a *gateway system*.

The originating system relies on three key ANSI-41 functions for normal call delivery:

- The location management functions (see Chapter 10) that keep track of the MS's location, i.e., the serving system identification
- The MS state management functions (see Chapter 10) that keep track of the MS's ability to receive incoming calls (i.e., the MS is either active or inactive)
- The call routing functions that provide the originating system with the routing information—at a minimum, a telephone number called the *temporary local directory number* (TLDN)—needed to get the call to the current serving system. These functions are provided by the ANSI-41 LocationRequest and RoutingRequest operations (see Figure 12.14).

ANSI-41 also allows an authorized subscriber to activate and deactivate the CD feature by using the single-step feature control mechanism.

Unsuccessful Call Delivery

There are a number of opportunities for failure in the call-delivery process (failures that can be mitigated with the other call-termination features, such as call forwarding, described in this chapter). In some cases, when the HLR receives the LocationRequest Invoke (LOCREQ) message, it can make an immediate decision to deny the call delivery attempt when:

- The directory number is not recognized.
- The MS's current location is not known.
- The MS is not authorized for call delivery.
- The MS is inactive.
- Termination to the MS is otherwise denied (e.g., due to a delinquent account).

In this case, the HLR immediately sends a LocationRequest Return Result (locreq) message to the originating system, including the ANSI-41

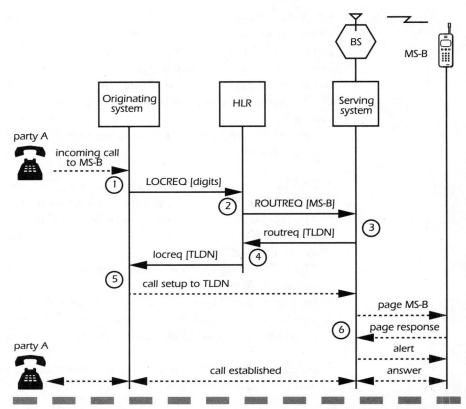

Figure 12.14   Use of the LocationRequest and RoutingRequest operations for call delivery: (1) Party A dials a phone number (i.e., MS-B's mobile directory number) that is routed via the PSTN to the originating system. The originating system launches a LOCREQ message to the HLR to determine how to route the call. (2) The HLR determines that the call is for MS-B and sends a ROUTREQ message to MS-B's current serving system. (3) The serving system allocates a TLDN for the call and returns it to the HLR in the routreq message. (4) The HLR relays the TLDN to the originating system in the locreq message. (5) The originating system routes the call to the TLDN. (6) The serving system associates the incoming call to the TLDN with the previous ROUTREQ message identifying MS-B; therefore, it pages MS-B. MS-B responds to the page, therefore, the serving system directs MS-B to alert the subscriber. When the subscriber answers, the call is established between the calling party A and MS-B.

AccessDeniedReason (ACCDEN) parameter with an indication of the reason for denying access to the called MS (see Figure 12.15). The message may also include an AnnouncementList parameter. This parameter specifies one or more tones or announcements that may be provided to the calling party by the originating system (e.g., a tone followed by "The customer you have called is not accessible. Please try your call again later").

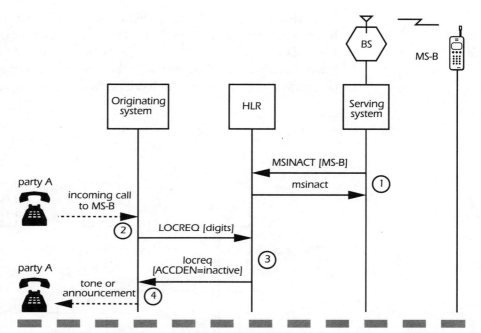

Figure 12.15   *An example of unsuccessful call delivery (MS is inactive): (1) The serving system declares MS-B inactive and notifies the HLR using the MSInactive Invoke (MSINACT) message. (2) Party A dials a phone number (i.e., MS-B's mobile directory number) that is routed via the PSTN to the originating system. The originating system launches a LOCREQ message to the HLR to determine how to route the call. (3) Since MS-B is inactive, the HLR denies the call delivery attempt. (4) The originating system provides a tone or announcement to the calling party.*

Otherwise, the HLR sends a RoutingRequest Invoke (ROUTREQ) message to the serving system. At this point, the serving system can make the decision to deny the call delivery attempt if:

- The MS is inactive, based on the serving system's determination.
- The MS is busy.
- The MS does not respond to paging (assuming the system performs paging at this point).
- The MS is otherwise unavailable[7] (e.g., turned off).

---

[7] Classifying the MS as *unavailable* versus *inactive* is a serving system decision. However, if the MS is labeled inactive, the HLR may suppress call delivery attempts to the MS; therefore, the serving system must explicitly notify the HLR when the MS becomes active again, using the RegistrationNotification operation (see Chapter 10). On the other hand, if the MS is labeled unavailable, the HLR will continue to attempt call delivery as it is notified of each new call to the MS.

In this case, the serving system sends a RoutingRequest Return Result (routreq) message to the HLR, including the ANSI-41 AccessDeniedReason parameter with the reason for denying access to the called MS. The HLR forwards the AccessDeniedReason parameter to the originating system in the locreq message (see Figure 12.16).

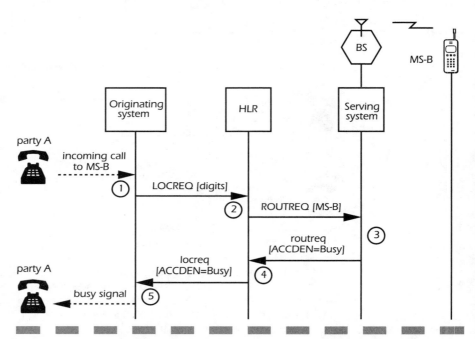

Figure 12.16 *An example of unsuccessful call delivery (MS is busy): (1) Party A dials a phone number (i.e., MS-B's mobile directory number) that is routed via the PSTN to the originating system. The originating system launches a LOCREQ message to the HLR to determine how to route the call. (2) The HLR determines that the call is for MS-B and sends a ROUTREQ message to MS-B's current serving system. (3) Since MS-B is busy, the serving system denies the call delivery attempt, returning the AccessDeniedReason (ACCDEN) parameter set to the value busy. (4) The HLR relays the ACCDEN parameter to the originating system in the locreq message. (5) The originating system provides a busy signal to the calling party.*

Finally, even when the MS is considered available, a TLDN is provided to the originating system, and a call is established to the serving system, call delivery can still fail under a number of circumstances, including:

- When the MS becomes inactive, busy, or unavailable after the serving system provides the TLDN to the HLR but before the call to the TLDN is received by the serving system.

- When the MS does not respond to paging or does not answer the incoming call alert.
- When the MS fails authentication.

In these cases, the calling party will be given the appropriate call treatment by the serving system—a busy signal, tone, or announcement (see Figure 12.17).

Figure 12.17
An example of unsuccessful call delivery (MS is busy when call arrives at the serving system): (1–4) Same as in Figure 12.14. (5) MS-B originates a call prior to the serving system receiving the call to the TLDN. (6) When the call to the TLDN arrives, the serving system provides a busy signal to the calling party.

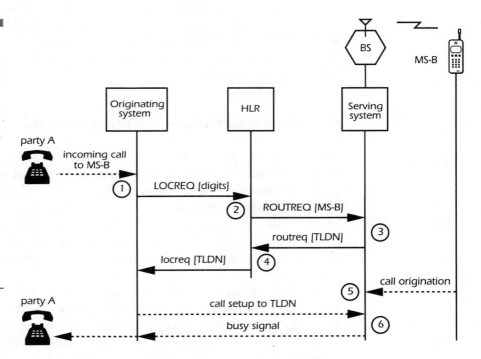

Call Delivery in Border Cell Situations

The border cell situations discussed in Chapter 10 may also affect call delivery if the MS sends a page response to a border system, even though the page was issued from the serving system. This situation can occur if:

1. The MS hears the page in the serving system, then rescans and chooses the border system as having the strongest control channel signal.
2. The MS has just registered in the border system, and the location update process has not yet been completed in the HLR.

3. The MS's location is otherwise incorrect in the HLR.

ANSI-41 includes a number of capabilities designed to alleviate these problems. The solutions fall under the categories of: *predelivery paging solutions* and *postdelivery paging solutions*. Predelivery paging solutions apply to serving systems that page the MS when the ROUTREQ message is received from the HLR; these methods use the ANSI-41 InterSystemPage or UnsolicitedResponse operations. The postdelivery solution can be used by serving systems that page the MS when the actual call to the TLDN is received from the originating system. We believe postdelivery paging is the easier approach and describe this border cell solution in Figure 12.18.

Call Forwarding—Unconditional

The call forwarding—unconditional feature enables the subscriber to have all incoming calls redirected either to a selected forward-to number or to the subscriber's voice mailbox. The basic CFU feature is supported by the ANSI-41 LocationRequest operation (see Figure 12.19). The subscriber may also be given an alert indication when CFU is invoked to forward a call, for example, to remind the subscriber that the feature is active. This reminder feature is enabled by the ANSI-41 InformationDirective operation.

ANSI-41 allows an authorized subscriber to activate and deactivate the CFU feature and to register a new CFU forward-to number by using the single-step feature control mechanism. A *courtesy call* may be provided to the newly registered CFU forward-to number, permitting the subscriber to verify the validity of the forward-to number, by using the feature control with call-routing procedures described in "Feature Control Services."

Call Forwarding—Busy and Call Forwarding—No Answer

The call forwarding—busy and call forwarding—no answer features enable the subscriber to have all incoming calls that encounter a busy or a no-answer condition, respectively, to be redirected either to a selected forward-to number or to the subscriber's voice mailbox. While the busy condition is easily defined—the subscriber is already engaged in a call

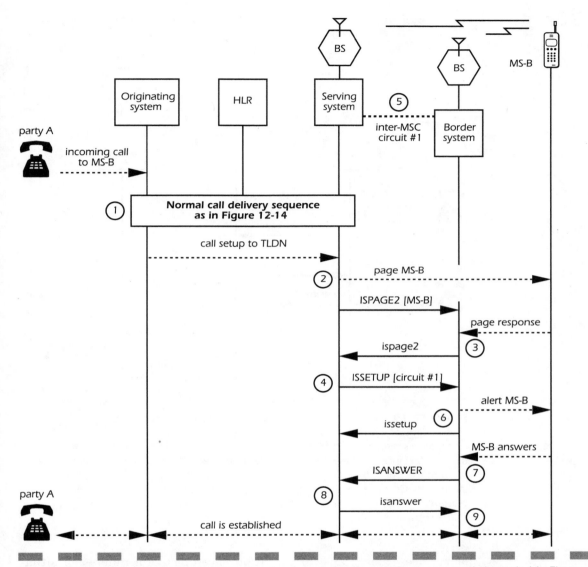

Figure 12.18 An example of call delivery in a border cell situation: (1) Same as steps 1 through 4 in Figure 12.14. (2) When the call to the TLDN arrives, the serving system pages MS-B, but also sends an InterSystemPage2 Invoke (ISPAGE2) message to one or more border system. The ISPAGE2 message informs the border system to either page MS-B and listen or just listen (as shown) for a page response from MS-B. (3) The border system receives the page response from MS-B; the border system notifies the serving system using the InterSystemPage2 Return Result (ispage2) message. (4) The serving system initiates call handoff to the border system using the InterSystemSetup Invoke (ISSETUP) message, specifying an inter-MSC circuit between the two systems for handoff purposes. (5) The inter-MSC circuit is established. (6) The border system alerts MS-B and responds to the serving system via the InterSystemSetup Return Result (issetup) message. (7) The subscriber answers the call; the border system notifies the serving system using the InterSystemAnswer Invoke (ISANSWER) message. (8) The serving (now anchor) system connects the call path from the originating system to the inter-MSC circuit and sends an InterSystemAnswer Return Result (isanswer) acknowledgment to the border system. (9) The call is now established between the calling party A and MS-B.

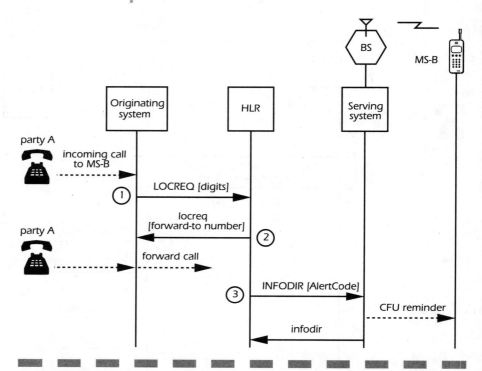

Figure 12.19 *An example of CFU: (1) Party A dials a phone number (i.e., MS-B's mobile directory number) that is routed via the PSTN to the originating system. The originating system launches a LOCREQ message to the HLR to determine how to route the call. (2) Since MS-B has CFU active, the HLR directs the originating system to forward the call to the CFU forward-to number. (3) The HLR may also send an InformationDirective Invoke (INFODIR) message to the serving system. The AlertCode parameter in the message directs the serving system to provide a CFU reminder alert to the subscriber.*

or service—the no-answer condition encompasses a number of situations, including these:

- The MS does not respond to paging.
- The subscriber does not respond to MS "alerting" (the phone ringing).
- The MS's current location is not known.
- The MS is inactive.
- The MS is otherwise inaccessible (e.g., the call delivery feature is inactive).

Checking the MS's busy status requires querying the serving system; however, in some cases the HLR can make an immediate CFNA invocation decision (e.g., if the MS is listed as inactive in the HLR's database). In these cases, the CFNA feature requires only the ANSI-41 LocationRequest operation, and the locreq message contains the CFNA forward-to number (see Figure 12.20).

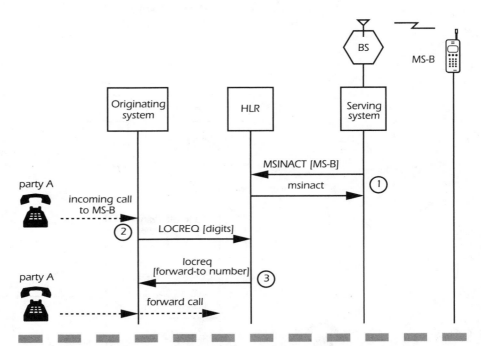

Figure 12.20 *An example of immediate CFNA by the HLR: (1) The serving system determines that MS-B is inactive; the HLR is notified using the MSInactive operation. (2) Party A dials a phone number (i.e., MS-B's mobile directory number) that is routed via the PSTN to the originating system. The originating system launches a LOCREQ message to the HLR to determine how to route the call. (3) Since MS-B is inactive and has CFNA active, the HLR immediately directs the originating system to forward the call to the CFNA forward-to number.*

Otherwise, the HLR sends a ROUTREQ message to the serving system, attempting normal call delivery. The serving system may respond with a routreq message, including the ANSI-41 AccessDeniedReason parameter with the reason for denying access to the called MS. The HLR directs the originating system to forward the call (1) to the CFB forward-to number if the AccessDeniedReason parameter value is busy (see Figure 12.21) or (2) to the CFNA forward-to number if the value of the parameter is any one of these:

- Unavailable
- No page response
- Inactive
- Unassigned directory number
- Termination denied

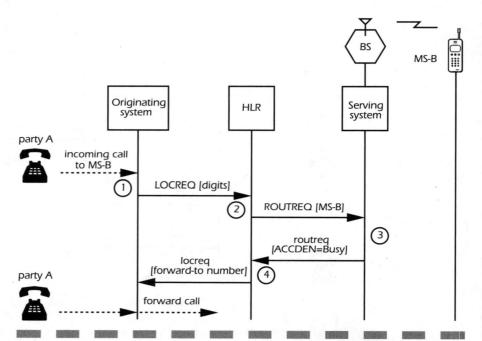

Figure 12.21 An example of CFB by the HLR: (1) Party A dials a phone number (i.e., MS-B's mobile directory number) that is routed via the PSTN to the originating system. The originating system launches a LOCREQ message to the HLR to determine how to route the call. (2) The HLR determines that the call is for MS-B and sends a ROUTREQ message to MS-B's current serving system. (3) Since MS-B is busy, the serving system denies the call delivery attempt, returning the AccessDeniedReason (ACCDEN) parameter set to the value "busy." (4) Since MS-B is busy and has CFB active, the HLR directs the originating system to forward the call to the CFB forward-to number.

Finally, even when the MS is considered available, a TLDN is provided to the originating system, and a call established to the serving system, CFB, or CFNA can still be invoked under a number of circumstances, including:

- When the MS becomes busy after the serving system provides the TLDN to the HLR but before the call to the TLDN is received by the serving system (this case is referred to as a *call collision* in ANSI-41).
- When the MS becomes inactive or unavailable after the serving system provides the TLDN to the HLR.
- When the MS does not respond to paging or does not answer the incoming call alert.
- When the MS fails authentication (ANSI-41 does not explicitly allow or disallow this case).

In these cases, the serving system initiates a set of call redirection processes that normally involve the steps shown in Figure 12.22.[8]

Call Forwarding—Default

The call forwarding—default feature enables a called subscriber to have the network send all incoming calls addressed to the subscriber's directory number to another directory number (i.e., the forward-to number) when a busy or no-answer condition (described in the previous section) is encountered.

From the ANSI-41 protocol perspective, a subscriber activation of the CFD feature is essentially equivalent to activating both the CFNA and CFB features simultaneously; registration of a CFD forward-to number registers a single forward-to number for both CFNA and CFB features. The only reference to the CFD feature in an ANSI-41 message or parameter is in the DMH_Redirection Indicator parameter, which is primarily used for call-recording purposes. Therefore, refer to the previous descriptions of the CFB and CFNA features for an understanding of ANSI-41's support of CFD.

The CFD feature is generally considered to be the basic form of call forwarding that would be provided to each subscriber. It enables call completion, typically to a voice mail system, when the subscriber cannot be reached, for just about any reason. With this basic—or default—forwarding feature in effect, the subscriber can add the CFB and CFNA features to provide special handling of these conditions. For example, a

---

[8] If the originating system does not accept the redirection request, ANSI-41 specifies that the serving system may query the HLR for the appropriate forward-to number (via the TransferToNumberRequest operation) and forward the call itself. However, this approach is less efficient in terms of the number of trunks involved since it includes the path between the originating and serving systems.

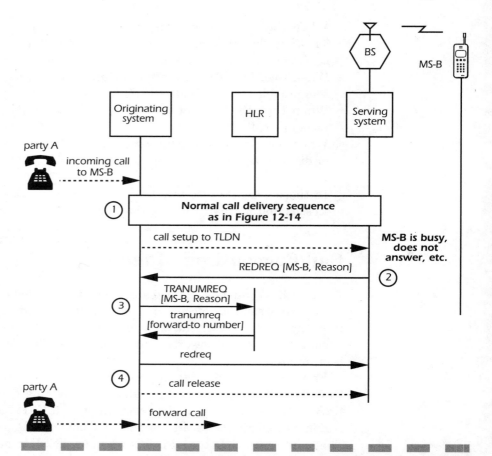

Figure 12.22   An example of CFB or CFNA after call delivery to the serving system: (1) Same as steps 1 through 4 in Figure 12.14. (2) When the call to the TLDN arrives, the serving system determines that CFB or CFNA should be invoked (see text for reasons). The serving system sends a RedirectionRequest Invoke (REDREQ) message to the originating system, requesting that the call be redirected. (3) The originating system uses the TransferToNumberRequest operation to obtain the forward-to number for the given redirection reason (i.e., busy or no answer) from the HLR. (4) The originating system notifies the serving system that it accepts the redirection request, releases the call leg to the serving system, and then forwards the call to the forward-to number provided by the HLR.

subscriber has CFD to voice mail but wants calls temporarily redirected to a coworker if the subscriber is engaged in a call, e.g., to handle particularly important calls. For this purpose, the subscriber activates the CFB feature and registers the coworker's telephone number as the CFB forward-to number.

Call Waiting

The call-waiting feature allows an MS already engaged in a call to be informed of an additional incoming call and to "toggle" between the two calls—placing one call on hold and getting connected to the other—by pressing the SEND key.

ANSI-41 protocol support for the CW feature has three components:

- *Conveying the subscriber's CW feature activity status to the serving system.* The ANSI-41 CallingFeaturesIndicator parameter contains this information; the HLR sends this to the serving system during the service qualification process.
- *Feature control.* The CW feature may be activated and deactivated by an authorized MS by using the single-step feature control procedure and may be deactivated on a per-call basis—either at call origination time or during an active call—by using the per-call feature control procedures described in "Feature Control Services."
- *Feature invocation.* This is described below.

Feature invocation begins in the anchor system when a ROUTREQ message is received from the HLR for an MS that is already engaged in a call (we refer to this as call-A). Rather than return the AccessDeniedReason parameter set to indicate that the MS is busy, the anchor system (the call may have been handed off to a new serving system) returns a normal TLDN. When the anchor system receives the incoming call (we refer to this as *call-B*) to the TLDN, it attempts to notify the MS. Depending on the anchor system's and serving system's capabilities and on those of the called MS, a number of CW notification methods are possible:

1. If the MS is capable of only an *in-band notification* (i.e., a tone injected momentarily into the voice path to the called MS), then this is provided by the anchor system, even if call-A has been handed off. The tone is applied once and then again in 15 seconds if call-B is not answered (i.e., by the subscriber's pressing the SEND key).
2. If call-A has not been handed off and the MS and anchor system are capable of an *out-of-band notification* (i.e., a signal sent outside the voice path to the called MS that directs the MS to generate a CW alert), then the anchor system orders the MS to apply the CW tone.
3. If call-A has been handed off and the MS, anchor, and serving systems are capable of an out-of-band notification, then the anchor sys-

tem sends an InformationForward Invoke (INFOFWD) message to the serving system. The INFOFWD message includes the AnnouncementList parameter that directs the serving system to order the MS to apply the CW tone. The subscriber may then press the SEND key to place call-A on hold and answer call-B. This triggers the serving system to send a FLASHREQ message to the anchor system. The anchor system places call-A on hold and connects the subscriber to call-B (see Figure 12.23).

Figure 12.23
An example of call waiting alert after handoff: (1) The subscriber is involved in a call that has been handed off from the anchor system to a non-anchor serving system. A new call arrives for the subscriber from party B. (2) Since the subscriber has CW active, the anchor system sends an INFOFWD message to the serving system, directing the serving system to order the MS to apply the CW tone. (3) The subscriber presses the SEND key. The serving system sends a FLASHREQ message to the anchor system. The anchor system places party A on hold and connects the subscriber to party B.

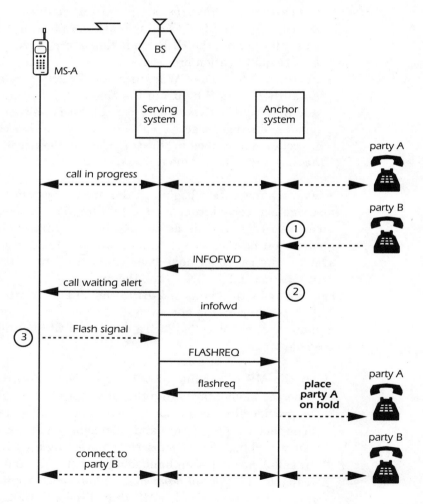

A little-known feature supported by the ANSI-41 protocol and procedures is the priority call waiting (PCW) feature. This is actually a call origination feature although it affects the CW feature: It allows calls originated by a PCW subscriber to be assigned a higher-than-normal priority level (this is indicated in the "Call waiting for incoming call" field of the OneTimeFeatureIndicator parameter). Once a call with this priority level is completed to a CW subscriber (MS-A), CW will be invoked only if another PCW subscriber attempts to reach MS-A; calls from subscribers without PCW are given busy treatment.

Do Not Disturb

When a subscriber activates the do-not-disturb (DND) feature, the MS is essentially placed in an inactive state for call delivery purposes. Incoming calls are provided the treatment described in Figure 12.15; that is, the calling party receives a call refusal tone, announcement, or both. The feature also blocks all other audible alerting of the called party, such as the message-waiting notification indications and the CFU alert.

The DND feature may be activated and deactivated by an authorized MS by using the single-step feature control procedure.

Calling Number Identification Presentation

The calling number identification presentation (CNIP) feature allows the subscriber to obtain—and usually view in some manner—the calling number identification information from the serving system when the incoming call is received and the subscriber is alerted. The CNI information can include up to two calling party numbers, a redirecting number, as well as calling and redirecting subaddress information. The ability to present two calling party numbers is optional. A user-provided calling party number is intended to represent the actual address of the calling party, useful when the calling party is served by a private branch exchange (PBX). The network-provided calling party number is intended to provide another number representing network address information that may be helpful to the called party. For example, the (user-provided) calling party's *direct inward dialing* (DID) number can be provided as well as the (network-provided) *listed directory number* for a given corporate office.

With normal call delivery, the CNI information gets to the serving system in the following manner (refer to Figure 12.14, which illustrates call delivery):

1. The originating system must receive the CNI information from the PSTN or other source of the call.
2. The originating system includes the CNI information in the LOCREQ message it sends to the HLR. The CNI information is carried in the following ANSI-41 parameters in the LOCREQ message:
   (a) CallingPartyNumberDigits1 (i.e., the network-provided number)
   (b) CallingPartyNumberDigits2 (i.e., the user-provided number)
   (c) CallingPartySubaddress
   (d) RedirectingNumberDigits
   (e) RedirectingSubaddress.
3. The HLR provides the CNI information to the anchor system in the ROUTREQ message. The CNI information is carried in the following ANSI-41 parameters in the ROUTREQ message, mapped from the corresponding parameters in the LOCREQ message:
   (a) CallingPartyNumberString1
   (b) CallingPartyNumberString2
   (c) CallingPartySubaddress
   (d) RedirectingNumberString
   (e) RedirectingSubaddress

The numbers—but not the subaddresses—are converted from one format in the LOCREQ message (e.g., CallingPartyNumberDigits1) to another format in the ROUTREQ message (e.g., CallingPartyNumberString1) for the following reasons:

1. The numbers are generally received from the PSTN in binary-coded decimal (BCD) format, so that is the format that is conveyed to the HLR in the LOCREQ message.
2. The numbers are generally delivered to the MS over the radio channel in the International Alphabet No. 5 (IA5) format, so that is the format that the HLR conveys to the anchor system in the ROUTREQ message.
3. Using the IA5 format in the ROUTREQ message to the anchor system gives the HLR the option to substitute letters (e.g., a name) for the numeric information.

At the serving system, the number information is analyzed. If the presentation indicator within a number parameter indicates that presentation to the called party is restricted (i.e., transmission of the number over the radio channel is not allowed or the number is not available), then the serving system shall not provide the number to the MS. Otherwise, the CNI information is transmitted to the MS.

Calling Name Presentation

The calling name presentation (CNAP) feature was initially defined in *TIA/EIA/IS-764* and will be incorporated in ANSI-41-E. CNAP allows the subscriber to obtain—and usually view in some manner—the calling name identification (CNA) information from the serving system when the incoming call is received and the subscriber is alerted. The CNA may take many forms; for example, a personal name, a company name, or a label like "restricted" or "not available."

Redirecting Name Delivery (RND) is a CNAP subscription option. When CNAP with RND is active and a call has been forwarded, CNAP provides the CNA of the last redirecting party and the original calling party to the called subscriber.

The development of the CNAP standard was originally undertaken as part of the wireless intelligent network (WIN) effort (see Chapter 19). However, the TR-45.2 subcommittee came to the conclusion that CNAP required little or no WIN-specific functionality and the CNAP feature was eventually defined in terms of straightforward extensions to ANSI-41-D. The only vestige of the WIN legacy is in the use of a Service Control Point (SCP) entity in the network, which handles queries from the HLR for CNA information, effectively doing number-to-name lookups. For interaction between the HLR and the SCP, a new ANSI-41 operation is defined, called ServiceRequest.[9] The use of this message for CNAP is illustrated in Figure 12.24.

The service also supports delivery of CNA information to the subscriber as part of the CFU alert indication, previously described in "Call forwarding—unconditional" and illustrated in Figure 12.19.

Selective Call Acceptance

The selective call acceptance (SCA) feature also makes use of the CNI information received from the PSTN or other source of the call to the subscriber. The SCA feature uses the CNI information to perform call screening at the HLR; that is, checking the calling party number against a list of allowed numbers. Calls from numbers on the list are allowed to proceed as normal call-delivery attempts (see Figure 12.14).

---

[9] According to IS-764, "the HLR may obtain the text to present to the called subscriber from an alternative source in lieu of performing the ServiceRequest operation." These *alternative sources* are not specified in IS-764.

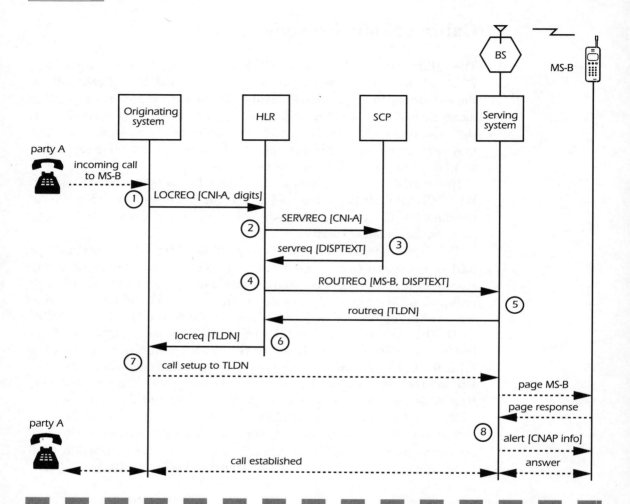

Figure 12.24   An example of CNAP: (1) Party A dials MS-B's mobile directory number and the call is routed to the originating system. The originating system launches a LOCREQ message to the HLR to determine how to route the call, including the caller's CNI information. (2) The HLR determines that CNAP should be invoked and sends a SERVREQ message to the SCP system capable of performing the name lookup. (3) The SCP retrieves the caller's name, based on the CNI information, and returns it to the HLR in the DisplayText (DISPTEXT) parameter within the servreq message. (4) The HLR sends a ROUTREQ message to MS-B's current serving system, including the DISPTEXT parameter. (5) The serving system stores the DISPTEXT information, allocates a TLDN for the call, and returns it to the HLR in the routreq message. (6) The HLR relays the TLDN to the originating system in the locreq message. (7) The originating system routes the call to the TLDN. (8) The serving system pages MS-B and then directs MS-B to alert the subscriber. In the alert command the serving system includes the text it received in the DISPTEXT parameter. When the subscriber answers, the call is established between the calling party A and MS-B.

Incoming calls from parties not on the screening list—including calls that do not provide CNI information—are given one of two possible call-refusal treatments:

1. If the subscriber has registered an SCA call-refusal treatment, by using the single-step feature control procedure described in "Feature Control Services," then refused calls are redirected either to the selected SCA forward-to number (see Figure 12.25) or to the subscriber's voice mailbox.
2. Otherwise, the call is given the default system call-refusal treatment; this could include a tone and announcement to the calling party (e.g., "The number you have dialed does not accept incoming calls").

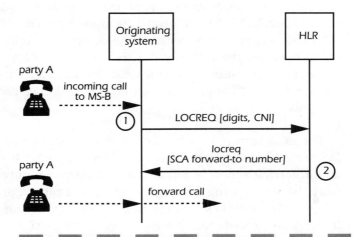

Figure 12.25 *An example of SCA when the caller is not on the screening list: (1) Party A dials a phone number (i.e., MS-B's mobile directory number) that is routed via the PSTN to the originating system. The originating system launches a LOCREQ message to the HLR to determine how to route the call. The LOCREQ message includes party A's CNI information. (2) Since MS-B has SCA active and the party A's calling number is not on the SCA screening list, the HLR directs the originating system to forward the call to the SCA forward-to number.*

The SCA feature may be activated and deactivated by an authorized MS using the single-step feature control procedure. It is also possible for an authorized subscriber to add and remove numbers from the screening list, again using the single-step feature control procedure.

Password Call Acceptance

The password call acceptance (PCA) feature can be used by itself or to augment the SCA feature just described. For example, if the calling party's regular number is on the SCA screening list but the party is calling from a hotel room, the SCA feature will refuse the call. However, if PCA is active, the calling party will be prompted for a password and—if the entered password is verified—call delivery to the subscriber will proceed as usual.

Incoming calls from parties not able to provide a valid password are given one of two possible call-refusal treatments:

1. If the subscriber has registered a PCA call-refusal treatment, using the single-step feature control procedure described in "Feature Control Services," then refused calls are redirected either to the selected PCA forward-to number or to the subscriber's voice mailbox.
2. Otherwise, the call is given the default system call refusal treatment; this could include a tone and announcement to the calling party (e.g., "The number you have dialed does not accept incoming calls").

The PCA feature may be activated and deactivated by an authorized MS using the single-step feature control procedure. It is also possible for an authorized subscriber to add to and remove numbers from the password list, again using the single-step feature control procedure.

ANSI-41 describes two methods of implementing the PCA feature invocation procedures, which we refer to as the *remote interaction* approach and the *home-based interaction* approach. The remote interaction approach uses the ANSI-41 RemoteUserInteractionDirective operation to allow the HLR remote control of the user interaction (i.e., the prompt for the password and digit collection) that is provided by the originating system (see Figure 12.26). The home-based interaction approach has the originating system route the call to the home system where, presumably, a more customized user interaction "experience" is provided; for example, voice entry of the password, rather than digit entry via the telephone keypad (see Figure 12.27).

In fact, the two approaches can coexist. If the originating system is not capable of interacting with the user via the remote interaction approach—this is indicated in the ANSI-41 TransactionCapability parameter supplied by the originating system in the LOCREQ message—the HLR can fall back on the home-based interaction approach.

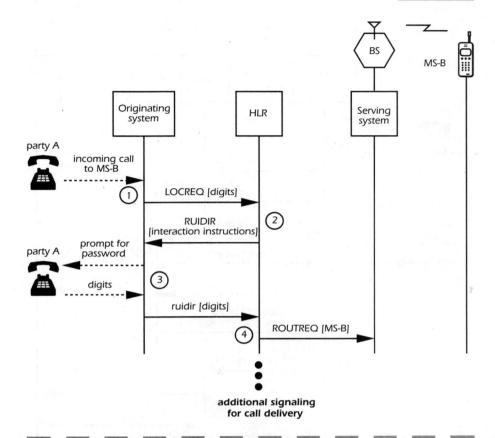

Figure 12.26 An example of the remote interaction method for PCA: (1) Party A dials a phone number (i.e., MS-B's mobile directory number) that is routed via the PSTN to the originating system. The originating system launches a LOCREQ message to the HLR to determine how to route the call. (2) Since MS-B has PCA active, the HLR must query the caller for a valid password before allowing the call to proceed; therefore, it sends interaction instructions to the originating system in the RUIDIR message. (3) The originating system connects the call to a system capable of user interaction (e.g., an interactive voice response unit), prompts party A for a PIN, collects the digits entered, and sends them to the HLR in the ruidir message. (4) The HLR verifies that the PIN is valid and then proceeds with normal call delivery.

In current practice, the originating system is normally the home system and is often colocated with the HLR; therefore, the two approaches appear identical (i.e., messages between the originating system and the HLR are internal). However, it is quite likely that in the future the HLR and originating system functions will reside in physically separate

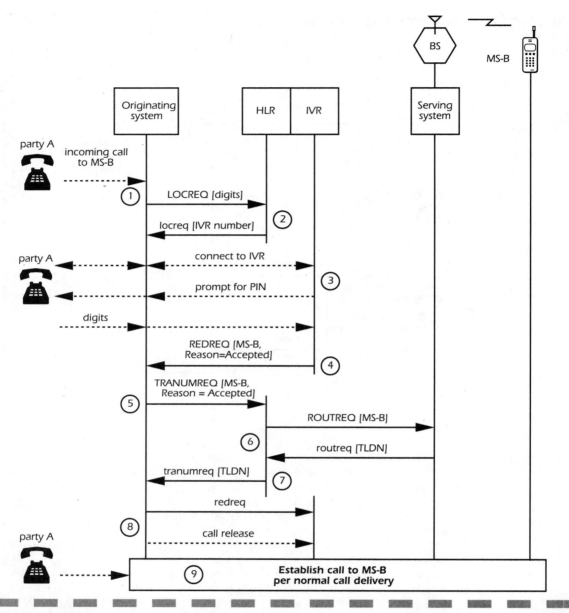

Figure 12.27 An example of the home-based interaction method for PCA: (1) Party A dials a phone number (i.e., MS-B's mobile directory number) that is routed via the PSTN to the originating system. The originating system launches a LOCREQ message to the HLR to determine how to route the call. (2) Since MS-B has PCA active, the HLR must query the subscriber for a valid password before allowing the call to proceed; therefore, it sends instructions to the originating system to route the call to a home-based interactive voice response (IVR) unit. The HLR communicates with the IVR to prepare for the incoming call. (3) The originating system routes the call to an IVR unit. The IVR unit prompts party A for a PIN, collects the digits entered, and verifies that the PIN is valid. (4) The IVR sends a REDREQ message to the originating system, requesting that the call to MS-B be redirected; the RedirectionReason (Reason) parameter is set to the value Call accepted. (5) The originating

equipment and locations, particularly if wire-line switches develop ANSI-41 originating system functionality; that is, when the wire-line calling party's central office switch can detect a call to a mobile subscriber and launch the LOCREQ message to the HLR. In this case, the remote interaction approach appears more efficient than routing the call home only to find that the call is refused.

Mobile Access Hunting

The call flows and procedures in ANSI-41 that define the mobile access hunting (MAH) feature may look complex, but the feature has a simple objective: to complete a call to one of a group of telephone numbers by proceeding sequentially through the group, one member at a time, until one of the group members answers. The MAH process is initiated when a party calls the MAH pilot directory number.

Two types of MAH groups are defined that affect the MAH feature processing:

- *Single-user type*. This type of group is considered busy when any one of the group members is detected as busy. It is useful for the individual who wants to set up a group including an MS, an office phone, and a business phone.
- *Multiple-user type*. This type of group is considered busy when all the group members are detected as busy. It is useful for the company that wants to set up a group including multiple employees.

Authorized MS members of the MAH group can enter and leave the group and change their position in the group (e.g., go to either first or last) by using the single-step feature control procedure.

Figure 12.28 illustrates the basic MAH call completion process for a single user group composed of mobile stations MS-A and MS-B and a land-line telephone.

system uses the TRANUMREQ message to obtain the forward-to number for the given redirection reason (i.e., Call accepted) from the HLR. (6) The HLR handling for the Call accepted forwarding case is to process the request as a normal call delivery attempt. Therefore, the HLR obtains a TLDN from the serving system using the RoutingRequest operation. (7) The HLR returns the TLDN to the originating system in the tranumreq message. (8) The originating system notifies the IVR that it accepts the redirection request, releases the call leg to the IVR, and then forwards the call to the TLDN provided by the HLR. (9) The call is then established to the subscriber as a normal call delivery.

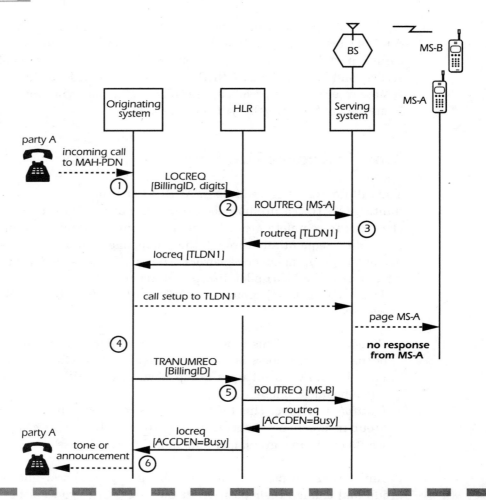

Figure 12.28 An example of MAH feature invocation: (1) The call to the MAH pilot directory number starts out as a normal call delivery attempt; i.e., the originating system sends a LOCREQ message to the HLR. (2) The HLR initiates MAH processing and selects the first member of the group, MS-A, which is roaming in another network. The HLR sends a ROUTREQ message to the serving system and receives a TLDN in return. (3) The HLR sends the TLDN to the originating system and instructs the system that if it cannot complete the call (e.g., because the MS is busy or does not answer), it must notify the HLR by using the TRANUMREQ message. This capability is provided by the ANSI-41 TerminationTriggers (TERMTRIG) parameter and is critical in that it allows the HLR to maintain control of the call in case of unsuccessful termination to an MAH member. The HLR may also provide the AnnouncementList parameter, directing the originating system to inform the calling party that the call is being forwarded. (4) The originating system establishes a call to the serving system, but MS-A does not answer. The originating system waits a period of time indicated by the HLR in the NoAnswerTime parameter (for example, 10 seconds) and then sends a TRANUMREQ message to the HLR. The message includes either the BillingID parameter used in the original LOCREQ message for the call or the PilotNumber parameter, containing the MAH pilot directory number; either of these parameters allows the HLR to relate the

Flexible Alerting

Like MAH, flexible alerting (FA) is another complex-looking feature with a simple objective: to complete a call to one of a group of telephone numbers by attempting to establish a call to each member of the group—in parallel—until one of the group members answers. The FA process is initiated when a party calls the FA pilot directory number.

Two types of FA group are defined that affect the FA feature processing:

- *Single-user type*. This type of group is considered busy when any one of the group members is detected as busy. It is useful for the individual who wants to set up a group including an MS, an office phone, and a business phone.
- *Multiple-user type*. This type of group is considered busy when all the group members are detected as busy. It is useful for the company that wants to set up a group including multiple employees.

Authorized MS members of the FA group can enter and leave the group using the single-step feature control procedure.

Figure 12.29 illustrates the basic FA call completion process for a single-user group composed of two mobile stations, MS-A and MS-B, with numbers MDN-A and MDN-B, respectively, and a land-line telephone with directory number POTS1.

Feature Control Services

In general, ANSI-41 feature control services are the network functions that allow the subscriber to activate, deactivate, and (in some cases) invoke features as well as register and "deregister" (i.e., delete) the information that enables the subsequent operation of the feature (e.g.,

---

TRANUMREQ message to the MAH call. (5) The HLR selects the next member of the group, MS-B, which is also roaming in another network. The HLR sends a ROUTREQ message to the serving system and receives an AccessDeniedReason parameter, indicating that MS-B is busy. Since the MAH group is a single-user group and a member is busy, the entire group is considered busy. Therefore, the HLR returns the AccessDeniedReason parameter to the originating system in the tranumreq message and terminates the MAH processing, without attempting to complete the call to the last member of the group, the land line telephone. (6) The originating system notifies the calling party that the call attempt was unsuccessful and completes its processing of the call.

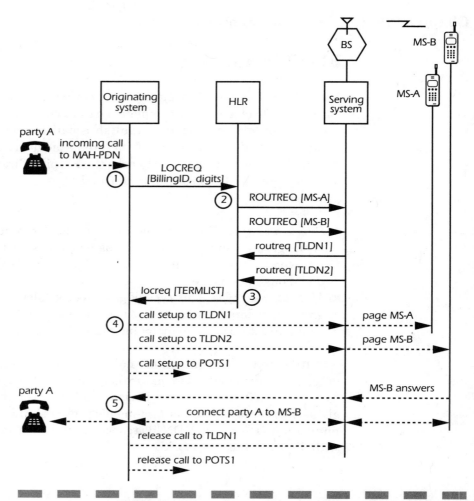

Figure 12.29 An example of the FA feature invocation: (1) The call to the FA pilot directory number starts out as a normal call delivery attempt—the originating system sends a LOCREQ message to the HLR. The originating system includes the TransactionCapability parameter, indicating whether it can support the multiple terminations required for the FA feature; this parameter is required in the ANSI-41 LOCREQ message. (2) The HLR initiates FA processing. Both mobile members of the group, MS-A and MS-B, are roaming in another network. The HLR sends two ROUTREQ messages to the serving system, one for each of the two MSs, and receives two temporary local directory numbers (TLDN1 and TLDN2) in return. (3) The HLR sends the two TLDNs and the land-line number POTS1 to the originating system in the ANSI-41 TerminationList parameter. The HLR may also provide the AnnouncementList parameter, directing the originating system to inform the calling party that the call is being forwarded. The HLR may then terminate its FA call processing. (4) The originating system establishes calls—in parallel—to each of TLDN1, TLDN2, and POTS1. (5) MS-B answers. Therefore, the originating system connects the calling party to MS-B and releases the calls to the other two members of the FA group. This completes the originating system's processing of the FA call.

320

forward-to numbers). These services effectively implement the feature control capabilities defined in the ANSI-664 standard. In particular, ANSI-41 seeks to standardize these services for subscribers who are roaming outside their home service areas.

ANSI-41 supports the following feature control services:

- Basic single-step feature control
- Feature control with call routing
- Per-call feature control
- Multistep feature control
- Remote feature control

We discuss these after introducing the feature code concept.

Feature Codes

Feature codes are most commonly used to modify, invoke, or cancel a cellular feature that has been subscribed to (refer to *ANSI/TIA/EIA-660* and *ANSI/TIA/EIA-664* for the standard specifications of feature codes).

Different feature code digit sequences are specified by the service provider for use with particular features. The sequence is known as a *feature code string* that consists of a preceding asterisk or double asterisk followed by a series of numeric digits (0 through 9); a pound sign (#) may be used as a delimiter between separate sequences of digits. For example, the feature code string: * 72, 4085550303 could mean that a call forwarding forward-to number is being registered. In this example, *72 indicates that the call-forwarding feature is being accessed. The digit sequence 4085550303 indicates the forward-to number. A # is not necessary since only one sequence of digits is used. The subscriber presses SEND (or some other key with similar functionality) on the mobile station to transmit the feature code string to the network.

One problem with feature codes is that they are not standardized. A feature code string that controls the use of a feature in one serving system may not be identical to that in another. The differences in feature codes are due to original disparate implementations in early cellular systems. Network operators are reluctant to change feature codes that have been in use for years since they have become familiar mechanisms for subscribers. However, ANSI-41 supports seamless operation by providing the protocol to convey the subscriber-entered feature code string to the home system for interpretation; in this manner, home-system feature codes may take precedence over local serving system codes. We describe these protocol mechanisms in the following sections.

Basic Single-step Feature Control

The *basic single-step feature control* service comprises the following steps (see Figure 12.30):

1. The subscriber enters a feature code string into the MS and presses the SEND key.
2. The serving system recognizes the received digits as a feature code string and sends them to the HLR in a FeatureRequest Invoke (FEATREQ) message.
3. The HLR processes the feature control request and updates its database, if necessary. The HLR returns a success or failure indication to the serving system in the FeatureRequest Return Result (featreq) message.
4. The serving system provides the appropriate indication to the MS.

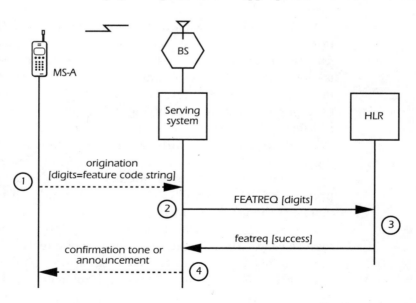

Figure 12.30
The single step feature control process (see text for the description of the steps).

Following completion of a successful feature control procedure, the HLR may invoke the service qualification process to convey the subscriber's new profile to the serving system.

When the call has been handed off and the serving system receives a feature code string from an MS, it uses the FLASHREQ message to forward the digits received from the MS to the anchor system (see Figure 12.31).

Figure 12.31
The single step feature control process under call handoff conditions. Note that the anchor system provides the confirmation tone or announcement.

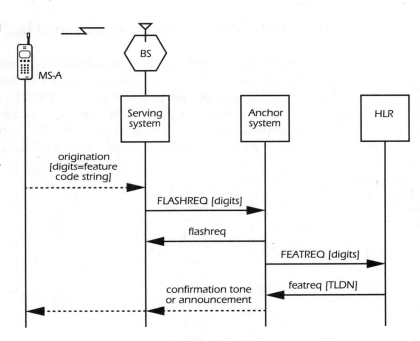

This process works flawlessly—as long as the serving system cooperates and sends the received feature code string to the HLR for processing. Unfortunately, because the serving system may not be able to distinguish a visiting MS from a home MS—or because the feature code string is interpreted as a local feature request—the feature code string may never leave the serving system. ANSI-41 procedures essentially direct the serving system to give the home system "first shot" at interpreting the feature code string. Then, if the HLR does not recognize the feature code string, the digits are returned to the serving system in the featreq message along with an indication that the HLR was unsuccessful in the FeatureResult parameter. At this point, the serving system may—if it is able—attempt to interpret the feature code string and provide the requested feature. The alternative method of accessing a local feature in a visited system, as defined in the ANSI-660 dialing plan standard, is for the subscriber to precede the feature code string with an additional * digit; that is, to access the local feature associated with the string *123, the visiting subscriber dials **123. This form of string should trigger immediate local processing in the serving system, though currently this is also not guaranteed!

Feature Control with Call Routing

The *feature control with call routing* service is an extension of the single-step feature control service. In addition to the feature result indication, the HLR includes routing information, like the Digits parameter, in the featreq message it sends to the serving system. The serving system gives the feature result indication to the MS and then attempts to establish a call to the number provided by the HLR (see Figure 12.32).

Figure 12.32
The feature control with call routing process.

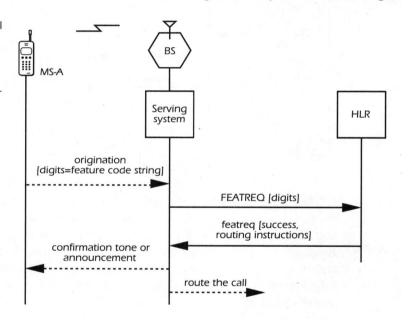

Per-call Feature Control

Some features allow activation or deactivation on a per-call basis:

- Calling number identification restriction may be activated or deactivated for a single call.
- Call waiting may be deactivated for a single call.

- Message-waiting notification may be deactivated for a single call.
- Priority access and channel assignment may be activated for a single call.

ANSI-41 supports this capability by supplementing the basic feature control procedures with the OneTimeFeatureIndicator (OTFI) parameter; this provides a *per-call feature control* service. An example of the process—for deactivating CW for a call—is illustrated in Figure 12.33.

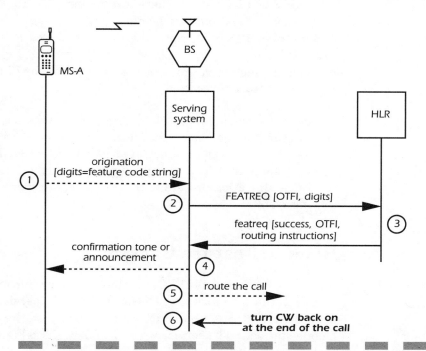

Figure 12.33 *The per-call feature control process: (1) The subscriber enters a feature code string, including the cancel-call-waiting feature code and a called directory number, into the MS and presses the SEND key. (2) The serving system recognizes the received digits as a feature code string and sends the string to the HLR in a FEATREQ message. ANSI-41 specifies that the serving system also initializes the OTFI parameter and includes it in the FEATREQ message. (3) The HLR processes the feature control request. The subscriber is authorized for demand CW cancellation; therefore, the HLR deactivates CW for a single call by setting the Call Waiting for Future Incoming Call (CWFI) field in the OTFI to the value No CW. The HLR returns a success indication to the serving system in the featreq message, along with the revised OTFI parameter and routing information corresponding to the called directory number received in the FEATREQ message. (4) The serving system provides the feature confirmation indication to the MS. (5) The serving system turns off CW for the call and establishes a call to the specified directory number. (6) At the end of the call, the serving system turns CW back on for the MS.*

Multistep Feature Control

The *multistep feature control* service is used when additional information—beyond the initial feature code string—is required to effect a feature control operation. The most common additional information is a password or a personal identification number. Currently, the only standardized features that require this capability are:

- Subscriber PIN access, where it is required for PIN entry during registration, activation, and deactivation.
- Subscriber PIN intercept, where it is required for PIN entry during registration.
- Voice message retrieval, where it is required for VMR password entry during feature invocation.

However, the capability is generic and may be used for future standard features as well as nonstandard feature control purposes.

The process works as illustrated in Figure 12.34. Following completion of a successful feature control procedure, the HLR may invoke the service qualification process to convey the subscriber's new profile to the serving system.

Remote Feature Control

Remote feature control (RFC) service is similar in concept to the multistep feature control service. However, the RFC service may be accessed from any land line or mobile station, not just the subscriber's own mobile station.[10] The service works as follows (see Figure 12.35):

1. The subscriber calls a special RFC directory number (say, 1-800-SYNACOM). The PSTN delivers the call to the RFC access system, which acts as the originating system for the call.
2. The RFC access system launches a LOCREQ message to the HLR.
3. The HLR sends a RUIDIR message to the RFC access system.
4. The RUIDIR message directs the RFC access system to
   (a) Connect the subscriber to a unit capable of voice prompts and digit collection

---

[10] Of course, the capability to access the service from other than the subscriber's (authenticated) mobile station opens the remote feature control service to potential fraud. It would appear advisable to deploy this feature with caution.

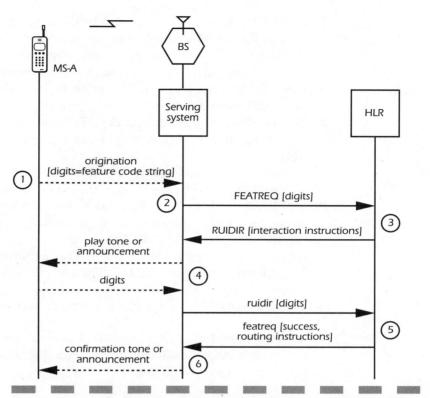

Figure 12.34 The multistep feature control process: (1) The subscriber enters a feature code string into the MS and presses the SEND key. (2) The serving system recognizes the received digits as a feature code string and sends them to the HLR in a FEATREQ message. (3) The HLR processes the feature control request. The HLR sends a RemoteUserInteractionDirective Invoke (RUIDIR) message to the serving system. (4) The RUIDIR message directs the serving system to (a) connect the caller to a unit capable of voice prompts and digit collection, (b) play the tone or announcement indicated in the AnnouncementList parameter, (c) collect digits entered by the subscriber; the digits may be sent as in-band DTMF signals or in an out-of-band signaling message, and (d) send the digits to the HLR in the RemoteUserInteractionDirective Return Result (ruidir) message. (5) The HLR processes the digits and updates its database, if necessary. Additional RemoteUserInteractionDirective exchanges may be executed. The HLR returns a success or failure indication to the serving system in the featreq message. (6) The serving system provides the appropriate indication to the MS and releases the call.

    (b) Prompt the subscriber for its mobile directory number
    (c) Collect digits entered by the subscriber
    (d) Send the digits to the HLR in the ruidir message
  5. The HLR processes the digits and sends another RUIDIR message to the RFC access system.

6. This RUIDIR message directs the RFC access system to
   (a) Prompt the subscriber for its personal identification number
   (b) Collect digits entered by the subscriber
   (c) Send the digits to the HLR in the ruidir message
7. The HLR processes the digits and sends another RUIDIR message to the RFC access system.
8. This RUIDIR message directs the RFC access system to
   (a) Prompt the subscriber for a feature request (i.e., a feature code string)
   (b) Collect digits entered by the subscriber
   (c) Send the digits to the HLR in the ruidir message
9. The HLR processes the digits as it would a normal feature code string, possibly updating its database. It then sends another RUIDIR message to the RFC access system, confirming the feature request and requesting any further feature control requests.
10. The subscriber releases the call. The serving system sends an empty ruidir message to the HLR.
11. The HLR responds with the locreq message, completing the transaction.

Following completion of a successful remote feature control procedure, the HLR may invoke the service qualification process to convey the subscriber's new profile to the serving system.

Summary of ANSI-41 Operations Used for Call Processing

Table 12.1 summarizes the ANSI-41 operations used by the call processing functions described in this chapter. Note that most functions are also supported by the service qualification operations described in Chapter 10.

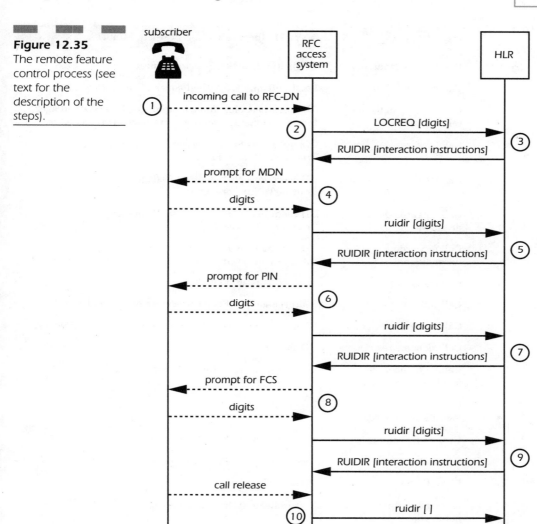

Figure 12.35
The remote feature control process (see text for the description of the steps).

TABLE 12.1

Use of ANSI-41 Operations for Call Processing

Function	ANSI-41 Operations Used for the Function
Basic call origination	OriginationRequest, FeatureRequest
Call delivery (CD)	LocationRequest, RoutingRequest, FeatureRequest
Call delivery in border cell situations	CD operations plus: InterSystemPage, InterSystemPage2, InterSystemSetup, InterSystemAnswer, UnsolicitedResponse
Call forwarding—busy	CD operations plus: RedirectionRequest, TransferToNumberRequest
Call forwarding—default	CD operations plus: RedirectionRequest, TransferToNumberRequest
Call forwarding—no answer	CD operations plus: RedirectionRequest, TransferToNumberRequest
Call forwarding—unconditional	CD operations plus: InformationDirective, InformationForward
Call transfer	FeatureRequest , FlashRequest
Call waiting	FeatureRequest , FlashRequest, InformationForward
Calling name presentation	CD operations plus: InformationForward, ServiceRequest
Calling name restriction	CD operations
Calling number identification presentation	CD operations plus: InformationForward
Calling number identification restriction	CD operations
Conference calling	FeatureRequest , FlashRequest
Do not disturb	LocationRequest, FeatureRequest
Flexible alerting	CD operations
Message waiting notification	FeatureRequest, InformationForward
Mobile access hunting	CD operations plus: RedirectionRequest, TransferToNumberRequest
Password call acceptance	CD operations plus: RemoteUserInteractionDirective or (RedirectionRequest and TransferToNumberRequest,)
Preferred language	FeatureRequest
Priority access and channel assignment	FeatureRequest
Selective call acceptance	CD operations

continued on next page

TABLE 12.1

Use of ANSI-41
Operations for Call
Processing
(continued)

Function	ANSI-41 operations used for the function
Subscriber PIN access	FeatureRequest, RemoteUserInteractionDirective
Subscriber PIN intercept	FeatureRequest, RemoteUserInteractionDirective, OriginationRequest
Three-way calling	FlashRequest
Voice message retrieval	CD operations plus: OriginationRequest
Voice privacy and signaling message encryption	AuthenticationRequest, FacilitiesDirective, FacilitiesDirective2, HandoffBack, HandoffBack2, HandoffToThird, HandoffToThird2, InterSystemSetup
Single-step feature control	FeatureRequest , FlashRequest
Feature control with call routing	FeatureRequest
Per-call feature control	FeatureRequest , FlashRequest
Multi-step feature control	FeatureRequest, FlashRequestplus, RemoteUserInteractionDirective
Remote feature control	LocationRequest, RemoteUserInteractionDirective

CHAPTER 13

Short-Message Service Functions

In this chapter, we discuss the ANSI-41 mobile telecommunications network functions related to short message service (SMS). These functions are divided into the following categories:[1]

- Short-message entity (SME[2]) service qualification
- SME location management
- SME state management
- Short message processing

We also examine the ANSI-41 application processes that support these network functions. In the course of describing the processes, we identify the ANSI-41 mobile application part (MAP) operations (e.g., SMSRequest) used to accomplish SMS process tasks. We summarize this information at the end of the chapter. Note that we employ the ANSI-41 convention for operation component acronyms; the Invoke component acronym is in all-capital letters (e.g., SMSREQ), while the Return Result component acronym is in all-lowercase letters (e.g., smsreq). Refer to the ANSI-41 standard for the descriptions of the individual ANSI-41 operations and parameters. Also, consult the Glossary for a description of general terms (e.g., serving system) not explicitly defined in this chapter.

Throughout this chapter we consider the serving system as a single entity encompassing the mobile switching center (MSC) and visitor location register (VLR) functional entities. This simplifies the descriptions of the ANSI-41 SMS functions and also is representative of a large percentage of the ANSI-41 implementations currently in service. However, keep in mind that the potential for separation of the MSC and VLR exists and is fully defined in ANSI-41.

What Is Short-message Service?

ANSI-41 specifies a set of data services collectively known as the short-message service (SMS). SMS includes services that are specially designed for the mobile environment. Most traditional data services are

---

[1] The list of SMS functions reflects the authors' subjective categorization of certain ANSI-41 functions. Likewise, the process descriptions are the authors' subjective interpretation of the ANSI-41 procedures.

[2] The short message entity acronym, SME, should not be confused with the same acronym used for signaling message encryption in Chapter 12.

not appropriate for this environment since they require terminals that are bulky compared to the size of a handheld mobile station (MS). That is, data services are generally not designed for an integrated implementation on mobile telephones since they usually require a keyboard and a reasonably large display screen. However, SMS supports the transmission and reception of simple messages that are convenient for display on mobile terminals.

The ANSI-41 SMS is designed as a generic short-message transport mechanism. The following design imperatives were applied to the SMS:

- Support a variety of teleservice applications
- Make use of commonly implemented transport protocols
- Incorporate a flexible addressing scheme
- Easily interwork with other packet-switched data networks
- Be compatible with electronic mail services, paging services, and other commonly used messaging services

The SMS is categorized into the following two types of services: SMS bearer service and SMS teleservices. These services require two functional entities in addition to those used for basic mobile telecommunications: the message center (MC) and the short-message entity (SME). These services and functional entities are described in detail in the following sections.

SMS Bearer Service

The SMS bearer service is the basic transport mechanism to convey a short message as a packet of data between two points on the network. The short message may have length up to 200 octets, although vendor implementations and service providers may further restrict this length.

The SMS bearer service does not contain many features itself. This allows for the flexibility of custom applications (i.e., teleservices) to use a simple transport mechanism. The bearer service supports the standardized teleservices described in the next section.

The SMS bearer service is designed to use any one of the following transport protocols:

- Signaling System No. 7 (SS7)
- X.25
- Internet protocol (IP)

In addition, ANSI-41 does not preclude the use of a proprietary data protocol for transport of the SMS bearer service messages (see Figure 13.1).

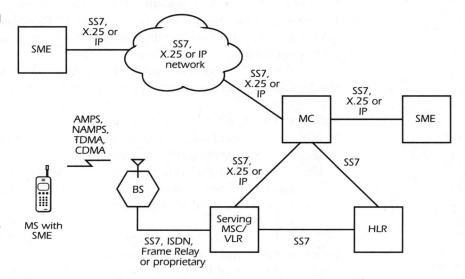

Figure 13.1
The basic network architecture supporting ANSI-41 SMS. SS7, X.25, and IP can be used for transport of short messages. If a packet-switch protocol other than SS7 is used, the serving MSC requires an interworking function to support that protocol.

As of December 2000, the current version of the SMS bearer service description—the *Short Message Delivery Point-to-Point Bearer Service*—is in ANSI/TIA/EIA/664-601-A, published in July 2000. The "points" are short-message entities, which are essentially applications that can both send and receive short messages. Thus, the bearer service is bidirectional and symmetrical, and there is no technical service differentiation (aside from signaling procedures) between mobile-originated SMS and mobile-terminated SMS.

The bearer service always attempts to deliver a short message to an MS-based SME whenever the MS is registered, even if the mobile station is engaged in a call. The network is informed whether a message has been received by the MS. This allows the messages to be held by the sender in cases of unsuccessful delivery and then retransmitted to the destination when possible.

SMS Teleservices

An *SMS teleservice* is defined as an SMS service that provides the complete application capability, including terminal equipment functions, for communication between SMS users (i.e., SMEs) according to application protocols established between the users. The SMS teleservices are designed as special applications that use the SMS bearer service as a transport mechanism (see Figure 13.2). There are two basic TIA-standardized teleservices:

- *Cellular paging teleservice* (CPT, now called the *wireless* paging teleservice in ANSI-664-A). The CPT specifies a short message-based teleservice to provide paging-like services to an MS. CPT is based upon a minimal character set, including uppercase letters A through Z, the digits 0 through 9, and various symbols and control characters. The maximum message length of 63 characters.
- *Cellular messaging teleservice* (CMT, now called the *wireless* messaging teleservice in ANSI-664-A). The CMT specifies a short message-based teleservice to provide a generic short-messaging application to an MS. CMT is based upon an extensive character set and a maximum message length of 200 characters.

Figure 13.2
Layered protocol model showing the SMS teleservices as applications supported by the SMS bearer service (i.e., SMS point-to-point delivery service). Note that X.25, IP, or SS7 can be used for network transport of short messages.

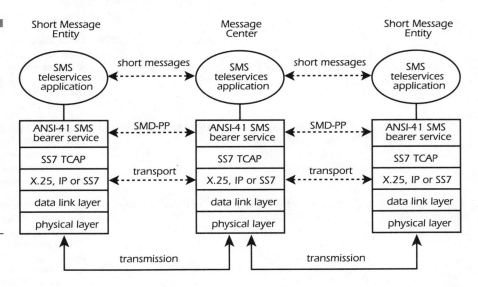

Although the stage 1 specifications of CMT (WMT) and CPT (WPT) are standardized in ANSI-664-602-A and ANSI-664-603-A, respectively, stages 2 and 3 are designed to be network-independent and are not specified in ANSI-41. The SMS stages 2 and 3 specifications in ANSI-41 are limited to the SMS bearer service.

However, the TR45.3 TDMA air-interface subcommittee has published a number of teleservice standards that include the stage 3 specifications. These standards specify the teleservice-specific protocols that *ride* on the SMS bearer service. References for some of these standards are listed in Table 13.1. For a complete listing, the reader should consult the main ANSI-136 document, ANSI/TIA/EIA-136-000. As of December 2000, the current version of this document is Revision C.

TABLE 13.1

Some TDMA Teleservice Standards

Teleservice Name	Document Number	Description
SMS Cellular Messaging Teleservice (CMT)	ANSI/TIA/EIA-136-710	See description in this section.
Over-the-air Activation Teleservice (OATS)	ANSI/TIA/EIA-136-720	OATS supports over-the-air activation, which is a method for provisioning a mobile user with the home service provider's network and programming the mobile's NAM with the mobile station identity, the home service provider identity, the A-key, and other information. See Chapter 15.
Over-the-air Programming Teleservice (OPTS)	ANSI/TIA/EIA-136-730	OPTS supports over-the-air programming, which is a method for programming non-NAM data into a mobile (e.g., the ANSI-136 intelligent roaming database data station). See Chapter 15.
System-Assisted Mobile Positioning through Satellite (SAMPS) Teleservice	ANSI/TIA/EIA-136-740	SAMPS utilizes the SMS capabilities of ANSI-136 TDMA networks to enhance the performance of Global Positioning System (GPS)-equipped MSs by providing GPS assistance.
General UDP Transport Service (GUTS)	ANSI/TIA/EIA-136-750	GUTS is a two-way teleservice that makes use of the SMS bearer service to transfer user datagram protocol (UDP) packets between transport endpoints. The UDP packet header, in turn, specifies the application endpoints (i.e., UDP ports).

Message Center

The message center (MC) provides the *store-and-forward* function for most (see below) mobile-originated short messages and for all mobile-terminated short messages. The MC is typically implemented as a physically separate network entity, but it may be combined with other functional entities. Many MCs may be connected on a single network, or a single MC may be connected to many networks.

SMS subscribers are associated with an MC, known as the *home* MC, in the MS's home system. The MC maintains the mobile station identity (MSID) address information of the MSs that it serves and is addressable by the directory number addresses of the MSs. In general, an MC provides the following capabilities:

- Forwarding messages to the addressed MS-based SME.
- Storing short messages for unavailable MS-based SMEs.
- Apply originating and terminating SMS supplementary services to short messages.
- As an option, providing interworking between different transport protocols.

ANSI-41 supports bypassing the MC for mobile-originated short messages, so that messages can be sent directly to the destination SME from the serving MSC; however, the authors are not aware of any SMS implementations that make use of this capability. Also, no originating supplementary services can be applied to these messages.

Short-message Entity

The short-message entity is a functional entity capable of composing (originating) and decomposing (receiving) a short message. An SME may be located in a fixed network (outside the ANSI-41 network), in a mobile station, or within the ANSI-41 network. An SME is generally considered to be an application entity that represents the originator and recipient of short messages provided via the point-to-point short-message delivery service. In general, an SME has the following capabilities:

- Composing short messages
- Disposing of, acting upon, or displaying short messages
- Requesting supplementary services
- Storing received short messages
- Managing stored messages

The methods for performing these tasks are implementation dependent and are not addressed by ANSI-41.

Where Are SMS Functions Specified in ANSI-41?

SMS functions are specified in three parts of ANSI-41:

- **ANSI-41.5** provides the required formats of all the ANSI-41 operation components, including those used for SMS. ANSI-41.5 defines both the messages (e.g., SMSRequest Invoke) and the message parameters (e.g., SMS_Address).
- **ANSI-41.6** provides algorithmic descriptions of the procedures associated with sending and receiving most ANSI-41 messages, including those used for SMS. This part of ANSI-41 also includes an example of an air-interface definition for SMS in the informative (i.e., not technically a requirement of the standard) Annex D. This information is intended to illustrate the assumptions regarding the air-interface SMS support that were made in the design of the ANSI-41 SMS network protocols and procedures.
- **ANSI-41.3** steps back from the protocol details contained in parts 5 and 6 and attempts to explain, using information flow diagrams and step-by-step descriptions, how the operations are used individually and together to accomplish SMS application process tasks.

Issues Associated with SMS

The following issues need to be addressed in the implementation of SMS:

- *Delivering short messages to roaming subscribers*. Location, status, and address information needs to be obtained from the serving system to deliver a mobile-terminated short message.
- *Delivering short messages during intersystem handoff*. Short messages being delivered during intersystem handoff must be delivered intact to the subscriber.
- *Methods for originating short messages from an MS*. Typical mobile stations do not have an adequate keypad for entering message information to be transmitted. In a basic implementation, a message may be selected from a predetermined set of messages stored in the MS; alternatively, operator assistance or portable computers can be used to generate messages.

■ *Subscribers may roam into areas where SMS is not supported.* The subscriber may not have access to SMS in all wireless coverage areas. Subscribers may not be aware that they have lost service.

■ *Different air-interface protocols support different message lengths.* For example, NAMPS supports transmission of only 14 characters at a time, whereas the digital protocols support short messages of at least 140 characters. Different systems may also support varying maximum message lengths, possibly limited by the service provider to conserve bandwidth.

■ *SMS interconnection to other data networks.* A primary consideration in the design of SMS is support of standardized transport protocols for easy access to packet data networks such as the Internet or X.25, in addition to SS7.

SME Service Qualification

The ANSI-41 SME service qualification function encompasses the processes that establish an SME's financial accountability and service capabilities in a serving system. There are two types of SMEs defined for SMS: MS-based and fixed. ANSI-41 defines the service qualification procedures for MS-based SMEs only; qualification procedures for fixed SMEs are determined by individual service providers.

The ANSI-41 MS-based SME service qualification processes are tightly coupled to the service qualification of the host MS itself; that is, if the MS is qualified, generally the SME is also qualified. These procedures are described in Chapter 10 and use the following ANSI-41 operations:

■ RegistrationNotification
■ QualificationRequest
■ QualificationDirective

The only significant SME-related addition to MS service qualification is that the HLR sends the SME service profile information—in the form of the ANSI-41 SMS_OriginationRestrictions and SMS_Termination-Restrictions parameters—to the SMS-capable serving system, along with the other MS-related profile parameters (see Figure 13.3).

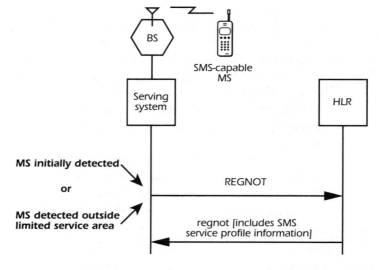

Figure 13.3
An example of SMS
service qualification
using the
RegistrationNotification
operation.

SME Location Management

Like the associated MS function described in Chapter 10, the SME location management function comprises two components:

- *SME location update processes* have the effect of creating or modifying the SME-related location information in an MS's temporary record in a visited system and updating the SME location information in an MS's record in the HLR.
- *SME location cancellation processes* have the effect of deleting SME-related location information in an MS's temporary record in a visited system and updating the SME location information in an MS's record in the HLR.

The ANSI-41 MS-based SME location management processes are tightly coupled to the associated processes of the host MS itself; the location of the MS is the location of the SME as well. These procedures are described in Chapter 10 and use the following ANSI-41 operations:

- To update location:
  - RegistrationNotification
- To cancel location:
  - MSInactive, including the DeregistrationType parameter

- BulkDeregistration
- RegistrationCancellation
- UnreliableRoamerDataDirective

Note that the serving system also includes a routing address—the SMS_Address parameter—in the RegistrationNotification Invoke message. This is used for routing short-messages to the serving system on their way to the visiting SME. However, we consider this address to be functionally associated with short message termination (described later in this chapter) rather than the SME location update process.

Note also that the MC is not responsible, nor is it enabled by ANSI-41, to maintain up-to-date location and status for the SME; this is the HLR's job. For example, ANSI-41 does not specify that the HLR automatically notify the MC when an MS-based SME moves from one serving system to another. Nor does the HLR notify the MC when the MS-based SME becomes unavailable for short-message termination (although the converse—the HLR notifies the MC when the MS-based SME becomes *available* for short message termination—is supported, as described in "SMS Delivery Pending Flag Management" on page 351).

SME State Management

The ANSI-41 SME state management function encompasses the processes by which the short-message termination availability state of an MS-based SME is maintained in the HLR.

Note that there is no formal definition of an *SME state* in ANSI-41. There are, however, procedures that define how the HLR responds to the MC's requests for short-message termination routing information. We use the concept of an SME state, having a value of either *available* or *unavailable*, to aid our explanation of these procedures. Generally, if the SME's state is *unavailable*, the HLR will indicate this to the MC when it receives a request for SMS routing information; otherwise, the HLR either will provide the MC with the routing information currently stored in its database or will attempt to obtain more up-to-date routing information from the serving system. The HLR may consider an MS-based SME *unavailable* because:

■ The MS is not "registered." When the HLR does not have a valid location for the MS, the associated MS-based SME is considered *unavailable*.

- The MS is registered on an SMS-incapable system.
- The SME is not authorized for SMS service on the current serving system, although the SME is generally authorized for SMS service on other systems.
- The host MS is out of radio contact. The MS may have missed autonomous registration events due to a loss of radio contact and been designated *inactive* by the serving system. The SME state in this case is system-dependent.
- The host MS is intentionally inaccessible. The MS may have gone into a mode (e.g., sleep mode) whereby it is intentionally inaccessible. The SME state in this case is also system dependent.

Note that the SME state is not necessarily the same as the MS state described in Chapter 10; an *inactive* MS may be *available* for short-message service deliveries, based on system-dependent algorithms.

The HLR SME state management process makes use of information provided by the serving system and HLR location management processes. Generally, the SME state is set to *available* by the HLR after a successful location update for the MS. Likewise, after a successful location cancellation, the HLR sets the SME state to *unavailable*; the state may be reset to *available* if the location cancellation is nested within a location update process. In this sense, the HLR SME state management process makes use of the same operations as the location update and location cancellation processes—RegistrationNotification, RegistrationCancellation, MSInactive, BulkDeregistration, and UnreliableRoamerDataDirective. The SME state may also be explicitly set to *unavailable* in the HLR in the following ways:

- When the HLR receives a valid ANSI-41 REGNOT message including a valid AvailabilityType parameter from the serving system, it may set the SME's state to *unavailable* (see Figure 13.4).
- When the HLR receives a valid ANSI-41 MSINACT message from the serving system, it may set the SME's state to *unavailable* (see Figure 13.5). Note that this message may also result in location cancellation for the MS if the DeregistrationType parameter is included.
- When the HLR receives a valid ANSI-41 RoutingRequest RETURN RESULT (routreq) message from the serving system, including the AccessDeniedReason parameter set to *inactive*, the HLR may set the SME's state to *unavailable* (see Figure 13.6).

Once set, the SME's state remains unavailable until a serving system—for which the SME is SMS-authorized—sends a valid REGNOT message to the HLR that includes the SMS_Address parameter but does not include the AvailabilityType parameter.

Figure 13.4
Setting the SME's state to unavailable using the RegistrationNotification operation.

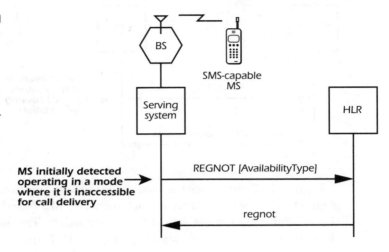

Figure 13.5
Setting the SME's state to unavailable using the MSInactive operation.

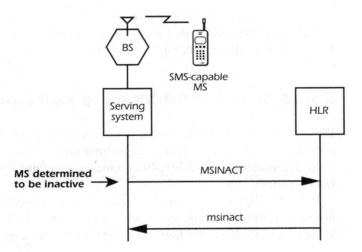

Short-message Processing

The ANSI-41 short-message processing functions encompass the processes that enable, restrict, supplement, or otherwise impact an SME's ability to originate and terminate a short message.

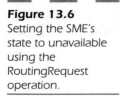

Figure 13.6
Setting the SME's state to unavailable using the RoutingRequest operation.

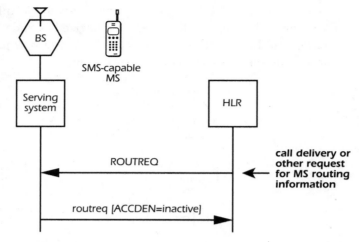

Figure 13.7 illustrates a basic message origination and termination sequence for message transfer between two MS-based SMEs; SME-A is the originator, and SME-B is the destination. The key SMS elements that apply to the scenario shown in Figure 13.7 are:

- Short-message addressing and routing
- Short-message barring, i.e., enforcing messaging restrictions
- Applying SMS supplementary services

Short-message Addressing and Routing

Because of the store-and-forward nature of the short-message transfer process, messages may take a circuitous path from the originator to the final destination. The addressing mechanisms defined in ANSI-41 provide for this.

Refer to Figure 13.7. ANSI-41 allows the originating SME (SME-A) to provide up to four pieces of address information in the air-interface (e.g., TDMA or CDMA) equivalent of the SMD-REQ message:

1. OriginalOriginatingAddress
2. OriginalDestinationAddress
3. OriginalOriginatingSubaddress
4. OriginalDestinationSubaddress

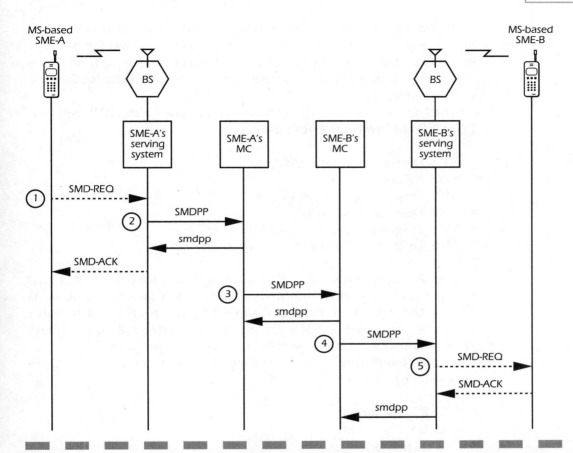

Figure 13.7 A basic message origination and termination sequence for short-message transfer between two MS-based SMEs: (1) MS-based SME-A sends an air-interface message, SMD-REQUEST (SMD-REQ), to the serving system. (2) The serving system routes the short message to SME-A's MC, using the ANSI-41 SMSDeliveryPointToPoint Invoke (SMDPP) message. Each of the SMDPP messages shown in this scenario may be routed by using the same SS7 signaling network as is used for routing other ANSI-41 messages; alternatively, a separate network, based on TCP/IP or some other network protocol, may be employed. When the smdpp message is received from the MC, the serving system converts it into an air-interface acknowledgment, the SMD-ACK message. (3) SME-A's MC may apply an originating supplementary service to the short message (this is not currently defined in ANSI-41); the SMDPP message is then routed to the destination SME's MC. (4) SME-B's MC may apply an terminating supplementary service to the short message (this is not currently defined in ANSI-41); the SMDPP message is then routed to the destination SME's serving system. (5) The serving system forwards the short message toward the destination SME using the air-interface SMD_REQ message. SME-B responds with an automatic acknowledgment (SMD-ACK) to signal acceptance of the SMD-REQ message.

In general, the OriginalOriginatingAddress information is required for message termination but is not necessarily included for message origination; likewise, the OriginalDestinationAddress information is required for message origination but is not necessarily included for message termination.

ANSI-41 defines six SMS address parameters for the SMSDelivery-PointToPoint Invoke (SMDPP) message:

- SMS_OriginalOriginatingAddress
- SMS_OriginatingAddress
- SMS_OriginalDestinationAddress
- SMS_DestinationAddress
- SMS_OriginalOriginatingSubaddress
- SMS_OriginalDestinationSubaddress

The air-interface OriginalOriginatingSubaddress and Original-DestinationSubaddress information is optional and, if provided, is passed transparently from end to end by the SMS point-to-point bearer service in the SMS_OriginalOriginatingSubaddress and SMS_OriginalDestinationSubaddress parameters, respectively.

Various numbering formats are supported for the other SMS address parameters, including:

- ITU-T E.164 format
- ITU-T X.121 format
- Private numbering plan formats
- Internet protocol (IP) address format

One possible scenario is for each MS-based SME in Figure 13.7 to be addressed by its host MS's MSID. In this example, we use the MIN form of the MSID; MIN-A (for SME-A) and MIN-B (for SME-B). The MIN-based addresses are encoded by using the E.164 format. Table 13.2 summarizes the relationship between the air-interface address values in the SMD-REQ messages and the ANSI-41 SMS address values in the SMDPP messages for each step identified in Figure 13.7. For the purposes of illustration, we assume that all possible address parameters—with the exception of the subaddress parameters—are included in each message.

Most of the mappings between the air interface and ANSI-41 address elements are straightforward:

TABLE 13.2

Relationship
Between Air-
interface and
ANSI-41 SMS
Address
Information

STEP	Original-Originating-Address	Original-Destination-Address	SMS_Original-Originating-Address	SMS_Originating-Address	SMS_Original-Destination-Address	SMS_Destination-Address
1	MIN-A	MIN-B				
2			MIN-A	MIN-A	MIN-B	MIN-A
3			MIN-A	MIN-A	MIN-B	MIN-B
4			MIN-A	MIN-B	MIN-B	MIN-B
5	MIN-A	MIN-B				

- The air interface OriginalOriginatingAddress parameter maps to the ANSI-41 SMS_OriginalOriginatingAddress parameter.
- The air interface OriginalDestinationAddress parameter maps to the ANSI-41 SMS_OriginalDestinationAddress parameter.

However, the values of the SMS_OriginatingAddress and SMS_DestinationAddress parameters vary depending on the particular information flow in Figure 13.7 and are set to ensure correct routing of the message:

- In step 2, the SMS_DestinationAddress parameter is set to MIN-A, rather than to MIN-B, to route the SMDPP message to SME-A's MC.
- In step 3, the SMS_OriginatingAddress parameter is set to MIN-A, to identify the source of the message as SME-A's MC; the SMS_DestinationAddress parameter is set to MIN-B, to route the SMDPP message to SME-B's MC.
- In step 4, the SMS_OriginatingAddress parameter is set to MIN-B, rather than to MIN-A, to identify the source of the message as SME-B's MC.

To route the message to SME-A's MC in step 2, the serving system can maintain a table of MIN-to-MC addresses (e.g., MIN to SS7 destination point codes), as is often done today in ANSI-41 networks for routing ANSI-41 messages to an MS's HLR. If messages are transported using an SS7 signaling network, the serving system can use the SS7 global title translation (GTT) capability. In this case, the serving system creates a global title address containing the originating SME's MIN value and requests a MIN-to-MC global title translation.[3]

These same routing possibilities exist for SME-A's MC in step 3; that is, to route the message to SME-B's MC, either a fixed MIN-to-MC address table or GTT on MIN-B may be employed, although the latter approach is more likely (i.e., it is far easier to maintain the MIN-to-MC routing information in the SS7 network than in each and every accessible MC).

Terminating Short Messages to MS-based SMEs

Steps 2 and 3 in Figure 13.7 involve message routing between fixed points. In step 4, the destination SME (SME-B) is mobile; therefore, if it does not already have one, SME-B's MC must get a valid routing address for the system currently serving SME-B. ANSI-41 provides the SMSRequest operation specifically for this purpose; this routing requirement also has an impact on the RegistrationNotification operation, as explained below (see Figure 13.8):

1. When a new SMS-capable MS is detected, the serving system sends a RegistrationNotification Invoke (REGNOT) message to the HLR. If the serving system is SMS capable, the message includes the SMS_Address parameter used to route short messages to the serving system for delivery to the MS-based SME. For example, if the short-message transport network is SS7-based, the SMS_Address parameter may contain an SS7 point code and subsystem number; when the serving system receives an SMDPP message addressed to this point code and subsystem number, it assumes the message is intended for a visiting MS-based SME. The specific destination MS is identified by the address parameters within the SMDPP message.

2. When the MC requires a current routing address for an MS-based SME, it sends an SMSRequest Invoke (SMSREQ) message to the MS's HLR.

3. The HLR then makes a decision: Is the SMS_Address received in step 1 sufficiently current for SMS delivery purposes, or should the HLR attempt to obtain a new address? If the SMS_Address is considered valid, the HLR returns it to the MC in the SMSRequest Return Result (smsreq) message; otherwise (not shown in Figure

---

[3] A third possibility exists in ANSI-136, the TDMA standard. It defines the *Teleservice Server Address* parameter, which is included in the TDMA version of the SMD-REQ message, called R-DATA. SME-A can populate the Teleservice Server Address parameter with its MC's point code and subsystem number values, or with an IMSI value assigned to its MC, which is used as the global title address in a IMSI-to-MC global title translation. Of course, this assumes that SME-A has been programmed with this Teleservice Server Address information.

13.8), the HLR may send an SMSREQ message to the serving system, requesting an update on the MS's accessibility and routing information.

4. The MC uses the SMS_Address to route the SMDPP message to the serving system.
5. The serving system sends the message to the MS-based SME identified in the SMDPP message.

Figure 13.8
Obtaining an SMS routing address for short-message termination.

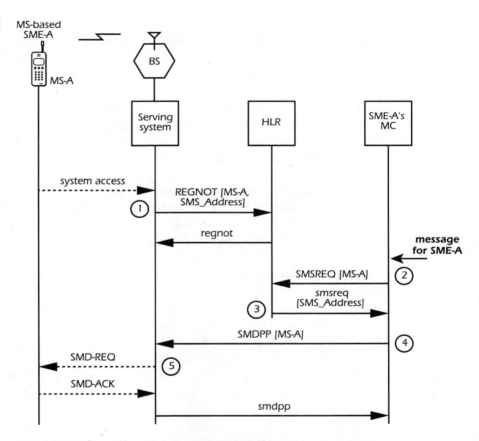

SMS Delivery Pending Flag (SMSDPF) Management

The MC may become aware of an MS-based SME's unavailability under three circumstances:

1. The HLR returns the SMS_AccessDeniedReason (SMSACCDEN) parameter in the smsreq message in response to the MC's request for a current routing address for the MS-based SME.

2. The serving system returns the SMSACCDEN parameter in the smdpp message in response to the MC's request for delivery of a short message to the MS-based SME.
3. The HLR or serving system sends the SMSACCDEN parameter in an SMSNotification Invoke message to the MC.

The MC can handle the MS-based SME unavailability condition by taking one of two approaches:

- *The polled approach.* The MC periodically resends the SMSREQ message to the HLR or the SMDPP message to the serving system until it is successful. This approach is entirely MC controlled and requires no special ANSI-41 protocol support.
- *The interrupt-driven approach.* The MC requests notification from the HLR or serving system when the MS-based SME becomes *available*. For this approach, the MC includes the SMS_NotificationIndicator (SMSNOTIND) parameter in the SMSREQ or SMDPP message, set to the value *notify when available*. If (1) the notification request in the SMSREQ message is accepted by the HLR (or serving system if the SMSREQ is relayed to the serving system) or (2) the notification request in the SMDPP message is accepted by the serving system, then the SMSACCDEN parameter is returned to the MC with value *postponed*; otherwise, the SMSACCDEN parameter is returned to the MC with value *denied*.

Once the notification request is accepted by a system (either the serving system or the HLR), the system sets the SMS delivery pending flag (SMSDPF). If the state of the MS-based SME changes, the system sends an SMSNotification Invoke (SMSNOT) message to the MC. The SMSNOT message is the "interrupt" signal for the MC, at which point it evaluates whether the pending message is still valid or has already been discarded.

Figure 13.9 illustrates an example of the interrupt-driven approach when the HLR provides the notification to the MC, while Figure 13.10 illustrates the case when the serving system provides the notification.

Once the SMSDPF is set in the serving system, special procedures are necessary to ensure that the short-message pending status is preserved when, if, for example, the serving system changes. The SMS_Message-WaitingIndicator (SMSMWI) parameter is designed for this purpose (see Figures 13.11 and 13.12).

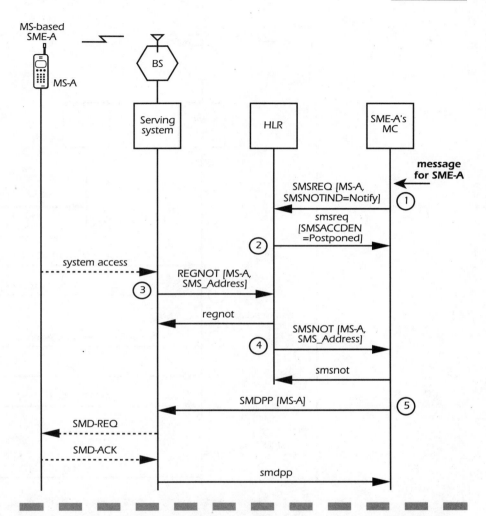

Figure 13.9 An example of the interrupt-driven approach to SME availability notification by the HLR: (1) When MS-based SME-A's MC requires a current routing address for SME-A, it sends an SMSREQ message to the MS's HLR. It includes the SMSNOTIND parameter set to the value "Notify when available." (2) The HLR considers SME-A to be unavailable. The HLR accepts the notification request by including the SMSACCDEN parameter set to the value "Postponed." The HLR sets the SMSDPF. (3) MS-A accesses the serving system. The HLR marks SME-A as available and stores the current SMS routing address (SMS_Address). (4) Since the SMSDPF is set, the HLR launches an SMSNOT message to SME-A's MC, informing the MC of the routing address for SME-A. (5) The MC uses this address to route the short message to SME-A.

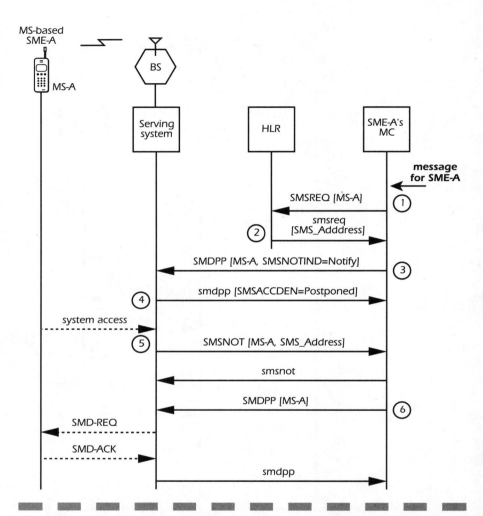

Figure 13.10   An example of the interrupt-driven approach to SME availability notification by the serving system: (1) When MS-based SME-A's MC requires a current routing address for SME-A, it sends an SMSREQ message to the MS's HLR. (2) The HLR considers SME-A to be available, and provides the current routing address for the SME in the SMS_Address parameter. (3) The MC uses this address to route the short message to SME-A. It includes the SMSNOTIND parameter set to the value "Notify when available." (4) SME-A is temporarily unavailable for short message delivery. The serving system accepts the notification request by including the SMSACCDEN parameter set to the value "Postponed." The serving system sets the SMSDPF. (5) MS-A accesses the serving system. Since the SMSDPF is set, the serving system launches an SMSNOT message to SME-A's MC, informing the MC of the routing address for SME-A. (6) The MC uses this address to route the short message to SME-A.

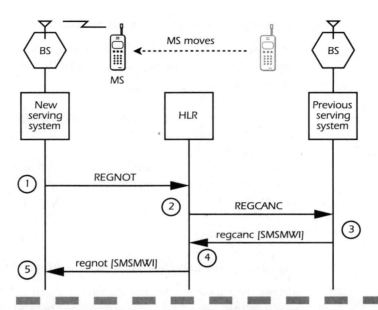

Figure 13.11 Transferring the state of the SMSDPF by using the SMSMWI parameter during the location update and location cancellation processes: (1) A new MS moves into its service area; therefore, the serving MSC requests location update in the HLR. (2) The HLR has stored location information for the MS; therefore, it requests location cancellation from the previous serving system. (3) The previous serving MSC deletes the temporary MS record and sends an acknowledgment to the HLR. It includes the SMSMWI parameter if the SMSDPF is set. (4) The HLR acknowledges the location update and relays the SMSMWI parameter to the new serving system. (5) When the new serving system receives the SMSMWI parameter, it sets the SMSDPF.

Figure 13.12
Transferring the state of the SMSDPF by using the SMSMWI parameter during location cancellation via the MSInactive operation. The HLR sets its SMSDPF when it receives the SMSMWI parameter.

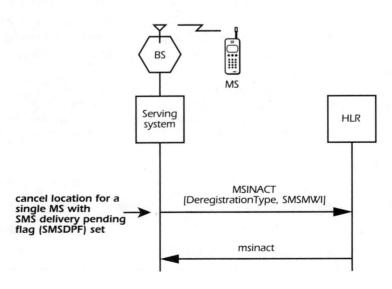

Short-message Routing with Call Handoff

One thing not shown in Figure 13.7 is ANSI-41's support for short-message origination and termination under call handoff conditions. Two operations are defined:

- *SMSDeliveryBackward* conveys the mobile-originated short message from the serving system, possibly via one or more tandem systems, to the anchor system.
- *SMSDeliveryForward* conveys the mobile-terminated short message from the anchor system, possibly via one or more tandem systems, to the serving system.

These operations are illustrated in Figures 13.13 and 13.14.

More on Short-message Addressing and Routing

The addressing and routing mechanisms in ANSI-41 provide an extremely flexible set of tools for short-message exchange. Part of the design requirement was to support multiple transport technologies—SS7, TCP/IP, and X.25—and to allow interworking between these technologies. Another requirement was to support the various air interfaces. The messages, parameters, and procedures in ANSI-41 support these requirements.

For example, Table 13.3 provides another example of the relationship between the air-interface address values in the SMD-REQ messages and the ANSI-41 SMS address values in the SMDPP messages for each step identified in Figure 13.15. This time, SME-B is addressed by an IP address, IP-B. For the purposes of illustration, we assume that no sub-address information is provided for SME-A or SME-B.

TABLE 13.3

Another Example of ANSI-41 SMS Address Information

STEP	Original-Originating-Address	Original-Destination-Address	SMS_Original-Originating-Address	SMS_Originating-Address	SMS_Original-Destination-Address	SMS_Destination-Address
1	MIN-A	IP-B				
2			MIN-A	MIN-A	IP-B	MIN-A
3			MIN-A	MIN-A	IP-B	IP-B
4			MIN-A	IP-B	IP-B	MIN-B
5	MIN-A	IP-B				

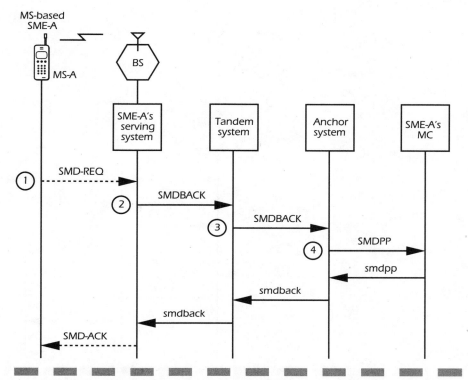

Figure 13.13 *MS-based SME message origination under call handoff conditions: (1) MS-based SME-A sends an air-interface message, SMD-REQ, to the serving system. (2) The call has previously been handed off from the anchor system to a tandem system and then to the current serving system; therefore, the serving system routes the short message towards the anchor system using the SMSDeliveryBackward Invoke (SMDBACK) message. (3) The tandem system forwards the SMDBACK message to the anchor system. (4) The anchor system then sends the message to SME-A's MC, using the SMDPP message. Acknowledgment messages are relayed forward through the handoff chain to SME-A.*

As we see in Table 13.3, routing proceeds per our first example in Table 13.2 until step 3. At this point, SME-A's MC must send the SMDPP message to SME-B's MC. Since SME-B is identified with an IP address, SME-A's MC sends the SMDPP message via a TCP/IP network, where it terminates on SME-B's MC. Noting that the message is intended for IP-B, and that IP-B is currently associated with MS-B, SME-B's MC sends the SMDPP message to MS-B's current serving system (e.g., via an SS7 transport network using the SMS_Address provided by the serving system). SME-B's MC informs the serving system that the origi-

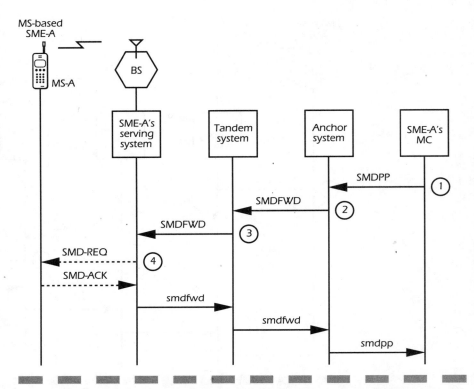

Figure 13.14 *MS-based SME message termination under call handoff conditions: (1) A message destined for MS-based SME-A is sent by SME-A's MC to the anchor system. (2) The call has previously been handed off from the anchor system to a tandem system and then to the current serving system; therefore, the anchor system routes the short message towards the serving system using the SMSDeliveryForward Invoke (SMDFWD) message. (3) The tandem system forwards the SMDFWD message to the serving system. (4) The serving system sends the air-interface message, SMD-REQ, to SME-A. Acknowledgment messages are relayed back through the handoff chain to the MC.*

nal destination address is IP-B, while the "intermediate" destination address is MIN-B (the MIN associated with MS-B). The serving system should, at this point, send the message to MS-B, identifying the destination address as IP-B.

A Final Word on Short-message Addressing and Routing

Unfortunately, while they are flexible, the ANSI-41 short-message addressing and routing mechanisms are also complex and subject to interpretation. The fact that ANSI-41 refers to idealized air-interface

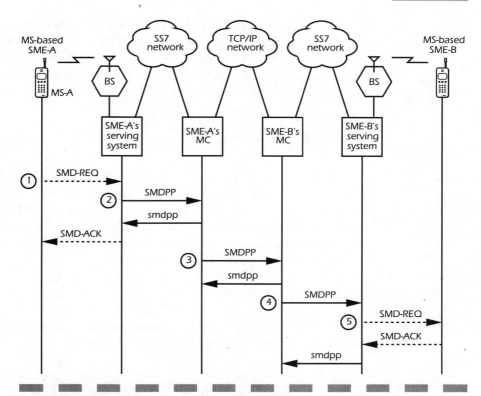

Figure 13.15 Another basic message origination and termination sequence, involving different network technologies: (1) MS-based SME-A sends an air interface message, SMD-REQ, to the serving system. (2) The serving system routes the short message to SME-A's MC, using the ANSI-41 SMDPP message. The SMDPP message is routed by using the same SS7 signaling network as is used for routing other ANSI-41 messages. When the smdpp message is received from the MC, the serving system converts it into an air-interface acknowledgment, the SMD-ACK message. (3) SME-A's MC may apply an originating supplementary service to the short message (this is not currently defined in ANSI-41); the SMDPP message is then routed to the destination SME's MC by using a TCP/IP network. (4) SME-B's MC may apply a terminating supplementary service to the short message (this is not currently defined in ANSI-41); the SMDPP message is then routed to the destination SME's serving system by using the SS7 signaling network. (5) The serving system forwards the short message toward the destination SME using the air-interface SMD_REQ message. SME-B responds with an automatic acknowledgment (SMD-ACK) to signal acceptance of the SMD-REQ message.

messages (e.g., the SMD-REQ message) rather than real air-interface messages makes things that much more difficult to decipher.

TR45.3, the subcommittee with responsibility for the TDMA air-interface standards, has attempted to limit the potential for confusion by producing a specification that explicitly identifies the mappings between the SMS air-interface message in TDMA, called R-DATA, and the corresponding ANSI-41 SMDPP message, for both mobile-originated and mobile-terminated SMS scenarios. If interested, the reader should see ANSI/TIA/EIA-136-610, *R-DATA / SMDPP Transport*.

SMS Barring

ANSI-41 provides a number of SMS origination and termination barring capabilities, conveyed to the serving system during the service qualification process for the host MS in the ANSI-41 parameters SMS_OriginationRestrictions and SMS_TerminationRestrictions. The SMS_OriginationRestrictions parameter defines the type of short messages the MS is allowed to originate:

- No messages; block all messages
- All messages; allow all messages
- Block message if direct routing is requested (see below)
- Force indirect routing for all messages (see below)

In the examples of the previous section, we showed messages routed via the originating SME's MC; this is called *indirect routing*. ANSI-41 allows MS-based SMEs to request that originated messages bypass the originator's MC; this is called *direct routing*. Figure 13.16 and Table 13.4 illustrate how our first example from the previous section (i.e., in Figure 13.7 and Table 13.2) changes when the originating SME requests direct routing of the message. Note the change in the SMS_Destination-Address parameter in step 2.

TABLE 13.4

Addressing When Direct Routing is Provided

STEP	Original-Originating-Address	Original-Destination-Address	SMS_Original-Originating-Address	SMS_Originating-Address	SMS_Original-Destination-Address	SMS_Destination-Address
1	MIN-A	MIN-B				
2			MIN-A	MIN-A	MIN-B	MIN-B
3			MIN-A	MIN-B	MIN-B	MIN-B
4	MIN-A	MIN-B				

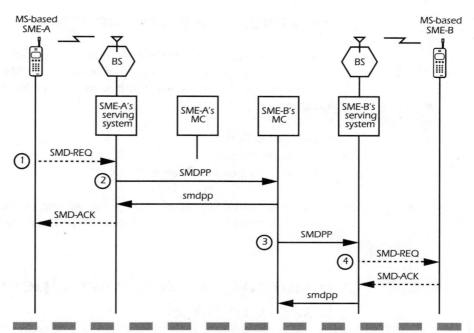

Figure 13.16   *Another basic message origination and termination sequence, involving direct routing: (1) MS-based SME-A sends an air-interface message, SMD-REQ, to the serving system, requesting direct routing. (2) The serving system routes the short message to SME-A's MC, using the ANSI-41 SMDPP message. When the smdpp message is received from the MC, the serving system converts it into an air-interface acknowledgment, the SMD-ACK message. (3) SME-A's MC may apply an originating supplementary service to the short message (this is not currently defined in ANSI-41); the SMDPP message is then routed to the destination SME's MC. (4) SME-B's MC may apply a terminating supplementary service to the short message (this is not currently defined in ANSI-41); the SMDPP message is then routed to the destination SME's serving system. (5) The serving system forwards the short message toward the destination SME using the air interface SMD-REQ message. SME-B responds with an automatic acknowledgment (SMD-ACK) to signal acceptance of the SMD-REQ message.*

The SMS_TerminationRestrictions parameter defines the type of short messages the MS is allowed to receive:

- No messages; block all messages.
- All messages; allow all messages.
- Block message if it is charged to destination (see below)

The SMS_ChargeIndicator parameter, when it is included in the SMDPP message, indicates whether the originator of the message has requested that the destination be charged for the message delivery.

SMS Supplementary Services

Routing messages through the originator's and destination's MCs allows supplementary services to be applied to the basic SMS bearer service at one or both of the originating and terminating MCs. Possible services include:

- Delayed delivery
- Repeated delivery
- Delivery to a distribution list

These potential supplementary services, and many others, are left as system implementation issues in the ANSI-41 specification.

Summary of ANSI-41 Operations Used for SMS

Table 13.5 summarizes the ANSI-41 operations used by the SMS functions described in this chapter.

TABLE 13.5

Use of ANSI-41 Operations for SMS

Function	ANSI-41 Operations Used for the Function
SME service qualification	RegistrationNotification, QualificationRequest, QualificationDirective
SME location management	RegistrationNotification, MSInactive, BulkDeregistration, RegistrationCancellation, UnreliableRoamerDataDirective
SME state management	SME location management operations plus: RoutingRequest
Short message processing	RegistrationNotification, MSInactive, RegistrationCancellation, SMSDeliveryBackward, SMSDeliveryForward, SMSDeliveryPointToPoint, SMSNotification, SMSRequest

CHAPTER 14

Operations, Administration, and Maintenance Functions

In this chapter, we discuss the ANSI-41 mobile telecommunications network functions related to operations, administration, and maintenance (OA&M). The subset of OA&M functions supported by ANSI-41 is limited to inter-MSC circuit management. Circuit management can be considered a subset of network management functionality. There are many functions used to manage and control inter-MSC circuits; however, the ANSI-41 intersystem operations only support the most primary functions:

- Inter-MSC circuit blocking (and unblocking)
- Inter-MSC circuit reset
- Inter-MSC circuit testing

These functions are necessary to remove trunks from service, place them back into service, test them for transmission, and reset them to an idle state when necessary. Other functions such as traffic load control and collection of traffic statistics are beyond the scope of ANSI-41.

We describe each of the functions and provide examples of how the ANSI-41 mobile application part (MAP) operations (e.g., TrunkTest) are used to accomplish inter-MSC circuit management tasks. We summarize this information at the end of the chapter. Note that we employ the ANSI-41 convention for operation component acronyms; the Invoke component acronym is in all-capital letters (e.g., TRUNKTEST), while the Return Result component acronym is in all-lowercase letters (e.g., trunktest). Refer to the ANSI-41 standard for the descriptions of the individual ANSI-41 operations and parameters. Also, the reader should consult the Glossary for the description of general terms (e.g., serving system) not explicitly defined in this chapter.

Throughout this chapter we consider the serving system as a single entity encompassing the mobile switching center (MSC) and visitor location register (VLR) functional entities. This simplifies the descriptions of the ANSI-41 OA&M functions and also is representative of a large percentage of the ANSI-41 implementations currently in service. However, the reader should keep in mind that the potential for separation of the MSC and VLR exists and is fully defined in ANSI-41.

Where Are OA&M Functions Specified in ANSI-41?

OA&M functions are specified in two parts of ANSI-41:

- **ANSI-41.5** provides the required formats—the bit-by-bit encoding—of all the ANSI-41 operation components, including those used for OA&M. ANSI-41.5 defines both the messages (e.g., TrunkTest Invoke) and the message parameters (e.g., InterMSCCircuitID).
- **ANSI-41.4** provides algorithmic descriptions of the procedures that are associated with sending and receiving the ANSI-41 OA&M-related messages; these procedures were considered sufficiently unique to warrant separating them from the main procedures in ANSI-41.6. ANSI-41.4 also includes stage 2–style information for OA&M.

Inter-MSC Circuit Blocking

The ANSI-41 circuit blocking function—which also encompasses circuit unblocking—allows the MSCs at each end of an inter-MSC circuit to coordinate:

1. Removal of the inter-MSC circuit from service (*blocking*).
2. Reinstatement of the inter-MSC circuit into service (*unblocking*).

Circuit blocking may be used for fault management or testing purposes. Since the inter-MSC circuits are used solely for handoff, circuit blocking has important consequences for the intersystem handoff functions described in Chapter 9. ANSI-41 defines the concept of a *blocking state* to describe the impact on handoff of circuit blocking. We describe the ANSI-41 blocking state in terms of the inter-MSC circuit model shown in Figure 14.1.

Figure 14.1
An inter-MSC circuit model.

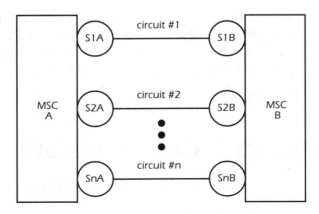

Each MSC maintains a record of the blocking state for each inter-MSC circuit. For our modeling purposes, *Sxy* is the blocking state of circuit x at MSC-y. In general, SxA is not necessarily equal to SxB. The blocking state for each circuit can take on one of four values at any given point in time. Table 14.1 describes the possible blocking state values for circuit 1 at MSC-A.

TABLE 14.1

Possible Values for S1A, the Blocking State of Circuit #1 at MSC-A

Blocking State, S1A	Symbol	Meaning
Active	ACT	Handoff can be initiated on circuit #1 by either MSC-A or MSC-B.
Locally blocked	LB	Handoff can be initiated on circuit #1 only by MSC-A.
Remotely blocked	RB	Handoff can be initiated on circuit #1 only by MSC-B.
Locally and remotely blocked	LRB	Circuit #1 cannot be used for handoff.

The blocking state for a circuit makes a transition from one state value to another based on *events*. Events include the sending or receiving of certain ANSI-41 messages. ANSI-41.4 specifies dozens of events, including:

- Sending and receiving blocking and unblocking messages
- Sending and receiving intersystem handoff messages
- Sending and receiving other OA&M messages (e.g., ResetCircuit Invoke)

Figure 14.2 illustrates a small portion of the overall *state transition diagram* for S1A, the blocking state of circuit 1 at MSC-A, when we consider only the events associated with receiving and transmitting the Blocking Invoke (BLOCKING) and Unblocking Invoke (UNBLOCKING) messages.

Inter-MSC Circuit Reset

The ANSI-41 inter-MSC circuit reset function may be used for two related purposes:

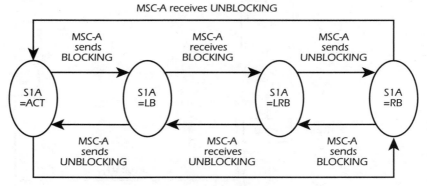

Figure 14.2
A portion of the overall state transition diagram for S1A, blocking state of circuit #1 at MSC-A (see Figure 14.1).

1. To initialize an inter-MSC circuit to a known state when it is placed into service.
2. To reinitialize an inter-MSC circuit to a known state after a failure that may have resulted in a loss of circuit state information.

The latter function addresses the sticky issue that arises when something in the "real world" is modeled as an object with a limited set of states and events that cause transitions between states—what do you do when the object's current state information is lost? The ANSI-41 Reset-Circuit operation solves this problem, as illustrated in Figure 14.3 and described below:

1. MSC-A and MSC-B have exchanged blocking messages and the state of inter-MSC circuit 1 is set to *locally and remotely blocked* at both ends.
2. MSC-A suffers a data failure.
3. When MSC-A recovers from the fault, it does not know the current state of inter-MSC circuit 1. It sends a ResetCircuit Invoke (RESET) message to MSC-B; effectively, this message tells MSC-B: "I consider circuit 1 to be in the *active* state. What is the state from your perspective?"
4. MSC-B responds with the ResetCircuit Return Result (reset) message, indicating that it considers the circuit to be blocked. This is conveyed in the ANSI-41 TrunkStatus parameter. MSC-B then changes the circuit blocking state to *locally blocked*, since the circuit is blocked from MSC-B's end and active from MSC-A's end.
5. When MSC-A receives the reset message, it sets the blocking state to *remotely blocked*. The circuit state is now reinitialized.

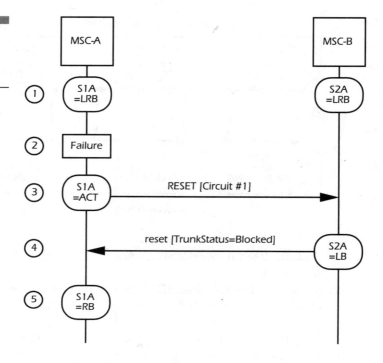

Figure 14.3
An example of the
use of the ANSI-41
Reset operation.

Inter-MSC Circuit Testing

The ANSI-41 inter-MSC circuit testing function supports *loop-back* testing of inter-MSC circuits. This capability is important to verify end-to-end path continuity and quality, especially when analog trunks are used to provide inter-MSC circuits. Otherwise, the only time a signal appears on the circuit is when it is used for handoff purposes—not the time to find out that the trunk is defective. With digital trunks—such as those based on a T-1 carrier system—there are alternate methods of monitoring the integrity of the physical facility but loop-back testing is still effective.

The circuit testing process is illustrated in Figure 14.4 and described below:

1. MSC-A initiates circuit testing by sending a TrunkTest Invoke (TTEST) message to MSC-B.

2. MSC-B decides to accept the test request. It places the selected circuit in loop-back mode; this involves connecting the circuit's receive-

audio path to the transmit-audio path, with appropriate signal level adjustment.

3. MSC-B then sends a TrunkTest Return Result (ttest) message to MSC-A.

4. MSC-A applies either a test tone or a termination to the circuit to test transmission level or noise level on the circuit, respectively.

5. When MSC-A wishes to end testing, it sends a TrunkTestDisconnect Invoke (TTESTDISC) message to MSC-B.

6. MSC-B ends testing, removes the loop-back, and sends a TrunkTestDisconnect Return Result (ttestdisc) message to MSC-A.

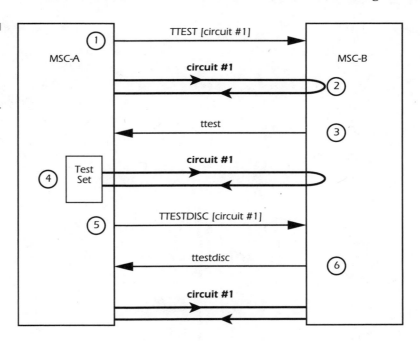

Figure 14.4
An example of the ANSI-41 inter-MSC circuit testing process.

Summary of ANSI-41 Operations Used for OA&M

Table 14.2 summarizes the ANSI-41 operations used by the OA&M functions described in this chapter.

TABLE 14.2

Use of ANSI-41
Operations for
OA&M

Function	ANSI-41 Operations Used for the Function
Inter-MSC circuit management	Blocking, Unblocking
Inter-MSC circuit reset	ResetCircuit
Inter-MSC circuit testing	TrunkTest, TrunkTestDisconnect

CHAPTER 15

Over-the-Air
Service
Provisioning

In this chapter, we discuss the basic ANSI-41 wireless telecommunications network functions related to over-the-air service provisioning (OTASP). Note that the OTASP functions are standardized in *TIA/EIA/IS-725 Revision A—Cellular Radiotelecommunications Intersystem Operations—Over-the-Air Service Provisioning (OTASP) & Parameter Administration (OTAPA)*. This interim standard provides modular extensions and functions defined as additions to ANSI-41. This is why we still refer to the OTASP operations as ANSI-41 operations.

This chapter is divided into the following categories:

- OTASP for ANSI/TIA/EIA-136 TDMA-based systems
- OTASP for ANSI/TIA/EIA-95 CDMA-based systems

For each category specific to CDMA and TDMA, the functions are subdivided into the following categories:

- OTASP automatic roaming functions, including call origination, registration, and authentication
- OTASP parameter transfer functions
- Over-the-air parameter activation (OTAPA) parameter transfer functions

We also examine the application processes associated with these functions. While many of the functions associated with OTASP have already been described in this book, we address the functions that have been modified for application to OTASP, such as authentication and the use of short-message service (SMS) transport.

In the course of describing the processes, we identify the ANSI-41 mobile application part (MAP) operations (e.g., OTASPRequest) used to accomplish basic OTASP process tasks. We summarize this information at the end of the chapter. Note that we employ the ANSI-41 convention for operation component acronyms; the Invoke component acronym is in all-capital letters (e.g., OTASPREQ), while the Return Result component acronym is in all-lowercase letters (e.g., otaspreq). Refer to the TIA/EIA/IS-725-A standard for the descriptions of the individual operations and parameters. Also, the reader should consult the Glossary for the description of general terms (e.g., serving system) not explicitly defined in this chapter.

What is Over-the-Air Service Provisioning?

Over-the-air service provisioning (OTASP) is sometimes informally called *over-the-air activation*. OTASP encompasses a set of network functions that allow a subscriber to obtain initial wireless telecommunications service remotely via the air interface. The concept of OTASP is a simple one: allow wireless service providers to offer initial service to potential subscribers without any intervention of a third party (e.g., a qualified dealer or representative). The feature consists of over-the-air programming of the number assignment module (NAM) within the mobile station (MS) equipment to authorize wireless telecommunications service with a particular service provider. Generally, customers desiring wireless service are required to visit a retail establishment where they choose their MS equipment. A customer service representative of the wireless service provider performs a credit check on the customer and manually programs the NAM of the MS with the parameters necessary to support the wireless subscription. These parameters include the mobile identification number (MIN) and the system ID (SID). This process can be tedious and prone to error since the programming sequence is typically designed so that it is difficult for subscribers to reprogram these parameters accidentally. If the customer service representative makes an error in the manual entry of these parameters, service will not be afforded and the customer's initial experience with the wireless service provider is an unpleasant one. There are many advantages to *properly* implemented OTASP:

- It makes the programming and activation process quick and easy.
- It minimizes opportunities for unauthorized use of the network.
- It enables better automation of the wireless service provider's provisioning systems.
- It avoids programming errors caused by manual programming.
- It enables wireless service providers to sell mobile stations at any retail distribution center without the need for authorized dealers and qualified personnel to perform the activation.
- It reduces churn by providing an easier method for customers to obtain wireless service.
- It reduces costs associated with mobile phone–center stores operated by the wireless service providers.

■ It reduces costs by avoiding the need to pay independent dealers for each subscriber activation.

The key to these advantages is that OTASP must be properly implemented. The average consumer can have a difficult time choosing a service provider, learning about wireless services, and learning how to operate the mobile station. For OTASP to be successful, wireless service providers must have highly trained customer service representatives to interface with potential subscribers during the OTASP process. Also, service providers must offer clear and simple documentation to enable potential subscribers to initiate the OTASP process.

One of the primary requirements of OTASP is the ability to provide an authentication key (A-key) to a mobile station to enable authentication. *Authentication* is the process by which information is exchanged between an MS and the network to confirm legitimate use of the MS. The transmission of information over the air interface to generate the A-key in the MS must be a very secure process to eliminate cloning fraud. Figure 15.1 shows an example flow chart for a typical OTASP scenario.

What is Over-the-Air Parameter Administration?

Over-the-air parameter administration (OTAPA) is similar to OTASP since the function involves the programming of NAM parameters into the mobile station over the air interface. OTAPA enables remote programming of the mobile station parameters subsequent to initial activation. OTAPA is used to program preferred roaming lists (PRLs), intelligent roaming databases (IRDBs), and changes to the MIN due to area code changes. Note that OTAPA requires no customer service intervention and these NAM parameters can be programmed into the mobile station whenever it is powered on and without the knowledge of the subscriber. OTAPA sessions are initiated autonomously by the network and do not limit the subscriber's ability to make or receive calls. In fact, OTAPA can occur while a call is in progress. Figure 15.2 shows an example flow chart for a typical OTAPA session.

Figure 15.1
Example flow chart of the typical sequence of procedures for a successful OTASP session.

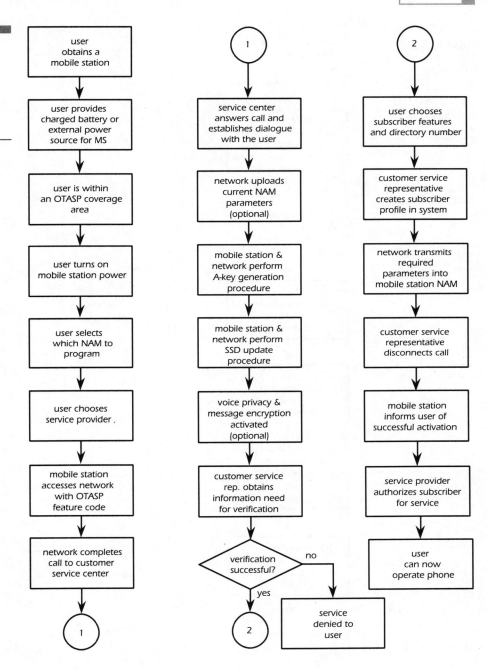

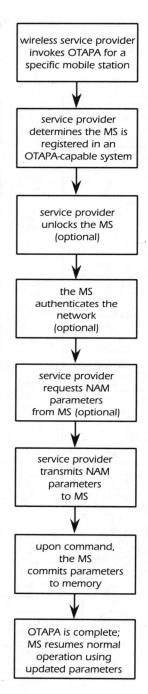

Figure 15.2
Example flow chart of the typical sequence of procedures for a successful OTAPA session.

wireless service provider invokes OTAPA for a specific mobile station

service provider determines the MS is registered in an OTAPA-capable system

service provider unlocks the MS (optional)

the MS authenticates the network (optional)

service provider requests NAM parameters from MS (optional)

service provider transmits NAM parameters to MS

upon command, the MS commits parameters to memory

OTAPA is complete; MS resumes normal operation using updated parameters

Where are OTASP and OTAPA Functions Specified?

OTASP functions are specified in three general sections of TIA/EIA/IS-725-A:

- Sections 1, 2, and 3 specify the functional overview and stage 1 description of OTASP and OTAPA.
- Sections 4T–9T (for TDMA) specify the detailed protocol (stages 2 and 3) and procedures for TDMA network OTASP operation.
- Sections 4C–9C (for CDMA) specify the detailed protocol (stages 2 and 3) and procedures for CDMA network OTASP operation.

Each of the CDMA and TDMA sections is analogous to each other. Note that each section of TIA/EIA/IS-725-A specifies the section of ANSI-41 where this information should be added.

- Sections 4C and 4T provide the automatic roaming functions supporting OTASP. Existing intersystem operations used for new scenarios and new intersystem operations along with the message parameters are described. This section is designed as an addendum to part 3 of ANSI-41 (ANSI/TIA/EIA-41.3).
- Sections 5C and 5T are sections within TIA/EIA/IS-725-A that provide new sections 8C, 9C and 8T, 9T for part 3 of ANSI-41 (ANSI/TIA/EIA-41.3). Sections 8C and 8T specify the individual OTASP call scenarios for all the automatic roaming functions. Sections 9C and 9T specify the individual OTAPA MS programming scenarios.
- Sections 6C and 6T provide the detailed signaling protocols for OTASP. This section is designed as an addendum to part 5 of ANSI-41 (ANSI/TIA/EIA-41.5).
- Sections 7C and 7T provide the detailed procedures for OTASP. This section is designed as an addendum to part 6 of ANSI-41 (ANSI/TIA/EIA-41.6).

Issues Associated with OTASP

The following list provides issues associated with the implementation and deployment of OTASP. Many of these need to be resolved before

OTASP becomes the primary means that wireless service providers use to activate new subscribers:

- Although OTASP has been introduced in a few markets, it will probably be several more years before it is the preferred and primary method for activating new subscribers. The main reason for this is the cost and complexity of properly implementing OTASP. OTASP requires wireless service providers to deploy a new functional entity, the over-the-air function (OTAF) and deploy modified existing functional entities. For example, OTASP requires upgrades to the air interface, MSC, VLR, HLR, AC, provisioning systems, and customer care systems.

- Radio anomalies may affect proper and complete programming of the mobile station. Subscribers cannot simply activate the phone anywhere; they must be in a location where the OTASP function is supported and a strong enough radio signal exists.

- A variety of problems can affect the wireless service provider's perceived quality of service, such as blocked calls due to congestion and enough available customer service representatives.

- Currently, OTASP is only defined for digital radio systems. Subscribers using dual-mode digital and analog mobile stations need to activate the phone where digital coverage is available, although the MS will appear to operate normally in analog service areas.

- Many digital mobile stations currently in operation do not support OTASP.

- Although the home system of a subscriber may support OTASP, a visited system while the subscriber is roaming may not, even when a roaming agreement is in place. This can occur when the subscriber obtains a new mobile station while roaming and discovers there is no means of activating it.

- Some wireless service providers may observe a significant increase in subscription and credit card fraud due to OTASP phones. Many service providers will delay the introduction of OTASP until better subscription fraud detection mechanisms are in place. The problem is that with OTASP, there is never face-to-face contact between a new subscriber and the service provider and no exchange of official identification documentation, such as a driver's license, credit card, passport, etc. Because of this, it is much easier to fake identification information during a phone call.

- Even when OTASP is the preferred and primary method for activating new subscribers, there will always be the need for manual programming at a customer service center or phone repair center when OTASP is not sufficient (that is, there will be occasions when OTASP fails to some extent and the subscriber requires a backup means to activate the phone).

■ ■ OTASP Functional Entities

OTASP requires the addition of two new functional entities in the ANSI-41 network: over-the-air function (OTAF) and customer service center (CSC).

The OTAF is a functional entity that supports proprietary interfaces to the CSC to support wireless service provisioning. It also supports standardized interfaces to the MSC, HLR, and VLR to complete service provisioning requests and pass the appropriate NAM parameters to the mobile stations. The OTAF may be a physically separate network entity, or may be implemented as a function within an HLR or a message center (MC). The OTAF does not explicitly interface with an MC; however, short-message service (SMS) transport is required to perform the OTASP and OTAPA functions. This is why the MC may be significant in the physical deployment of the over-the-air functions. The OTAF also contains the short-message entity (SME) application that uses short-message services (SMS) to communicate with the mobile station to be programmed. The MS contains a peer SME application that communicates with the OTAF SME for the actual transfer of NAM parameters using SMS as the transport protocol. A single OTAF functional entity can interface with multiple serving systems and with multiple customer service centers (CSCs).

The CSC is a functional entity that interfaces with customer service representatives receiving phone calls from potential subscribers who are activating initial service or making changes to their existing service. The CSC also supports proprietary interfaces to the OTAF to support wireless service provisioning and mobile station parameter modifications. The CSC is also an integral function that maintains proprietary interfaces with the service provider's back-end provisioning and billing systems. Due to the nature of internal marketing and business practices of the wireless service providers, the CSC is usually completely proprietary and customized to the service provider's particular needs. A single CSC entity can interface with multiple OTAF functional entities.

OTASP Automatic Roaming Functions for TDMA-based Networks

The automatic roaming functions required for TDMA-based OTASP are as follows:

1. Call origination from the MS to the network to initiate the OTASP process.
2. MS registration as part of call origination or initial MS power-on.
3. Authentication as part of A-key generation and SSD update during the OTASP process.

These functions support the basic scenario for activating an MS via OTASP. This basic scenario is as follows:

1. The subscriber calls the wireless service provider.
2. The MS registers on the wireless service provider's network.
3. The serving MSC becomes *attached to* (i.e., establishes a communications dialogue with) the OTAF functional entity.
4. The authentication function is performed.
5. MS activation parameters are downloaded to the MS from the OTAF.

MS Call Origination for OTASP

The first step in the network OTASP process is for the potential subscriber to originate a call to the wireless service provider. This function performs origination activation, to attach the serving MSC to the OTAF. Complexities occur for this function since the MS may be completely *unprogrammed*, meaning that the MS is attempting an original activation with no existing MS parameters. The MS may also be *preprogrammed*, meaning that the MS requires *reprogramming* of an existing MIN or IMSI and a profile in the HLR. An added twist is that the MS may be preprogrammed, but the MIN or IMSI is unknown and the profile cannot be found (e.g., when a subscriber changes wireless service providers or parameters are programmed by the phone manufacturer).

For preprogrammed MSs, the MIN or IMSI stored in the phone is used as the identity of the MS for the call. If the MS is unprogrammed, a special *activation MIN* is used. This MIN is a standard 10-digit identifier of the format NPA-NXX-XXXX, where NPA is defined as 000 and NXX-XXXX is assigned using a decimal representation of the lower digits of the ESN.

The MSC determines the address of the appropriate OTAF using the mobile country code (MCC) of the IMSI plus the SID or SOC, or the address is determined by the feature code sequence, which includes the dialed directory number. Figure 15.3 shows the call origination for OTASP scenario.

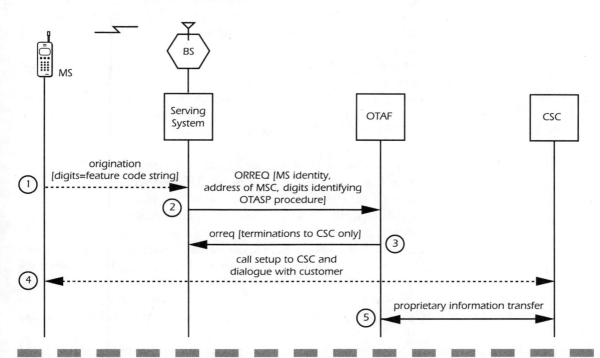

Figure 15.3 OTASP call origination for activation: (1) The potential subscriber makes a call to the wireless service provider. The subscriber enters a feature code and presses SEND. (2) The serving system (MSC) launches an ORREQ message to the OTAF. This message includes the MIN and ESN of the MS. If the MS is preprogrammed, the existing MIN is used. If the MS is an original activation, a special activation MIN is used. The address of the current serving MSC using the SMS protocol for activation, and digits identifying this as an OTASP procedure, are also sent. (3) The OTAF responds back to the MSC using the orreq message. This message acts as an acknowledgment and specifies that the MS can currently only make a call to the OTAF and to no other party. (4) The call is established between the potential subscriber and a customer service representative at the CSC. (5) The CSC informs the OTAF of the initial contact with the MS using a proprietary protocol. The OTAF then associates the CSC information with the data stored from the ORREQ message.

MS Registration with the Network for OTASP

For both unprogrammed and preprogrammed MSs, there are alternate means the network uses to initiate the activation process and to attach the MSC to the OTAF to begin this process. There are five distinct scenarios used for this process:

- Registration activation on call origination (unprogrammed MS has no real MIN).
- Registration activation on MS power on (unprogrammed MS has no real MIN).
- Deferred registration activation (when a call is made to the CSC from a phone other than the one to be programmed and the unprogrammed MS has no real MIN).
- Registration activation on call origination (preprogrammed MS has a MIN or IMSI).
- Deferred registration activation (when a call is made to the CSC from a phone other than the one to be programmed and the preprogrammed MS has a MIN or IMSI).

Registration activation on call origination for an unprogrammed MS is similar to the basic call origination for OTASP scenario, except that the initial step to attach the MSC to the OTAF is accomplished using the RegistrationNotification message (see Figure 15.4).

Registration activation on MS power-on for an unprogrammed MS is very similar to registration activation on call origination. The only difference is that when the MS is powered on, the registration is implicitly invoked prior to the potential subscriber's originating a call. The scenario is the same as is shown in Figure 15.4, except that step 1 comes after step 5, and the MSC is already attached to the OTAF before the activation call is made.

Deferred registration activation for an unprogrammed MS is used when the potential subscriber calls the CSC from a wire-line or other phone, not the MS to be programmed. This scenario is similar to the registration activation on power-on scenario, except that the call is made to the CSC before the MS is powered on. The CSC then transfers information to the OTAF using a proprietary protocol. When the MS is eventually powered on, the registration is implicitly invoked to attach the MSC to the OTAF and no subsequent call to the CSC is needed.

Registration activation on call origination for a preprogrammed MS is only used for an MS performing a reprogramming procedure. This is

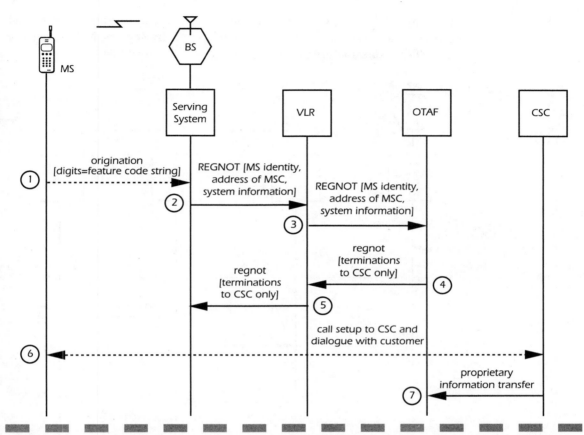

Figure 15.4 *OTASP registration activation on call origination for an unprogrammed MS: (1) The potential subscriber makes a call to the wireless service provider. The subscriber enters a feature code and presses SEND. (2) The serving system (MSC) launches a REGNOT message to the VLR. This message includes the activation MIN of the MS, the address of the current serving MSC using the SMS protocol for activation, and system information specifying that this is an initial OTASP activation. (3) The VLR forwards the REGNOT message to the OTAF. (4) The OTAF responds back to the VLR using the regnot message. This message acts as an acknowledgment and specifies that the MS can currently only make a call to the OTAF and to no other party. (5) The VLR forwards the regnot message to the serving MSC. (6) The call is established between the potential subscriber and a customer service representative at the CSC. (7) The CSC informs the OTAF of the initial contact with the MS using a proprietary protocol. The OTAF then associates the CSC information with the data stored from the REGNOT message.*

desired, for example, when a subscriber wishes to change the directory number of the MS and the MIN serves as the directory number. In this scenario, the subscriber originates a call to the CSC and normal registration occurs (either on MS power-on prior to the call, or during call origination). The CSC typically informs the OTAF that this is a reprogramming procedure. The OTAF then uses the short-message service SMSRequest message to query the HLR for the address of the serving MSC receiving the OTASP parameters for programming (see Figure 15.5).

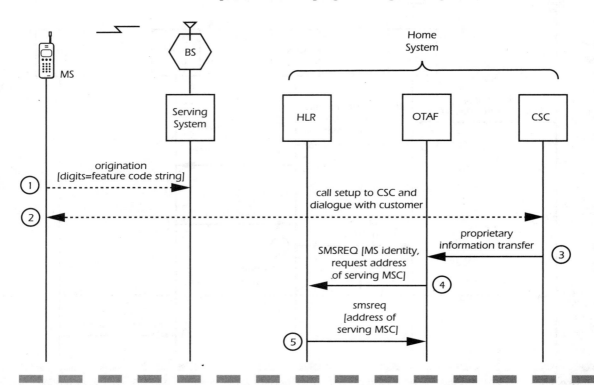

Figure 15.5 Registration activation on call origination for an MS requiring reprogramming. (1) The potential subscriber makes a call to the wireless service provider. The subscriber enters a feature code and presses SEND. (2) Since the MS is already programmed and active, the call is established between the subscriber and a customer service representative at the CSC. (3) The CSC informs the OTAF of the initial contact with the MS using a proprietary protocol. Included with the information may be an indication that this is a reprogramming procedure for the MS. (4) The OTAF launches an SMSREQ message to the HLR associated with the MIN or IMSI received from the MS. The message includes parameters requesting the address of the serving MSC used for the reprogramming. If the OTAF does not know the address of the serving MSC to be used, the SMSREQ message is then forwarded to the serving system to obtain this information from the serving system. (5) The HLR responds back to the OTAF using the smsreq message. This message acts as an acknowledgment and specifies the address of the serving MSC to be used for reprogramming the MS.

Deferred registration activation for a preprogrammed MS is used when an existing subscriber calls the CSC from a wire-line or other phone, not the MS to be programmed. This scenario is similar to the registration activation on call origination for a preprogrammed MS, except that the call is made to the CSC before the MS is powered on. Since this is a reprogramming procedure, a subscriber profile exists within the HLR. The CSC informs the OTAF that this is a reprogramming procedure and the OTAF uses the short-message service SMSRequest message to query the HLR for the address of the OTASP SME. At some later time when the MS is powered on, normal registration will occur and the network will be aware of the location of the MS to be programmed (see Figure 15.6).

In some cases, the MS requires reprogramming, but the preprogrammed parameters within the NAM are unknown to the network. In this case, registration of the MS will fail, and OTASP can continue as if the MS were unprogrammed.

Authentication Procedures for OTASP

After the serving MSC is *attached* to the OTAF using call origination or registration procedures, the A-key for subscriber authentication must be generated. This procedure is quite different from the typical method for entering an A-key into the mobile station (see Chapter 11). Typically the entire 26-digit A-key is entered manually via the MS keypad, preprogrammed internally by the MS manufacturer, or programmed through the data port of the MS equipment. Since OTASP must be a secure process, it makes no sense to send the full A-key as an over-the-air parameter to be programmed into the MS. This could lead to fraudulent access of the A-key and enable cloning fraud. Therefore, a new procedure has been developed to *generate* the 26-digit (i.e., 64-bit) A-key within the MS based on the Diffie-Hellman Key Agreement Standard procedure.

Diffie-Hellman Key Agreement Standard Procedure

The Diffie-Hellman Key Agreement Standard procedure is a cryptography algorithm that supports the distribution of temporary keys to generate identical authentication keys (A-keys) on both the mobile station and the authentication center (AC). The Diffie-Hellman procedure is patented in the United States and Canada under patents #4,200,770 and #4,218,582. The patent expired on April 29, 1997 and is available for public use by any party, with no obligation to pay royalties. Patent # 5,930,362 is in effect for applying the Diffie-Hellman Key Agreement Standard to A-key generation in wireless networks.

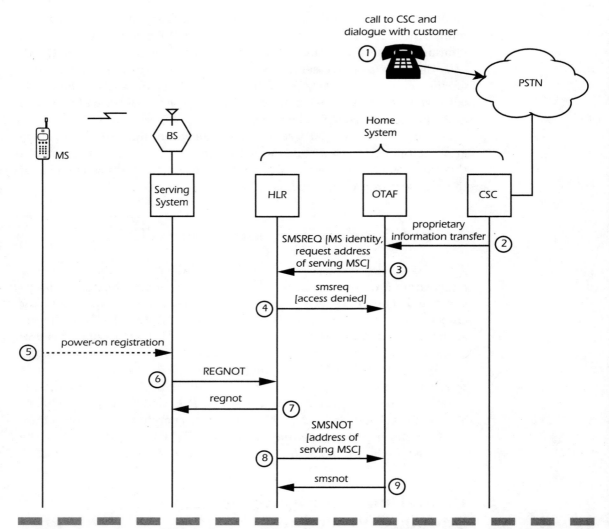

Figure 15.6 Deferred registration activation for an MS requiring reprogramming. (1) The subscriber makes a call to the wireless service provider using a phone other than the one to be programmed. (2) The CSC transfers information to the OTAF, including the identity of the MS, and informing it that this is a reprogramming procedure. (3) The OTAF launches an SMSREQ message to the HLR associated with the MIN or IMSI of the MS to request the address of the last known serving MSC for the MS. (4) Since the MS is not registered and is considered unavailable, the HLR responds back to the OTAF using the smsreq message specifying that access is denied. This causes the OTAF and the HLR to mark the procedure as pending, so that when the MS is available, the OTASP procedure can continue. (5) At some later time, the MS is powered-on and initiates the registration procedure with the network. (6) The serving system performs normal registration with the HLR using the REGNOT message. (7) The HLR responds back to the serving system using the regnot message. This message acts as an acknowledgment that registration was successful. (8) The HLR now knows the identity of the serving MSC to continue with the OTASP procedure, and launches the SMSNOT message to the OTAF informing it of the serving MSC to be used. (9) The OTAF responds back to the HLR using the smsnot message informing it that the OTASP procedure can continue.

The main advantage of the Diffie-Hellman Key Agreement Standard algorithm over other forms of public-key cryptography is that all keys are temporary. There are no public keys to publish and no private keys to store securely. All parties look the same, which is ideal for mass-produced systems such as mobile stations.

The example below shows the Diffie-Hellman algorithm used by the network and the MS to generate the secret A-key value:

Let X_A be a large random number that is private to the network.

Let X_B be a large random number that is private to the MS.

Let p be a large prime number that is public.

(p = parameter PRIMVAL sent from the network to the MS for OTASP.)

Let g be a large random number base that is public ($0 < g < p$).

(g = parameter MODVAL sent from the network to the MS for OTASP.)

The network calculates the value Y_A using the following equation:

$$Y_A = g^{XA} \bmod p, 0 < Y_A < p$$

The network sends the values g, p, and Y_A (the partial network key value) to the MS.

The MS calculates the value Y_B using the following equation:

$$Y_B = g^{XB} \bmod p, 0 < Y_B < p$$

The MS sends the calculated value Y_B to the network (the partial MS key value).

The network calculates the agreed secret value K (key) using the following equation:

$$Key = Y_B X^A \bmod p \text{ (which equals } g^{XA}X^B \bmod p)$$

The MS calculates the agreed secret value K (key) using the following equation:

$$Key = Y_A X^B \bmod p \text{ (which equals } g^{XA}X^B \bmod p)$$

Each party stores the least-significant 64 bits of Key as the A-key for authentication. Note that for OTASP, p is a 512-bit random prime number and g is a 160-bit number calculated by the authentication center (AC).

This method is considered to be secure for generating identical 64-bit A-keys in the mobile station and in the network. Since only the values Y^A and Y^B are transmitted over the air interface, eavesdroppers cannot derive the secret value key except by computing a discrete logarithm, mod p, which is computationally infeasible for properly chosen values of g and p.

One of the drawbacks of this method, however, is the time it takes both the MS and the network to complete these calculations. Typically, mobile stations do not support the computing power to calculate the A-key quickly using the Diffie-Hellman method. The computation can take several minutes to complete and this causes an inconvenience to the potential subscriber. The computation time can be shortened by choosing smaller values of g and p, but these smaller values may reduce security and increase the ability of a hacker to derive the A-key using brute-force computational methods.

Initiate A-key Generation

The A-key generation procedure uses short-message services (SMS) and the OTASPRequest operation to initiate the process of passing the proper parameters to securely generate the A-key for use in authentication. The Diffie-Hellman Key Agreement Standard is used by both the AC and the MS to calculate the A-key based on the parameters generated by the AC and delivered to the MS. When SMS is used to deliver the parameters from the network to the MS, both the MS and the network need to specify the SMS teleservice identifier. Since there are many possible applications that can use SMS transport services, the teleservice ID is used to identify the SMS peer applications specifically used for OTASP. This teleservice is known as the *over-the-air activation teleservice* (OATS). Figure 15.7 shows the procedure for initiating A-key generation in OTASP for authentication.

1. The OTAF determines that an A-key must be generated via OTASP and launches an OTASPREQ message to the HLR. The message contains the ID of the mobile station, ID of the MSC serving the MS, and a request to initiate the A-key generation process.
2. The HLR forwards the OTASPREQ message to the AC to generate the Diffie-Hellman parameters used to generate the A-key.
3. The AC calculates the 160-bit modulus (MODVAL), 512-bit prime number, and the partial network key value and responds back to the HLR using the otaspreq message containing these parameters.
4. The HLR forwards the otaspreq message to the OTAF for transport to the MS.
5. The OTAF uses the SMDPP message to send the A-key generation parameters to the serving system (and subsequently to the MS via

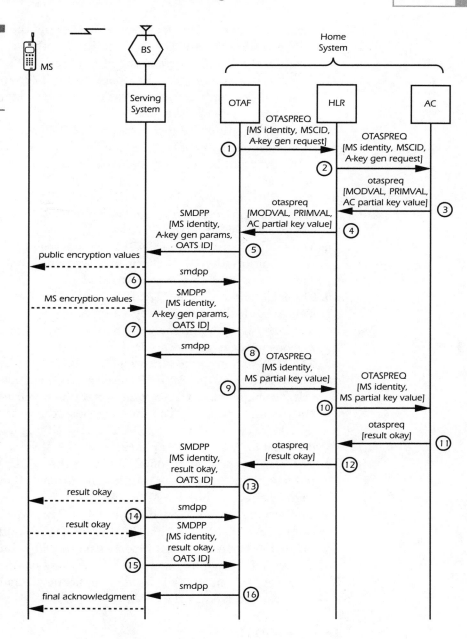

Figure 15.7
A-key generation
procedure. See text
for a description of
the scenario.

the air interface). This message also contains the ID of the MS (i.e., MIN or IMSI) and the OATS ID specifying that this message is destined for the OTASP application at the MS.

6. The serving system responds back to the OTAF using the smdpp message. This message is used as an acknowledgment that the message was received by the MS.

7. The serving system launches the SMDPP message to the OTAF containing the identity of the MS, the OATS ID, and the partial MS key value that will be used by the network to generate the actual A-key at the AC.

8:. The OTAF responds back to the serving system using the smdpp message. This message is used as an acknowledgment that the message was received by the OTAF.

9. The OTAF launches an OTASPREQ message to the HLR. The message contains the ID of the mobile station, ID of the MSC serving the MS, and the MS partial-key value.

10. The HLR forwards the OTASPREQ message to the AC.

11. The AC calculates the A-key value and responds back to the HLR using the otaspreq message with an acknowledgment that the A-key has been properly generated.

12. The HLR forwards the otaspreq message to the OTAF.

13. The OTAF uses the SMDPP message to send the A-key generation acknowledgment to the serving system (and subsequently to the MS via the air interface). This message also contains the ID of the MS (i.e., MIN or IMSI) and the OATS ID specifying that this message is destined for the OTASP application at the MS.

14. The serving system responds back to the OTAF using the smdpp message. This message is used as an acknowledgment that the message was received by the MS.

15. The serving system launches the SMDPP message to the OTAF containing the identity of the MS, the OATS ID, and the acknowledgment that the A-key was properly generated by the MS.

16. The OTAF responds back to the serving system using the smdpp message. This message is used as an acknowledgment that the message was received by the OTAF.

Initiate SSD Update

After the A-key is properly generated at both the AC and the MS, the shared secret data (SSD) value must be updated to enable authentication. This procedure is invoked using the OTASPRequest message. Subsequent to this invocation, the authentication procedures are performed as described in detail in Chapter 11. These procedures are defined when SSD is shared with the serving system and when SSD is not shared. Figure 15.8 shows the SSD update and subsequent unique challenge and base-station challenge performed by the network when SSD is not shared to ensure the legitimacy of the MS being activated via OTASP.

Reauthentication to Send Encryption Parameters

As an optional part of OTASP, the network can perform a reauthentication procedure for the purpose of invoking signaling message encryption and voice privacy. This is done after the A-key is generated and the SSD is updated for the MS. These procedures are identical to the procedures described in Chapter 11, except that they are initiated by the OTAF using the OTASPRequest message. The primary reason for this procedure is to encrypt the voice and signaling channels for the exchange of a subscriber's personal information to activate the subscription. For example, voice privacy should be used if the subscriber is providing a credit card number to the customer service representative at the CSC during an activation conversation. Alternatively, this process can be automated, since the subscriber may enter the credit card number via the keypad on the MS. This would require signaling message encryption to keep this information secure. Other subscriber information that should be kept secure includes the social security number, driver's license number, and home address information, if required to activate the subscription.

Committing the A-key

The initial A-key generation procedure performed during OTASP is primarily used to support authentication, signaling message encryption, and voice privacy during the actual OTASP process to keep this process secure. However, the newly generated A-key can be committed to the AC database for subsequent authentications of a subscriber after OTASP is completed. Figure 15.9 shows the procedure for committing this new A-key to the AC's subscriber database.

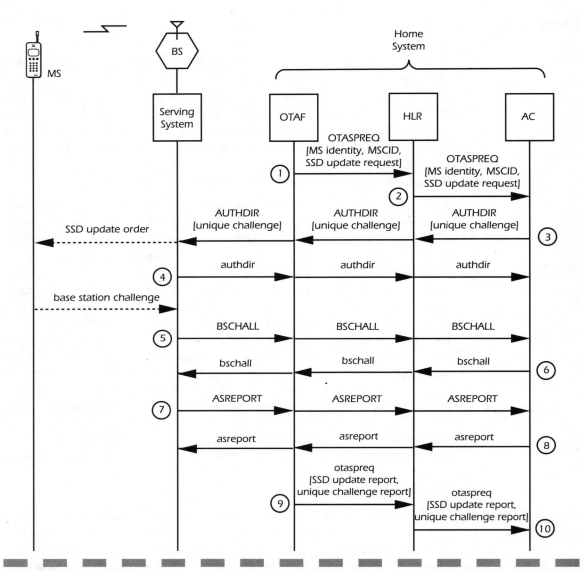

Figure 15.8 SSD update after A-key generation. (1) The OTAF launches an OTASPREQ message to the HLR associated with the MIN or IMSI of the MS to request an SSD update for the MS. (2) The HLR forwards the OTASPREQ message to the AC. (3) The AC launches an AUTHDIR message through the network to the MS to invoke the unique challenge procedure. (4) The AUTHDIR message is acknowledged through the network via the authdir message. (5) The serving system forwards the BSCHALL message through the network to the AC indicating that the MS is challenging the network for authenticity with the new SSD value. (6) The AC acknowledges the base station challenge and responds through the network with a bschall message containing the response to the challenge. (7) The serving system launches an ASREPORT message through the network to the AC indicating the success or failure of the SSD update and the unique challenge. (8) The AC acknowledges the indication from the MS using the asreport message. If all goes well, this message indicates that the MS can be provided service. (9) The OTAF acknowledges the original OTASPREQ message with an otaspreq message back to the HLR containing the result (success or failure) of both the SSD update and the unique challenge. (10) The HLR forwards the otaspreq message to the AC. See Chapter 11 for details of the unique challenge, base station challenge, and authentication status report procedures.

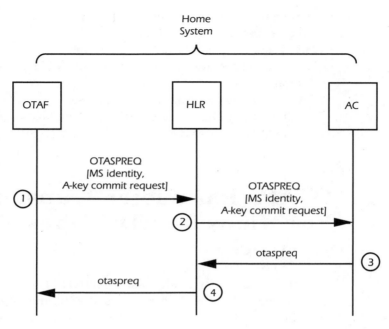

Figure 15.9
A-key commit procedure. (1) The OTAF launches an OTASPREQ message to the HLR associated with the MIN or IMSI of the MS to request that the previously generated A-key be committed to the AC's database. (2) The HLR forwards this request to the AC. (3) The AC responds back to the HLR acknowledging that the A-key has been committed to its database using the otaspreq message. (4) The HLR forwards this acknowledgment to the OTAF.

Over-the-air Activation Teleservice (OATS)

The over-the-air activation teleservice (OATS) is an application-layer protocol defined specifically for TDMA-based networks. OATS is the teleservice that uses the SMS transport protocol to send the specific NAM parameters to be programmed to the mobile station (see Chapter 13 for more details on SMS). OATS is identified as the peer-to-peer teleservice by a specific teleservice ID that is sent within the ShortMessageDeliveryPointToPoint (SMDPP) operation. The NAM parameters to be programmed are encapsulated within the SMS_BearerData parameter sent within the SMDPP message. The NAM parameters that can be sent via OATS include the following:

- Generic configuration data (e.g., MIN and IMSI)
- Public encryption values (for A-key generation, voice privacy, and signaling message encryption)
- Key-generation results (for A-key generation procedure)

- Parameter download requests
- System operator codes (SOCs)
- MS-specific requests
- NAM commit requests (to commit the NAM parameters to the MS database)
- CSC authentication challenge responses
- Intelligent roaming database (IRDB) downloads (i.e., SID lists)
- Abort OATS process (for failures during OTASP)

OTAPA Parameter Transfer Functions for TDMA-based Networks

Over-the-air parameter administration (OTAPA) transfer functions are used to modify MS NAM parameters at any time after initial activation, regardless of whether or not OTASP was used for activation. This function is used, for example, when a subscriber requires changes to the MIN or IMSI in the MS, or when the network requires updates to system operator codes (SOCs) or system IDs (SIDs). OTAPA requires no interaction by the subscriber or by the customer service center (CSC). When OTAPA is performed, the subscriber is completely unaware of the changes being made to the parameters within the MS. Figure 15.10 shows the basic successful OTAPA process.

If the OTAPA process fails because the MS rejects the programming of the new parameters, the serving system informs the OTAF in the response to the ShortMessageDeliveryPointToPoint message (i.e., smsnot). This message contains the parameter SMS_CauseCode, indicating a failure. If this happens, The OTAF can restart the entire process and attempt to reprogram the MS the next time it registers.

Over-the-air Programming Teleservice (OPTS)

The over-the-air programming teleservice (OPTS) is an application-layer protocol defined specifically for TDMA-based networks. OPTS is used for the procedures to modify NAM parameters in the MS subsequent to ini-

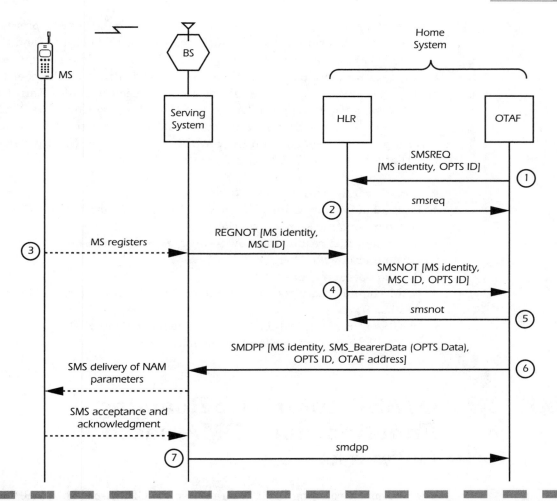

Figure 15.10 Basic successful OTAPA process using OPTS. (1) The OTAF launches an SMSREQ message to the HLR requesting to reprogram the NAM parameters for the identified MS via OPTS. (2) The HLR acknowledges this request by sending the smsreq message back to the OTAF. (3) At some later time the MS registers and is available to receive the new parameters. If the MS is already registered, it is currently available to receive these parameters. (4) The HLR launches an SMSNOT message to the OTAF indicating that the MS is registered and supplies the OTAF with the current serving system address of the MS. (5) The OTAF acknowledges that the MS is now available by responding to the HLR with the smsnot message. (6) This step represents the standard delivery mechanism of TDMA NAM parameters via OPTS. The OTAF delivers the parameters to the serving system via the SMDPP message. This message contains the MS identity, the OPTS NAM parameter data, the OPTS teleservice ID, and the address of the OTAF sending the data. (7) The NAM parameter data is sent to the MS via the air interface and the serving system acknowledges the successful programming of the MS by sending the smdpp message to the OTAF.

tial activation or the over-the-air service provisioning (OTASP) process. OPTS is the teleservice that uses the SMS transport protocol to send the specific NAM parameters to be programmed to the mobile station (see Chapter 13 for more details on SMS). OPTS is identified as the peer-to-peer teleservice, by a specific teleservice ID, which is sent within the ShortMessageDeliveryPointToPoint (SMDPP) operation. The NAM parameters to be programmed are encapsulated within the SMS_BearerData parameter sent within the SMDPP message. The NAM parameters that can be sent via OPTS include the following:

- Generic configuration data (e.g., MIN and IMSI)
- Public encryption values (for new A-key generation, voice privacy, and signaling message encryption)
- Key generation results (for new A-key generation procedure)
- Parameter download requests
- System operator codes (SOCs)
- NAM commit requests (to commit the NAM parameters to the MS database)
- Intelligent roaming database (IRDB) downloads (i.e., SID lists)
- Abort OPTS process (for failures during OTAPA)

OTASP Automatic Roaming Functions for CDMA-based Networks

The automatic roaming functions required for CDMA-based OTASP are as follows:

1. Call origination from the MS to the network to initiate the OTASP process.
2. Redirection from initial CSC to desired CSC.
3. Authentication as part of A-key generation and SSD update during the OTASP process.
4. Registration following successful OTASP.

These functions support the basic scenario for activating an MS via OTASP:

1. The subscriber calls the wireless service provider.
2. The serving MSC becomes *attached to* (i.e., establishes a communications dialogue with) the OTAF functional entity.
3. The authentication function is performed.
4. MS activation parameters are downloaded to the MS from the OTAF.

MS Call Origination for OTASP

The first step in the netork OTASP process is for the potential subscriber to originate a call to the wireless service provider. This function performs call origination and uses SMS to attach the serving MSC to the OTAF. The SMSDeliveryPointToPoint operation is used perform this attachment to associate the subscriber's call with the OTASP processes. A temporary reference number is used to create the association between the subscriber's call and the OTASP processes controlled at the OTAF. The temporary reference number is similar to a temporary local directory number (TLDN) and is assigned from a pool of numbers controlled by the CSC. This association is released prior to the performance of the authentication functions and the programming of activation parameters. Figure 15.11 shows the basic scenario to perform this function.

An alternative procedure is also defined for call origination to attach the OTAF to the serving system. During the subscriber's call origination, the serving system can launch a FeatureRequest message to the OTAF before the call is completed and the dialogue with a customer service representative takes place. The FeatureRequest message contains the MIN or IMSI of the MS, the MSC ID, and a billing ID. The billing ID is used to inform the OTAF to which CSC the call should be completed. When the serving system receives the response to the FeatureRequest message, the subscriber's call is completed to the CSC using the temporary reference number included in the digits (destination) parameter in the response from the OTAF. Following the call establishment to the CSC, the MS call origination scenario is identical to that shown in Figure 15.11.

Call Redirection from an Initial CSC to a Desired CSC

In CDMA-based networks, consideration was given to roamers who can have their mobile stations activated outside of their home systems,

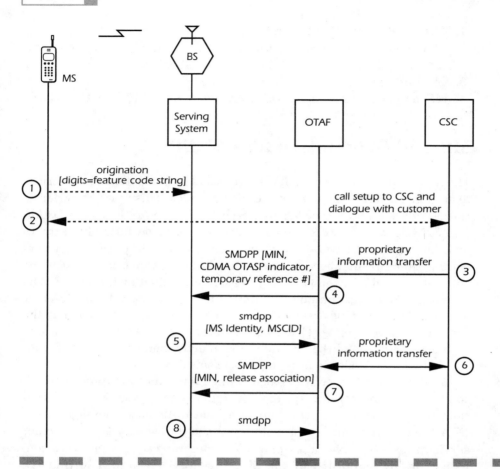

Figure 15.11 OTASP call origination for activation: (1) The potential subscriber makes a call to the wireless service provider. The subscriber enters a feature code and presses SEND. (2) The call is established between the potential subscriber and a customer service representative at the CSC. At this point the serving system may perform normal validation and authentication of the subscriber, if the MS requires reprogramming rather than initial activation. A temporary reference number is assigned by the serving system and sent as part of the feature code or as the calling or called party number, based on the signaling technique used. This temporary reference number is used to associate the OTASP processes and the CSC with the OTASP call. (3) The CSC informs the OTAF of the initial contact with the MS using a proprietary protocol to prepare the OTAF for activation of the MS. (4) The OTAF launches an SMDPP message to the serving system. This message includes the special activation MIN of the MS, a service indication specifying that this is an initial OTASP activation, and the temporary reference number that the serving system uses to associate the OTASP processes and the CSC with the OTASP call from the potential subscriber. (5) The serving MSC responds back to the OTAF using the smdpp message. This message includes the mobile station's identity (MIN or IMSI) and the MSC ID of the serving MSC. (6) The OTAF and CSC exchange information via a proprietary protocol. The OTAF informs the CSC that the attachment with the serving MSC has occurred. The CSC informs the OTAF to release the association of the OTASP processes from the OTASP call. (7) The OTAF launches a second SMDPP message to the serving system. This message includes the activation MIN and a code directing the MSC to release the association of the OTASP processes from the OTASP call. (8) The serving MSC responds back to the OTAF using the smdpp message as an acknowledgment.

where the call requires extension or redirection from the initial attached OTAF to the desired or appropriate OTAF. In this scenario, the desired OTAF is not served by the initial CSC to which the subscriber's call was completed. Rather, the desired OTAF is served by another CSC, based on a call extension from the initial CSC (see Figure 15.12).

An alternative procedure is also defined for call redirection from an initial CSC to a desired CSC. Instead of the call's being *extended* from the initial CSC to the desired CSC, the call is redirected; that is, the call to the initial CSC is released and redirected from the serving system to the desired CSC. The primary benefit of this alternate technique is that call-processing resources used to establish the call to the initial CSC are freed for use and the subscriber's call to the proper CSC requires fewer call legs (see Figure 15.13).

Authentication Procedures for OTASP

After the serving MSC is *attached* to the OTAF using call origination or registration procedures, the A-key for subscriber authentication must be generated. This procedure is quite different from the typical method for entering an A-key into the mobile station (see Chapter 11). Typically the entire 26-digit A-key is entered manually via the MS keypad, preprogrammed internally by the MS manufacturer, or programmed through the data port of the MS equipment. Since OTASP must be a secure process, it makes no sense to send the full A-key as an over-the-air parameter to be programmed into the MS. This could lead to fraudulent access of the A-key and enable cloning fraud. Therefore, a new procedure has been developed to *generate* the 26-digit (i.e., 64-bit) A-key within the MS based on the Diffie-Hellman Key Agreement Standard procedure (refer to "Diffie-Hellman Key Agreement Standard" procedure within the "OTASP Automatic Roaming Functions for TDMA-based Networks" section of this chapter). Note that the CDMA-based network procedures for A-key generation are very similar to those for TDMA-based networks.

Initiate A-key Generation

The A-key generation procedures for CDMA-based ANSI-41 networks are very similar to the TDMA procedures. The primary differences are as follows:

- CDMA systems do not use a specific SMS application teleservice (i.e., OATS) to perform A-key generation.

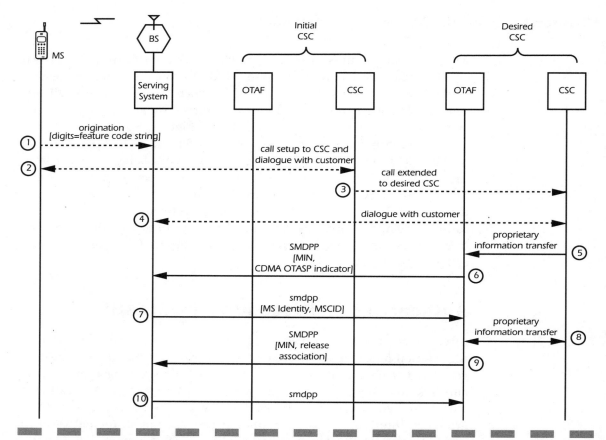

Figure 15.12 OTASP call origination from an initial CSC to a desired CSC. (1) The potential subscriber makes a call to the wireless service provider. The subscriber enters a feature code and presses SEND. (2) The call is established between the potential subscriber and a customer service representative at the initial CSC. At this point the serving system may perform normal validation and authentication of the subscriber, if the MS requires reprogramming rather than initial activation. A temporary reference number is assigned by the serving system and sent as part of the feature code or as the calling- or called-party number based on the signaling technique used. This temporary reference number is used to associate the OTASP processes and the CSC with the OTASP call. (3) Based on a dialogue with the user, the customer service representative at the initial CSC may determine that the desired OTAF is not associated with this CSC and extends the voice call, along with the temporary reference number, to the desired CSC. (4) A new customer service representative at the desired CSC begins a dialogue with the subscriber. (5) The desired CSC informs the desired OTAF of the initial contact with the MS using a proprietary protocol to prepare the OTAF for activation of the MS. (6) The OTAF launches an SMDPP message to the serving system. This message includes the special activation MIN of the MS, a service indication specifying that this is an initial OTASP activation, and the temporary reference number that the serving system uses to associate the OTASP processes and the CSC with the OTASP call from the potential subscriber. (7) The serving MSC responds back to the OTAF using the smdpp message. This message includes the mobile station's identity (MIN or IMSI) and the MSC ID of the serving MSC. (8) The OTAF and CSC exchange information via a proprietary protocol. The OTAF informs the CSC that the attachment with the serving MSC has occurred. The CSC informs the OTAF to release the association of the OTASP processes from the OTASP call. (9) The OTAF launches a second SMDPP message to the serving system. This message includes the activation MIN and a code directing the MSC to release the association of the OTASP processes from the OTASP call. (10) The serving MSC responds back to the OTAF using the smdpp message as an acknowledgment.

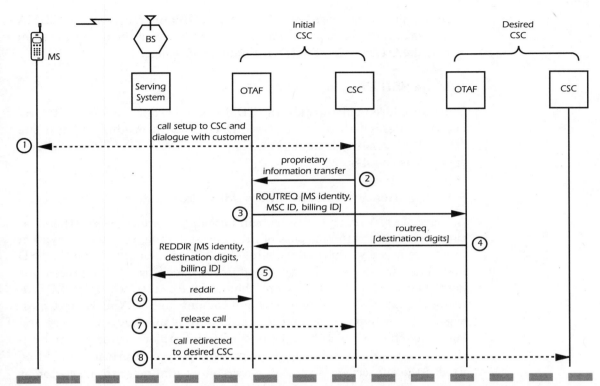

Figure 15.13 OTASP call redirection from an initial CSC to a desired CSC. (1) The subscriber has established the call to the initial CSC from the serving system. (2) Local address information is used to determine that the call should be attached to another CSC. This address information is transferred from the initial CSC to the initial OTAF using a proprietary protocol. (3) The initial OTAF launches a ROUTREQ message to the desired OTAF, to begin the call redirection process. This message contains the identity of the MS, the MSC ID, and the billing ID. The billing ID is used to inform the OTAF to which CSC the call should be completed. (4) The desired OTAF responds back to the initial OTAF using the routreq message containing the destination digits parameter. This parameter is used to redirect the call from the initial OTAF. This number will also serve as the temporary reference number once the call is redirected. (5) The initial OTAF launches a REDDIR message to the serving system containing the identity of the MS, the destination digits, and the billing ID. (6) The serving system responds back to the initial OTAF using the reddir message as an acknowledgment. (7) Based on the redirection request from the initial OTAF, the serving system releases the call to the initial CSC. (8) The call is redirected to the destination CSC.

- CDMA systems use the MS Key request/response and the Key Generation request/response parameters as SMS bearer data within the ShortMessageDeliveryPointToPoint message to pass the parameters required to perform A-key generation.
- In CDMA systems, the final acknowledgments indicating that the A-key was properly generated are sent between the AC and the OTAF

using the ShortMessageDeliveryPointToPoint message. In TDMA systems, these acknowledgments are sent between the serving system and the OTAF using the OTASPRequest message.

Initiate SSD Update

This procedure is identical to the TDMA-based OTASP SSD update procedures. Refer to the "Initiate SSD Update" procedure within the "OTASP Automatic Roaming Functions for TDMA-based Networks" section of this chapter.

Reauthentication to Send Encryption Parameters

As part of OTASP, the network can optionally perform a reauthentication procedure for the purpose of invoking signaling message encryption and voice privacy. This is done after the A-key is generated and the SSD is updated for the MS. These procedures are identical to the procedures described in Chapter 11, except that they are initiated by the OTAF using the ShortMessageDeliveryPointToPoint and OTASPRequest messages. The primary reason for this procedure is to encrypt the voice and signaling channels for the exchange of a subscriber's personal information to activate the subscription. For example, voice privacy should be used if the subscriber is providing a credit card number to the customer service representative at the CSC during an activation conversation. Alternatively, this process can be automated, since the subscriber may enter the credit card number via the keypad on the MS. This would require signaling message encryption to keep this information secure. Other subscriber information that should be kept secure includes the social security number, driver's license number, and home address information, if these are required to activate the subscription.

Committing the A-key

This procedure is identical to the TDMA-based OTASP commit A-key procedures. Refer to the "Committing the A-key" procedure within the "OTASP Automatic Roaming Functions for TDMA-based Networks" section of this chapter.

OTASP Parameter Transfer Functions for CDMA-based Networks

CDMA-based ANSI-41 networks do not specify a particular SMS application teleservice for OTASP. Instead, they use ShortMessageDelivery-PointToPoint message procedures to transfer the NAM activation parameters to the mobile station. This process is known as OTASP data exchange.

OTASP Data Exchange

This procedure is used to transfer the NAM programming and activation parameters to the mobile station to enable service. The procedure is performed subsequent to the OTASP authentication procedures to ensure security during the process. The NAM parameters are sent within the ShortMessageDeliveryPointToPoint (SMDPP) operation. The NAM parameters to be programmed are encapsulated within the SMS_BearerData parameter (using the OTASP data message information element) sent within the SMDPP message. The NAM parameters that can be programmed include the following:

- Generic configuration data (e.g., MIN and IMSI)
- Public encryption values (for A-key generation, voice privacy, and signaling message encryption)
- Key-generation results (for A-key generation procedure)
- Parameter download requests
- MS-specific requests
- NAM commit requests (to commit the NAM parameters to the MS database)
- CSC authentication challenge responses
- Preferred roaming list (PRL) downloads (i.e., SID lists)

Figure 15.14 shows the basic scenario for the exchange of OTASP data between the OTAF and the MS.

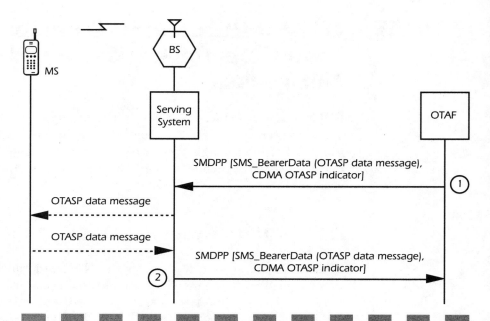

Figure 15.14 OTASP data exchange. (1) The OTAF launches the SMDPP message containing the SMS_BearerData parameter to the serving system. The SMS_BearerData parameter includes the OTASP data message information element. When the serving system receives this message, the OTASP data message containing NAM parameter information is sent over the air interface to the mobile station. (2) The serving system launches the SMDPP message containing the SMS_BearerData parameter to the OTAF. The SMS_BearerData parameter includes the OTASP data message information element. The OTASP data message containing NAM parameter information was received from the mobile station over the air interface.

Functions Following Successful CDMA OTASP

Two optional procedures can be invoked following successful OTASP of a CDMA subscriber: AC request to release resources and registration following OTASP.

The *AC request to release resources* procedure is used to release previously allocated resources at the authentication center, which were required for A-key generation and other authentication functions (see Figure 15.15).

The *registration following OTASP* procedure is used to test the registration function of a newly activated mobile station subsequent to the

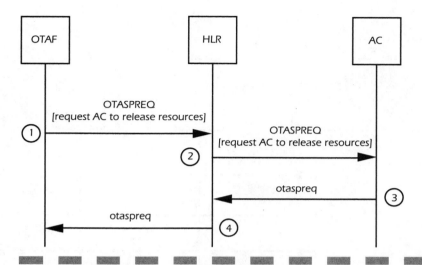

Figure 15.15 AC request to release resources. (1) The OTAF launches an OTASPREQ message to the HLR. The OTASPREQ message contains an action code parameter set to "release resources." (2) The HLR forwards the OTASPREQ message to the AC. (3) The AC releases any OTASP-associated resources and acknowledges this action by responding back to the HLR with the otaspreq message. (4) The HLR forwards the otaspreq message to the OTAF.

OTASP procedures. The mobile station's new MIN, IMSI, or both is sent to the serving system to be used for registration in the new home system associated with the MIN or IMSI (see Figure 15.16).

OTAPA Parameter Transfer Functions for CDMA-based Networks

Over-the-air parameter administration (OTAPA) transfer functions are used to modify MS NAM parameters at any time after initial activation, regardless of whether or not OTASP was used for activation. This function is used, for example, when a subscriber requires changes to the MIN or IMSI in the MS, or when the network requires updates to system IDs (SIDs). OTAPA requires no interaction by the subscriber. When OTAPA is performed, the subscriber is completely unaware of the changes being made to the parameters within the MS.

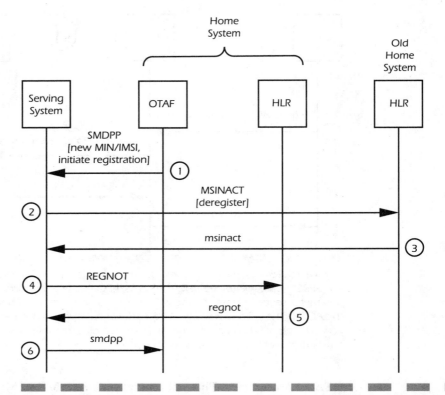

Figure 15.16 *Registration following OTASP. (1) The OTAF launches an SMDPP message to the serving system. This message contains the new MIN, IMSI, or both and an action code parameter set to "initiate registration." (2) The serving system launches an MSINACT message to the previously serving HLR to deregister the MS. (3) The previously serving HLR uses the msinact message to acknowledge the deregistration. (4) The serving system launches a REGNOT message to the new HLR to register the subscriber. (5) The new HLR uses the regnot message to acknowledge the registration. (6) The serving system sends the smdpp message to the OTAF to acknowledge successful registration of the MS.*

Registration for OTAPA

A mobile station (MS) must be registered with the network for OTAPA to occur. This is performed via standard registration functions using the RegistrationNotification message. However, the MS must be OTAPA capable; that is, the MS equipment must contain the software to accept programmable NAM parameters over the air. This capability is made known to the HLR via the Transaction Capabilities (TRANSCAP) parameter within the RegistrationNotification message sent from the serving system to the HLR. The OTAPA capability bit must be set within this parameter for OTAPA to occur.

OTAPA Data Exchange

Subsequent to registration, there are three scenarios defined for OTAPA data exchange with the MS:

- MS is available to receive OTAPA parameters.
- MS is unavailable to receive OTAPA parameters (OTAPA postponed at the HLR).
- MS becomes available after postponement at the HLR.

MS Is Available

This process occurs when the mobile station is registered and idle, or can occur while a call is in progress. There are basically six steps used for the procedure to download NAM parameters to the CDMA mobile station:

1. OTAF requests parameter download from the HLR.
2. OTAF starts the parameter download procedure (so that the serving system can assign a traffic channel for OTAPA).
3. MS performs a base-station authentication challenge (optional, if the NAM is in a protected mode).
4. OTAF downloads parameters to the MS.
5. OTAF commits the parameters.
6. OTAF stops the parameter download procedure (so that the serving system can release the traffic channel for OTAPA).

Figure 15.17 shows the OTAPA NAM parameter download to an available CDMA mobile station.

MS Is Unavailable and HLR Postponement

If the mobile station is unavailable to perform OTAPA when it is desired by the OTAF, the process is postponed at the HLR until the MS becomes available. The MS may be unavailable (not registered) due to being powered off, outside the coverage area, or otherwise inactive. Also, the MS may be in an insecure area (as determined by the HLR), thus preventing OTAPA from occurring at that location.

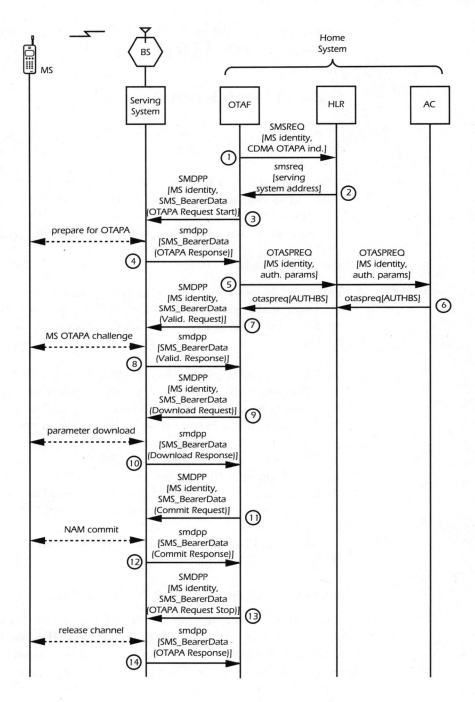

The OTAF starts the OTAPA process by sending the SMSRequest message to the HLR to determine the availability of the MS and address of the serving system (see Figure 15.17). If the MS is unavailable, or in an insecure location, the HLR responds back to the OTAF using the smsreq message containing an access-denied parameter indicating that the process is *postponed* until the MS is available. This causes an *OTAPA delivery pending flag* to be set at the HLR.

If the OTAPA process for a particular MS is marked as postponed at the HLR, the procedure can continue when the MS becomes available or moves into a location considered secure at some later time. When the MS registers and is considered active, the HLR sends an SMSNotification Invoke message to the OTAF, indicating that the MS is now available for OTAPA and providing the address of the MS's serving system. The OTAF acknowledges the message by sending the SMSNotification Return Result message to the OTAF. The OTAPA process then proceeds as specified in the previous section for an available MS (see Figure 15.17).

Figure 15.17 OTAPA data exchange with an available CDMA MS. (1) The OTAF launches an SMSREQ message to the HLR to determine the availability of the MS and address of the serving system. (2) The HLR responds to the OTAF using the smsreq message informing the OTAF of the OTAPA capabilities of the MS and the serving system. (3) The OTAF launches an SMDPP message to the serving system requesting the MS to start the OTAPA procedure. (4) After the serving system informs the MS to prepare for OTAPA, the serving system responds back to the OTAF using the smdpp message to begin a base station challenge. (5) The OTAF launches an OTASPREQ message to the AC (via the HLR), initiating the authentication process using authentication parameters received from the serving system. (6) The AC responds back to the OTAF (via the HLR) using the otaspreq message with the authentication calculation signature in the AUTHBS parameter. (7) The OTAF launches an SMDPP message to the serving system containing the authentication information in the Validation-Request parameter. (8) After the serving system successfully exchanges the authentication information with the MS, it responds back to the OTAF with a successful response to the OTAF using the smdpp message. (9) The OTAF launches an SMDPP message to the serving system to download the parameters to the MS. (10) After the serving system successfully exchanges the NAM parameters with the MS, it acknowledges the success using the smdpp message. (11) The OTAF launches an SMDPP message to the serving system to request the MS to commit the NAM parameters to its memory. (12) After the MS successfully commits the NAM parameters, the serving system acknowledges the success using the smdpp message. (13) The OTAF launches an SMDPP message using the serving system to complete (i.e., stop) the OTAPA procedure. (14) The serving system acknowledges completion of the OTAPA procedure using the smdpp message.

Summary of ANSI-41 Operations Used for OTASP and OTAPA

Table 15.1 summarizes the ANSI-41 operations used by the OTASP and OTAPA functions described in this chapter.

TABLE 15.1

Use of ANSI-41 Operations for OTASP and OTAPA

Function	ANSI-41 Operations Used for the Function
Call origination and registration (TDMA)	OriginationRequest, RegistrationNotification
Authentication (TDMA)	AuthenticationDirective, AuthenticationStatusReport, BaseStationChallenge, OTASPrequest, SMSDeliveryPointToPoint
OTASP parameter transfer (TDMA)	SMSNotification, SMSRequest, RegistrationNotification
OTAPA parameter transfer (TDMA)	SMSDeliveryPointToPoint, SMSNotification, SMSRequest
Call origination and registration (CDMA)	FeatureRequest, MSInactive, RedirectionDirective, RegistrationNotification, RouteRequest, SMSDeliveryPointToPoint
Authentication (CDMA)	AuthenticationDirective, AuthenticationStatusReport, BaseStationChallenge, OTASPrequest, SMSDeliveryPointToPoint
OTASP parameter transfer (CDMA)	SMSDeliveryPointToPoint
OTAPA parameter transfer (CDMA)	OTASPrequest, SMSDeliveryPointToPoint, SMSRequest

CHAPTER 16

Looking Ahead

ANSI-41 Revision D and its precursor, IS-41-C, represented a significant advance in the intersystem functionality available to roaming mobile telecommunications network subscribers—and more functionality has been added since 1997, which will eventually coalesce into ANSI-41 Revision E sometime in 2001 (it seems that moving between revisions of the standard is always an odyssey). The protocol provides ready tools for extending its capabilities, in both standard and nonstandard manners. Even as we write, the TIA standards committees are at work developing new capabilities that will drive the standard into the future.

Protocol Extension Mechanisms

ANSI-41 provides a number of mechanisms for extending the protocol. These mechanisms offer the potential for new operations, new parameters, and new parameter values. Furthermore, section 7 of ANSI-41.5 describes guidelines to be applied when the protocol is extended to maintain a level of backward and forward compatibility.

New Operations

Each of the 51 operations defined in ANSI-41-D (e.g., LocationRequest) is identified by a number known as the *operation specifier* (see Chapter 8). Operation specifier values are used as listed in Table 16.1.

As Table 16.1 shows, ANSI-41-D provides 32 operation specifier values for use in defining nonstandard protocol extensions. Individual vendors may use these values to identify new operations with the assurance that future revisions of the standard will not use these operation specifiers, thus forcing the vendor to modify its implementation to remain standard-compliant. These proprietary operations can be defined and implemented to provide differentiating capabilities for a vendor. Of course, protocol extension in this manner runs the risk of conflicts between two manufacturers' use of the same specifier.

New Parameters

Each of the 167 parameters defined in ANSI-41-D (e.g., MobileIdentificationNumber) is identified by a number known as the *parameter identi-*

TABLE 16.1

Use of Operation
Specifier Values in
ANSI-41-D

Operation Specifier Values	Use of Operation Specifiers
0	Not used
1 through 17	Used by operations in ANSI-41-D
18, 19	Reserved (previously used for ServiceProfile-Request and ServiceProfileDirective operations)
20	Used by operations in ANSI-41-D
21	Reserved (previously used for CallDataRequest operation)
22 through 40	Used by operations in ANSI-41-D
41	Reserved
42 through 55	Used by operations in ANSI-41-D
56 through 223	Reserved for future standard operations
224 through 255	Reserved for nonstandard protocol extensions

fier (see Chapter 8). Parameter identifier values are used as listed in Table 16.2.

TABLE 16.2

Use of Parameter
Identifier Values in
ANSI-41-D

Parameter Identifier Values	Use of Parameter Identifiers
0	Not used
1 through 26	Used by parameters in ANSI-41-D
27	Reserved
28 through 124	Used by parameters in ANSI-41-D
125 through 127	Reserved
128 through 169	Used by parameters in ANSI-41-D
170 through 16255	Reserved for future standard operations
16256 through 16383	Reserved for nonstandard protocol extensions

As Table 16.2 shows, ANSI-41-D provides 128 parameter identifier values for use in defining nonstandard protocol extensions. Individual vendors may use these values to identify new parameters with the assurance that future revisions of the standard will not use these parameter identifiers, thus forcing the vendor to modify its implementation to remain standard-compliant.

New Parameter Values

Many ANSI-41-D parameters have ranges of values (e.g., 224 to 255) explicitly reserved for private protocol extension purposes. For example, the values of the ANSI-41-D ActionCode parameter are listed in Table 16.3.

TABLE 16.3

Values of the
ANSI-41-D
ActionCode
Parameter

ActionCode Parameter Values	Meaning
0	Not used
1 through 7	Meaning is defined in ANSI-41-D
8 through 95	Reserved for future standard parameter values
96 through 127	Reserved for nonstandard parameter values
128 through 223	Reserved for future standard parameter values
224 through 255	Reserved for nonstandard parameter values

As Table 16.3 shows, ANSI-41-D provides the ranges from 96 to 127 and from 224 through 255 for use in defining nonstandard parameter values. Individual vendors may use these values to identify new actions associated with the ActionCode parameter with the assurance that future revisions of the standard will not use these parameter values, thus forcing the vendor to modify its implementation to remain standard-compliant.

Capabilities Planned for Future Versions of ANSI-41

As of the end of 2000, the current version of the ANSI-41 standard is Revision D; however, a number of interim standards and other documents have been developed since ANSI-41-D was published in 1997. The current TR-45.2 workplan calls for (at least) the following documents to be combined with ANSI-41-D to create the ANSI-41 Revision E standard planned for publication in 2001. We discuss most of these services and capabilities in this book.

- TSB-76 (PCS Multiband)
- IS-730 (DCCH features)
- IS-737 (TDMA Circuit Modes Data Services)
- IS-751 (IMSI)
- IS-735 (CDMA features)
- IS-756-A (Wireless Number Portability, WNP, Phase 1 and Phase II)
- IS-725-A (OTASP and OTAPA)
- IS-764 (CNAP/CNAR)
- J-STD-034 (Emergency Services)
- IS-771 (Wireless Intelligent Network)
- IS-778 (Authentication Enhancements)
- IS-807 (Internationalization of ANSI-41)
- IS-812 (ANSI-41 Message Segmentation)
- Miscellaneous Enhancements Document

Beyond ANSI-41-E, work continues (and, in many cases, has been completed) in the TR45.2 standards committee on enhancements to the standard that are planned to show up in Revision F of the ANSI standard (and probably future editions of this book). These capabilities include:

- Automatic Code Gapping (IS-786)
- Broadcast Teleservice Transport Capability (IS-824, published in 2000)
- WIN Phase II: Prepaid Charging (IS-826, published in 2000)
- Answer Hold (IS-837, published in 2000)
- User Selective Call Forwarding (IS-838, published in 2000)
- MDN Based Message Centers (formerly WNP Phase III, IS-841, published in 2000)
- Roamer Database Verification (IS-847)
- WIN Phase II (i.e., features other than Prepaid Charging, IS-848)
- IP Based Data Transfer Services (PN-4762)
- User Identity Module (IS-808, published in 2000)
- Emergency Services (J-STD-036, published in 2000)

The reader should refer to Chapter 5 for a discussion of the User Identity Module (UIM) and to Chapter 19 for a discussion of Emergency Services. What follows are descriptions extracted from the relevant stage 1 specifications.

Automatic Code Gapping

Automatic Code Gapping (ACG) is intended to provide a network entity, such as a service control point (SCP), the ability to throttle selected types of traffic from other network entities (such as MSCs), which may be passing through its domain of operation. The reason for ACG controls may be load related (e.g., from the SCP) or in response to a traffic engineering command from a Service Management System (SMS). The network entity can specify that ACG controls—which are generally of the form "wait for x seconds before sending another message, and keep this control on for y seconds"—be applied to query messages destined for a specific point code and subsystem number or for an SCCP global title address.

Broadcast Teleservice Transport Capability

The Broadcast Teleservice Transport Capability (BTTC) provides a method to deliver and manage broadcast SMS messages and other compatible teleservice messages.

WIN Phase II: Prepaid Charging

PPC allows the subscriber to pay for voice telecommunication services prior to usage. A PPC subscriber establishes an account with the service provider to access voice telecommunications services in home and roaming networks. Charges for voice telecommunication services are applied to the PPC account by debiting the account in real time. The PPC subscriber may be notified about the account information at the beginning, during, or at the end of the voice telecommunications service. When the account balance is low, the subscriber may be notified of the need to refill the account. When the account balance is below a predefined threshold, the subscriber's use of voice telecommunications services may be deauthorized.

PPC may be activated for all calls or on a single-call basis. For "All-Calls" activation, charges for all voice telecommunications services invoked are applied to the PPC account. For "Single-Call" activation, only charges for voice telecommunications services invoked in association with the call origination are applied to the PPC account.

Answer Hold

Answer Hold (AH) provides to a called subscriber the capability to answer the call, but selectively delay the conversation (e.g., for calls in the alerting or call-waiting state). The incoming call is provided an appropriate network announcement to request the calling party to hold. AH is also applicable to incoming calls being delivered to called AH authorized subscribers as call waiting (CW) calls. AH authorized subscribers, with AH active on an incoming call, must resolve the call in the AH state prior to originating a new outgoing call or call leg. New incoming calls to AH authorized subscribers, with AH active on an incoming call, receive the appropriate busy treatment (e.g., busy tone, redirected to voice mail, etc.) for the called AH subscriber.

User Selective Call Forwarding

User Selective Call Forwarding (USCF) provides a called subscriber the capability to selectively redirect unanswered, incoming calls, in the alerting or call waiting state, to an alternate destination; for example, to a voice-mail system, to a network registered USCF directory number (DN), or to a DN stored in the mobile station. USCF is applicable both to calls being offered to a subscriber via alerting and to calls being offered to a subscriber via call-waiting notification.

USCF subscribers may base a decision to invoke USCF on any consideration they deem appropriate (e.g., the time of day, their location, the activities in which they are engaged, etc.). USCF-called subscribers that are also authorized for calling party information presentation features (e.g., Calling Number Identification Presentation, Calling Name Presentation, etc.) may redirect incoming calls to different DNs based on the received calling party's identity. USCF subscribers are capable of redirecting incoming calls to an alternate destination DN independent of their authorization status for other redirection features; for example, Call Forwarding Busy (CFB), Call Forwarding Default (CFD), Call Forwarding No Answer (CFNA), etc.

MDN-based Message Centers

This standard identifies the ANSI-41-D technical enhancements required to support SMS delivery to Mobile Directory Number (MDN)-based Message Centers. The need for this capability arises with the introduction of the MDN parameter in ANSI-41-D and becomes pressing with the potential for Wireless Number Portability, when the subscriber's MIN is no longer the same as the MDN.

Roamer Database Verification

The objective of the Roamer Database Verification feature is to provide a secure method for a home carrier to verify that its MIN or IMSI ranges have been accurately provisioned in a roaming partner's VLR (see Chapter 17 for a discussion of roaming agreements). The focus is on verifying point code routing (i.e., that the proper HLR point code is provisioned for a given MSID range), with the response limited to *Yes* or *No*.

WIN Phase II: Other Features

Besides Prepaid Charging, WIN Phase II includes a number of features:

- Premium Rate Charging
- Freephone
- Rejection of Undesired Annoying Calls
- Advice of Charging

Premium Rate Charging (PRC) permits a wireless subscriber to be charged a premium rate for telecommunication services. The premium rate is different from the normal rate charged to the wireless subscriber; it may be higher or lower. The premium rate may be charged for the entire duration of the call or for any part thereof.

A number of criteria may be used to determine if a premium rate is to be applied. These criteria may be related to calling-party or to called-party characteristics. Examples of PRC criteria include:

- Calling MS directory number
- Called MS directory number
- Identity of the party calling the MS
- Digits dialed by the MS
- Time and date
- Location of the MS at the time of telecommunication service invocation
- Radio access characteristics such as priority access or type of bearer channel

Freephone (FPH) permits a wireless user to originate calls to a Freephone directory number and the called party to pay all charges associated with the call, including roaming charges, air-time charges, and interexchange charges. The calling party does not pay for any portion of the call.

Rejection of Undesired Annoying Calls (RUAC) is a call screening service that allows a subscriber to reject calls from parties whose Calling Party Numbers (CPNs) are in an Undesired Annoying Calls (UAC) screening list. Calls from CPNs in the UAC screening list are given call rejection treatment while RUAC is active, even if presentation of the calling party's CPN to the called party is restricted (CNIR). The called subscriber does not receive notification of a rejected call. The UAC screening list is a set of CPNs not permitted to terminate to the called subscriber. The subscriber can:

- Add a specific CPN to the list
- Add the "last calling party" CPN to the list
- Delete a specific CPN from the list
- Delete the most recent CPN added to the list
- Delete all the entries in the list.

Advice of Charging (AOC) permits a wireless subscriber to receive charging information for telecommunication services. AOC information is presented at the start of a call, or during a call, or at the end of a call. AOC information is conveyed to the subscriber within 5 seconds of the appropriate event (start of call, mid-call charging event, end of call).

AOC information may be presented using visual display, distinctive alerting, audible tone or announcement, or a combination of these.

AOC may convey the following types of information to the subscriber:

- Calling rate (display the charge per unit of time, e.g., $0.10 per minute)
- Change in calling rate (notify subscriber of a change in charge per unit of time, e.g., audible chirp)
- Accumulated usage or charge for the call just completed (e.g., display the last call usage = 4 minutes and 15 seconds, or last call charge = $0.40)
- Accumulated usage or charge for the billing period or calendar period (e.g., display the total billing period usage = 91 minutes, or total calendar period charge = $65.90)
- Remaining balance in an account (e.g., display the remaining balance = 33 minutes)

▪ Special charging is being applied (e.g., distinctive alerting for an incoming call).

IP-based Data Transfer Services

The IP-based Data Transfer Services standard will describe the enhancements to enable a wireless system to communicate with other ANSI-41 MAP signaling entities using IP-based transport. The document assumes that the introduction of IP-based transport into the SS7 ANSI-41 signaling network will be of a gradual, phased nature. Therefore, the standard intends to specify the means by which IP-based signaling nodes can interwork with SS7-based signaling nodes, and vice versa.

Of course, adding IP-based data transfer services to ANSI-41 is perhaps the tip of the iceberg with respect to the Internet's eventual impact on ANSI-41 and wireless networks in general. But that's the subject of another book.

PART 3

ANSI-41 Network Implementations

CHAPTER 17

ANSI-41 Interoperation with Other Networks

ANSI-41 wireless networks interoperate with each other to provide nationwide roaming services to wireless subscribers. They also interoperate with wire line networks to support calls between wireless and wire-line parties. These calls can be traditional voice calls or data calls that can convey circuit-switched or packet-switched data. ANSI-41 network interoperation involves the interconnection of trunks and signaling links between the networks as well as business agreements between the service providers. This chapter describes the interconnection of ANSI-41 networks with other wireless and wire-line networks.

Roaming Agreements

A roaming agreement is a legal and business contract entered into by two wireless network service providers. This contract defines the methods by which one wireless network provides service to subscribers from the other wireless network. It also defines the tariffs and the methods for billing roaming subscribers. Roaming agreements are also the basis for proper routing of signaling information between the two networks. Based on these agreements, provisionable tables in the network equipment are populated with information that enables subscribers to register and receive services from a visited system (see Figure 17.1).

Figure 17.1
The ability to route signaling information and provide mobility management functions (e.g., registration, location management, service qualification) is based on a roaming agreement. An MS from wireless network Y can only receive seamless roaming service from wireless network X if a roaming agreement exists between the two networks.

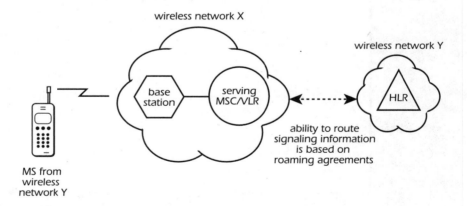

Roaming agreements permit the ANSI-41 signaling communications between wireless networks that provide the mobility management and call processing functions. Without these agreements, a subscriber may be prohibited from registering on a visited system and receiving any wireless service. Some networks support *indirect* roaming. This type of roaming occurs when there is no explicit roaming agreement between the visited system and the subscriber's home system. The subscriber can still be afforded service, but it may not be seamless. When a subscriber attempts to register on a visited system, that system may not be able to route the registration signaling information to the subscriber's home network (i.e., the HLR), since there is no roaming agreement. In this case, the visited system can perform one of three actions:

■ Deny service to the visiting subscriber.
■ Provide service to the visiting subscriber by obtaining credit information directly from the subscriber.
■ Route the signaling information about the visiting subscriber to a clearinghouse network, which may have a roaming agreement with the subscriber's home system.

The most undesirable action by the visited system is to deny service to the subscriber. The subscriber suffers and there is no revenue generated for either the home or visited network.

When a roaming subscriber attempts call origination and the visited system does not recognize the subscriber's MIN, it can direct the subscriber to a customer service center. This customer service center typically requests credit information from the subscriber so that calls can be originated and billed directly to the subscriber (see Figure 17.2). In this case, the location management and service qualification processes are not performed, and mobile-terminated calls cannot be delivered to the subscriber.

Another option is based on an agreement the visited system has with a *clearinghouse network* (see Figure 17.3). Signaling information for all unrecognized mobile subscribers is routed to the clearinghouse network. If the clearinghouse network has an agreement with the subscriber's home system, it can support the mobility management functions on behalf of the visited system. If the clearinghouse network does not have an agreement with the home system, it can obtain credit information directly from the subscriber.

Figure 17.2
If no roaming agreement exists, the serving system can request credit information from the subscriber and allow the subscriber to originate calls; however, mobile-terminated calls will not be delivered to the subscriber.

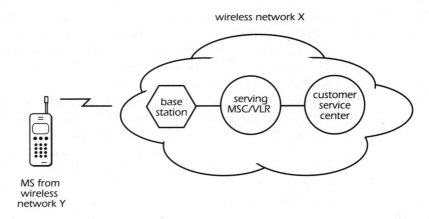

Figure 17.3
A clearinghouse network can provide roaming capabilities to an MS visiting a system that has no roaming agreement with the MS's home service provider system.

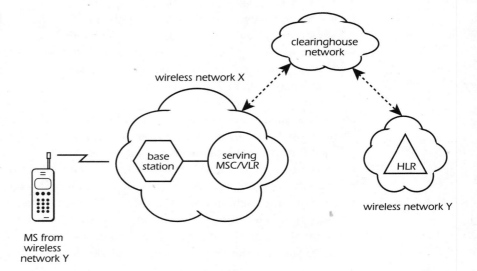

Roaming agreements define the interactions permitted between wireless systems. Examples of these interactions are:

- Methods for routing signaling information about roaming subscribers.
- Methods for the networks to settle accounting and billing information.
- Methods for recording call details to support accounting and billing.
- Methods for handling toll calls (i.e., use of interexchange carriers for calls).
- Methods for handling fraud (i.e., responsibility for and methods to avoid lost revenue due to roamer fraud).

- Methods for updating routing tables.
- Methods for sharing statistical data about roaming subscribers.

Roaming agreements also require that technical information be shared about the wireless networks involved. Examples of this information are:

- MSC NPA-NXX codes supported
- MSC identification codes supported
- SS7 point codes
- Switch makes and models
- Software generic versions
- Feature code values supported

Roaming agreements provide the basis for service providers to offer nationwide wireless service to subscribers. Without these agreements, service providers would need to operate their own wireless networks in every area to provide nationwide coverage for subscribers.

Network Interconnections

Whenever two telecommunications service providers connect to each other to pass signaling information, deliver calls, or do both, they enter into an interconnection agreement. The *interconnection agreement* defines methods used to provide communications between the two networks.

Wireless networks can interconnect with a variety of other networks including the following:

- Public switched telephone networks (PSTNs)
- Other wireless cellular networks
- Other wireless personal communications services (PCS) networks
- Public switched packet data networks (PPDN)
- Signaling System No. 7 (SS7) *backbone* networks

Wireless networks directly interconnect with the PSTN to support wire line-to-wireless and wireless-to-wire line calls. The PSTN may also be used to support wireless-to-wireless calls. The PSTN comprises local exchange carriers (LECs), interexchange carriers (IXCs), and other wire-line carrier networks. Some wireless networks directly interconnect with other wireless networks to support wireless-to-wireless calls. These interconnections are sometimes used in highly populated areas where there is a potential for many wireless-to-wireless calls between different wireless service providers serving the same area. Wireless cellular networks will also interconnect with wireless PCS networks. From a network perspective, there is no difference between PCS and cellular network operation if they're both based on ANSI-41 signaling. Wireless network interconnection with a PPDN is also possible if the wireless network supports packet data services.

Most wireless networks use SS7 to provide call control and ANSI-41 signaling. These wireless networks directly interconnect with an SS7 backbone network, which can be owned and operated by the wireless service provider itself or can be provided by a separate independent SS7 network provider.

The interconnection between an individual wireless network and other networks is primarily defined by three criteria:

- Physical communications facilities
- Communications protocol supported across the physical facilities
- The type of information transferred by the communications protocol

The physical communications facilities consist of the lines, trunks, and data links directly provided between the network nodes of the different service providers. The communications protocols supported range from the physical protocol layer through the application protocol layer. For example, a trunk interconnection may support T1 transmission, SS7, and ANSI-41 MAP signaling. The type of information transferred across an interconnection is limited by the protocol supported. The protocol, however, may support optional information that is provided only under agreement between the two networks.

A_i–D_i Interfaces

The interconnections between a wireless service provider and another network are specified in the standard *ANSI/TIA/EIA-93 Cellular Radio Telecommunications A_i–D_i Interfaces Standard*. This standard provides signaling protocol requirements for interfaces that interconnect a switching system in a wireless network with a switching system in another network. The A_i and D_i interfaces represent the interfaces from the MSC to the PSTN and ISDN, respectively, in the ANSI-41 network reference model. For all intents and purposes, these two interfaces are the same. The A and the D in the interface names stand for *analog* and *digital*. The distinction between these two interfaces has not been specified and is left over from an anticipated need to differentiate them. Since the PSTN supports both analog and digital protocols, there is no need to distinguish between interfaces to an analog or digital network.

The interconnection types specified in *ANSI/TIA/EIA-93* define generic interfaces that support a number of individual signaling protocols with a variety of signaling methods to provide telecommunications services. The standard specifies two subtle characteristics of the wireless-to-PSTN interface that distinguish it from other PSTN interface specifications:

- The interfaces are bidirectional and symmetric.
- The interfaces support the transport of automatic number identification (ANI) information into the wireless network.

Bidirectional and symmetrical interfaces mean that a given interface can support the same address and signaling information in both directions (i.e., either entering or exiting the wireless network). The transport of ANI information into the wireless network is an important business issue concerning network interconnections. ANI represents charge number information that can be received with a call from the originating or a mediating network. This information, when available to a terminating network, allows for control over billing and the ability to provide additional services that can be charged for by the terminating network. Note that the ANI information may or may not be the same as the calling party number. Typically, the calling party number represents the number to charge a call to; however, this may not be the case (e.g., calls from businesses with multiple calling party numbers, but one billing number). It should never be assumed that the ANI is always the same as the calling party number or vice versa.

Prior to the publication of *ANSI/TIA/EIA-93*, interconnections to the PSTN were specified by Bellcore (now Telcordia). In these specifications, ANI (or charge number) information was never included in the signaling information provided to the wireless networks for mobile-terminating calls. The reason for this is historical and to the advantage of the PSTN and, especially, the local exchange carriers that delivered calls to the wireless network. Wireless networks were traditionally treated as extensions to the PSTN, much as PBXs are treated.

The wireless service providers originally depended on the PSTN to carry wireless-to-wire line calls, which are still the most common type. This dependency came from the fact that wireless service providers needed only to provide MSC capabilities in their service areas and then use the existing nationwide switching infrastructure of the PSTN to carry the calls. Because of this dependency, wireless networks were considered subordinate to the PSTN and, hence, were never given charge number information that would allow them control over services and billing arrangements. The primary purpose of *ANSI / TIA / EIA-93* was to provide wireless service providers with a nationally standardized interface specification that supported the transport of charge number information into their networks. This important change enabled the wireless networks to gain a certain degree of "economic equality" with the PSTN, and to negotiate interconnection agreements favorable to both networks.

ANSI / TIA / EIA-93 specifies the logical interface reference model used for wireless-to-PSTN interconnections (see Figure 17.4). This model represents a generic interconnection between a wireless network element and any other network element. This enables the application of the interface types between a variety of network elements. The trunk point of interface (POI-T) is designed to convey trunk user traffic and in-band multifrequency (MF) signaling. The SS7 point of interface (POI-S) is used in conjunction with POI-T to represent out-of-band SS7 signaling used on the interface.

Figure 17.4
Generic interface reference model for wireless interconnection to the PSTN. The trunk point of interface (POI-T) specifies trunk interconnections to convey user traffic, while the SS7 point of interface (POI-S) specifies SS7 interconnections.

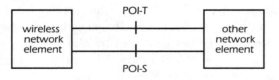

The interface types specified in *ANSI / TIA / EIA-93* are distinguished by the telecommunications protocol supported and the information conveyed with that protocol. The information defines the application of the interface type. For example, some interfaces support general trunk access for interexchange carrier calls, and others support emergency services calls. These interconnections require different information to be transferred to provide the appropriate service.

Of the interface types specified in *ANSI/TIA/EIA-93*, there are primarily only three in practice:

- Trunk with line treatment
- General trunk access
- Direct trunk access

These three interface types are the same as the original interface types specified by Telcordia (*Compatibility Information for Interconnection of a Wireless Services Provider and a Local Exchange Carrier Network, GR-145*), with one difference: They are bidirectional and symmetric. The other interface types specified in *ANSI/TIA/EIA-93* were added to provide various forms of flexibility to the wireless service providers while allowing them to negotiate and charge a tariff for each type of information terminated in their networks. The three primary interface types are really supersets of all the information required over a wireless-to-PSTN interconnection. Table 17.1 shows the mapping of these three interface types to the Telcordia-specified interface types.

The trunk with line-treatment interface type uses MF signaling to establish a connection to an end office to access valid directory number addresses, which are directly connected to that end office. The primary signaling information included in the address signaling sequence to obtain access to network addresses is the called party number. Charging information is not exchanged over this interface type (see Figure 17.5).

The general trunk access interface type uses MF or SS7 ISUP signaling to establish a connection to any common carrier switch to access any valid network endpoint accessible by that common carrier switch. The primary signaling information included in the address signaling sequence to obtain access to network addresses is the called-party number, charge number, originating line information, and a carrier identification code (see Figure 17.6).

TABLE 17.1

Mapping of ANSI/TIA/EIA-93 Interface Types to the Telcordia (GR-145) Defined Interface Types

Interface Name	ANSI/TIA/EIA-93 Type(s)	Telcordia (GR-145) Type
Trunk with line treatment	POI-T1	Type 1
General trunk access	POI-T4, POI-T5 & POI-S5	Type 2A
Direct trunk access	POI-T6, POI-T7 & POI-S7	Type 2B

Figure 17.5
The trunk with line treatment interface type. The interconnection allows access to directory numbers directly accessible by the interconnected switches only.

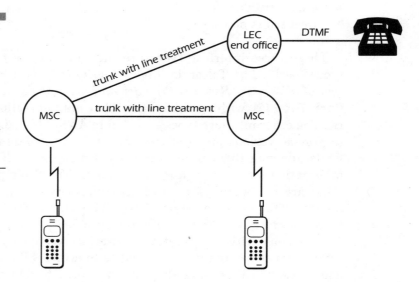

Figure 17.6
The general trunk access interface type. The interconnection allows access to any endpoint accessible by the connected switches.

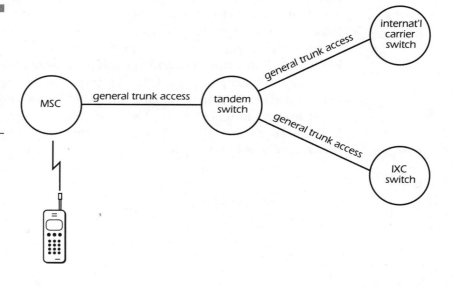

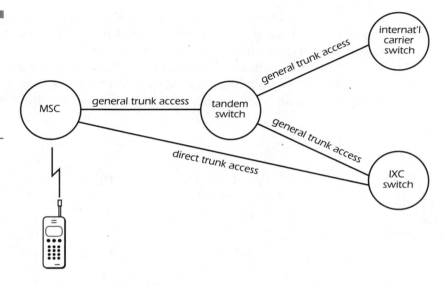

Figure 17.7
The direct trunk access interface type. The interconnection allows direct access to directory numbers served by the connected switch.

The direct trunk access interface type uses MF or SS7 ISUP signaling to establish a direct connection to any common carrier switch to access any valid directory numbers accessible by that common carrier switch. The directory number address space is restricted to local, national, and international directory numbers. This interface type is designed for high-volume traffic routes in conjunction with the general trunk access interface type only. For example, MSCs that carry a large volume of traffic to a single IXC may add this type of trunk specifically for the purpose of accessing directory numbers accessible by that IXC only. The primary signaling information included in the address signaling sequence to obtain access to network addresses is the called-party number (see Figure 17.7).

SS7 Network Interfaces

Although some older North American ANSI-41 networks may use the X.25 protocol for signaling message transport, the majority use SS7. SS7 is not just a signaling and transport protocol; SS7 connectivity defines an architectural strategy to provide a robust, highly reliable, and highly available signaling network.

Basic SS7 Network Configuration

The basic North American SS7 network configuration is shown in Figure 17.8. The end signaling points (SPs) represent the signaling capabilities of switches that can originate or terminate signaling traffic. The signaling transfer points (STPs) route signaling traffic through the network by transferring that traffic from one signaling link to another to direct it toward a particular destination. This configuration is known as *quasi-associated signaling*, since the signaling relationship between the end SPs is not a direct association (i.e., there are no direct links between end SPs).

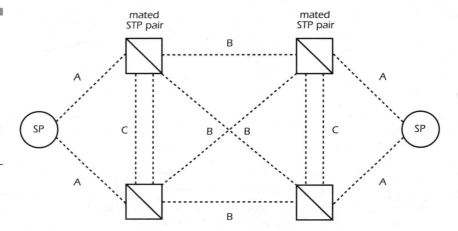

Figure 17.8
Basic SS7 configuration for North America. The architecture uses quasi-associated signaling to create a signaling relationship between end signaling points.

The STPs are deployed as a load-sharing mated pair, with each STP carrying half the traffic load. In case of a failure, the remaining STP is capable of carrying the full traffic load. An STP can be implemented as a standalone system or be integrated with a switch.

End signaling points are connected to the STP mated pair by a pair of *A links*. Each of these A links is a member of a *link set*. The pair of link sets containing the A links is known as a *combined link set*. The signaling traffic load is shared equally among the links of the combined link set. Mated pairs of STPs are connected by C links. Connections between STP pairs of different SS7 networks (i.e., gateway STPs) are made with *B links*. There is functionally no difference between the link types. They simply define different types of traffic that can traverse them since they connect different types of SS7 nodes. Some SS7 networks employ *D links*, which can define multiple levels of STPs, for example, local and regional STP mated pairs (see Figure 17.9). *E links* are used to avoid massive failures or outages of STP mated pairs. They connect an end signaling point to a remote STP pair (in addition to the local pair), which can be within the same SS7 network or in a different network (see Figure 17.10).

Figure 17.9
Local STP mated pairs can connect with regional STP mated pairs to create a hierarchy for the SS7 network topology.

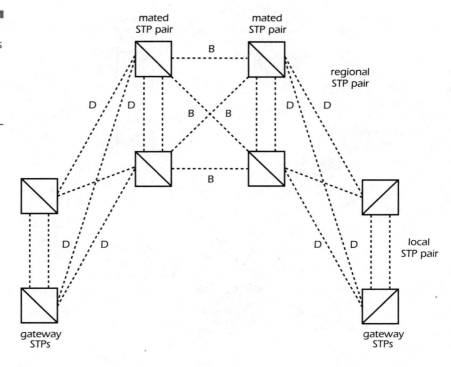

Figure 17.10
End SPs can connect with more than one STP mated pair. The remote pair is used to avoid failures of the local STP pair.

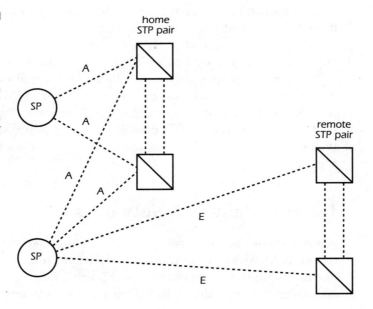

Figure 17.11
SCPs are usually connected to regional STP pairs for concentration of signaling traffic. The SCPs are connected by A links since they are considered to be end signaling points.

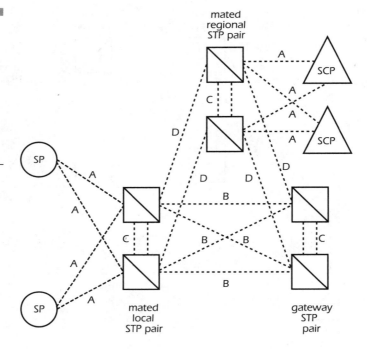

Service control points (SCPs) are special types of end signaling points that use the SS7 TCAP protocol to perform transaction processing of remote operations. In many cases, the SCP supports a database to perform the required operations. SCPs are typically connected to an STP mated pair by a pair of A links (see Figure 17.11). SCPs can also be deployed as a redundant pair for higher reliability.

MSCs connect with SS7 networks as end signaling points. An MSC may be part of the wireless service provider's own SS7 network, or may connect to another service provider's SS7 network. Typically MSCs connect to a local gateway STP mated pair using A links. Analog or digital trunks connect the MSC to common carrier switches for call traffic (see Figure 17.12).

Independent SS7 Networks

Many wireless service providers connect to independent SS7 networks to carry ANSI-41 signaling traffic, rather than operating their own backbone signaling networks (see Figure 17.13). The largest wireless service providers in the United States do own and operate their own SS7

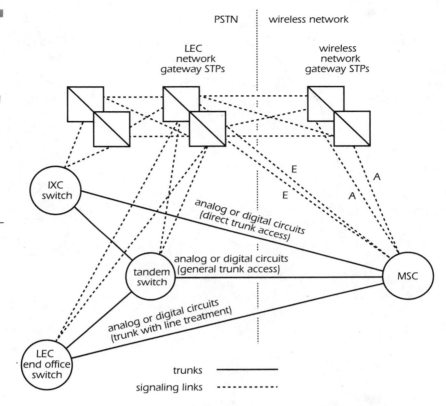

Figure 17.12
MSCs connect to the SS7 network as end SPs. The wireless SS7 backbone network connects to the PSTN SS7 backbone network through gateway STPs. Note the use of the ANSI/TIA/EIA-93 interface types for trunk connections to the PSTN.

networks, but this may not be advantageous for smaller wireless service providers. The independent SS7 network service providers coordinate and manage the SS7 network as well, allowing the wireless service provider to concentrate on managing the wireless infrastructure.

Wireless service providers can interconnect their MSCs directly to gateway STPs of the independent SS7 network (see Figure 17.13). The wireless service providers may also operate part of an SS7 network themselves and interconnect their local STP mated pair to the gateway STPs of the independent network (see Figure 17.14). This enables wireless service providers to access signaling destination points that may be inaccessible through their own networks, such as international signaling points.

In July, 1993, the Cellular Telecommunications & Internet Association (CTIA) named Independent Telecommunications Network, Inc. (ITN, now known as Illuminet) the official CTIA SS7/ANSI-41 backbone signaling network provider for wireless service providers. Of course, there is no requirement to use Illuminet's SS7 network, but it is among the largest SS7 network service providers in North America.

Figure 17.13
Wireless network
MSCs can connect to
an independent SS7
network directly via A
links to the gateway
STP.

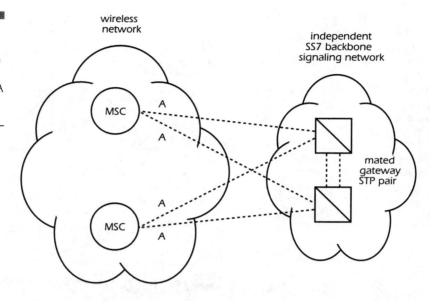

Figure 17.14
Wireless network
MSCs can connect to
an independent SS7
network via B links
between two sets of
gateway STPs, one
set operated by the
wireless network and
the other set
operated by the
independent SS7
network.

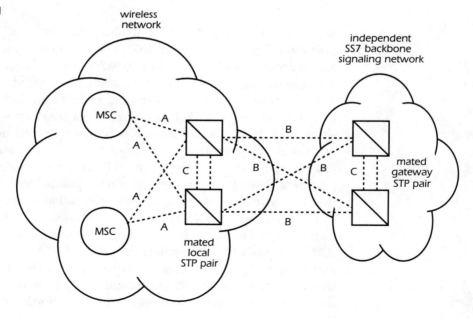

Wireless Resellers

Wireless resellers are wireless service providers that do not operate their own networks. Resellers make arrangements to prepurchase a percentage of the available capacity of a service provider's facilities in bulk and at a discount. These facilities generally include radio channels and switching systems, but they can also include any location registers and processing centers in the service provider's network. The reseller provides service under a different brand name using an infrastructure identical to the service provider's.

The reseller obtains a block of mobile identification numbers (MINs) from the wireless service provider. These are the numbers that the reseller provides to subscribers of the reseller service. Since the MINs fall within the range already supported by the original service provider, no routing changes are required to provide service to subscribers. The reseller's subscribers may also be supported by the service provider's HLR. The HLR can be partitioned to support the block of the reseller's subscribers. Alternative architectures may include a separate HLR operated by the reseller, but this would affect routing in the network and would require additional roaming agreements beyond those supported by the original service provider. From a network perspective, wireless resellers are typically distinguishable from the original service provider only by different customer service centers and different billing systems.

Wireless reselling is typically very advantageous to the service provider due to guaranteed revenue when the facilities may not otherwise have been in use. It is also advantageous to the reseller, which is afforded the ability to provide wireless service and reap profits without owning and operating a network. The reseller may even compete for subscribers with the service provider providing the facilities, although the agreement between the companies usually prohibits the direct luring of subscribers from each other's service.

International Roaming

With the explosive growth of the global wireless market it is very desirable to provide seamless automatic roaming internationally. International roaming, from a North American perspective, includes the ability of either a North American subscriber to roam seamlessly to other countries or subscribers from other countries to roam seamlessly in North

America. From a wireless network perspective, North America consists generally of the United States and Canada, since these countries both use common administration of numbering and addressing. International roaming in this context refers to roaming between ANSI-41 networks in different countries.

Numbering Problems

Mobile stations in North America are always identified by a 10-digit mobile identification number. The MIN happens to also be used as the 10-digit subscriber *directory number*, which follows the North American Numbering Plan (NANP). The NANP is a subset of the international ISDN numbering plan standardized in *ITU-T Recommendation E.164*. The international E.164 numbering plan includes a country code (CC), a national destination code (NDC), and a subscriber number (SN). Together the NDC and the SN comprise the national significant number (NSN). The 10-digit MIN complies with the international NSN format of E.164 (see Figure 17.15).

Figure 17.15
The 10-digit MIN complies with the format of the national significant number field of the E.164 international ISDN numbering plan.

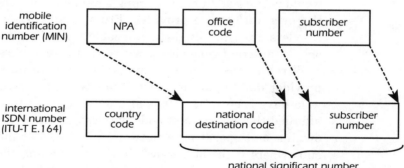

The primary problem arises because the 10-digit MIN follows only a national numbering format and contains no country code. It is impossible to distinguish some 10-digit MINs used in North American ANSI-41 networks from 10-digit MINs that include a country code used in other countries. Therefore, it becomes difficult to register an international subscriber in North American networks; the country code of the MIN could be construed as part of the numbering plan area (NPA) or area code and the wrong HLR could be queried for MS service qualification. Similarly, a North American subscriber roaming internationally would have the first few digits of the MIN construed as a country code and thus registration would also be attempted at the wrong HLR.

An excellent example of this problem is with Mexico. The country code in Mexico is 52. However, a valid area code in Arizona is 520. A Mexican roamer in Arizona would attempt a call and be denied, since service qualification would be attempted in Arizona instead of the sub-scriber's HLR in Mexico. In the short term, Mexico has agreed not to assign any MIN values beginning with 520; however, this problem exists for the country code values of dozens of countries.

This problem has been slightly alleviated by *TIA/EIA TSB29*, which defines a numbering plan for mobile stations outside of world zone 1, of which the United States and Canada are members. World zones are speci-fied by *ITU-T E.164*. They categorize each of the countries of the world into one of nine zones. The zones cover large contiguous geographic areas.

TIA/EIA TSB29 specifies that non–world zone 1 mobile stations be programmed with MINs formatted like 10-digit international mobile station identification (IMSI) numbers as specified in *ITU-T Recommen-dation E.212. TIA/EIA TSB29* specifies that the first three digits of the MIN represent a mobile country code, the fourth digit represent a mobile network code (specified as either 0 or 1), and the last six digits represent the subscriber number. Since the fourth digit is 0 or 1 (which is not allowed in the NANP), the MIN can have a unique address space distinct from all other NANP-based MIN values (see Figure 17.16).

Figure 17.16
TSB29 specifies a unique format for the MIN that can be recognized as international for roaming purposes.

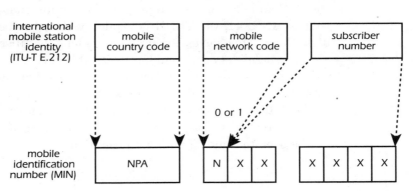

The best solution to the international numbering problem is to expand the length of the MIN to at least 12 digits to accommodate a country code value. This would enable true seamless automatic roaming using existing internationally standardized number formats. The prob-lem is that this requires major changes throughout the wireless networks and there are currently over 150 million existing subscribers worldwide with mobile stations containing 10-digit MINs.

Another solution that has been proposed is known as HLR *double dipping*. When an MS registers with a MIN that may be ambiguous, sequential queries are made from the serving system to the two HLRs that maintain the ambiguous MIN. This solution has drawbacks, though, since extraneous signaling traffic and delayed call setup times are never desirable.

International Call Delivery Problems

Another problem with international roaming involves the call delivery procedures used to deliver mobile-terminated calls to roaming subscribers. In IS-41-A and IS-41-B systems, the temporary local directory number (TLDN) used to redirect a call from an originating MSC to a serving MSC is commonly implemented as a 10-digit number (IS-41-A and IS-41-B do not specify minimum and maximum lengths for the TLDN digit string). This poses a problem when calls need to be redirected from the subscriber's home system to another country where the serving system resides. A 10-digit TLDN may not clearly identify the country for the call to be delivered to. Another problem for the originating MSC is knowing when to add an international dialing prefix code to the TLDN to deliver the call properly via an international gateway. Note that in ANSI-41, the TLDN is explicitly stated to be between 0 and 15 digits long. This solves the problem as long as all serving systems providing call delivery are compliant with this revision.

The best long-term solution to this and other numbering problems involving international roaming is the implementation and deployment of the IMSI. This value decouples the subscriber's dialable directory number from a unique internationally used identification number that can be used to route signaling information.

ANSI-41/GSM Interworking

With the growth of personal communications service (PCS) systems in North America, competing networking and radio technologies for systems based on wireless technology have emerged. In North America, the primary wireless telecommunications networking technology is ANSI-

41, supporting AMPS, TDMA, and CDMA radio technologies. The advent of PCS has provided the opportunity for the North American standardization of the Global System for Mobile (GSM) communications, an alternative international wireless technology standard. GSM comprises both networking and radio technologies not directly compatible with ANSI-41 and its associated radio technologies.

GSM was originally standardized in Europe by the European Telecommunications Standards Institute (ETSI). The following versions of GSM have been standardized:

- GSM 900 (GSM at 900 MHz, international)
- GSM 1800 (GSM at 1800 MHz, international)
- GSM 1900 (GSM at 1900 MHz, network and radio technology for North America)

For North American PCS, the GSM 1900 network and radio technology has been selected for use by many PCS service providers. However, ANSI-41 technology, along with its associated radio technologies, has also been selected for use by many other PCS service providers. ANSI-41 networking for PCS has the advantage of being fully compatible with existing North American wireless cellular networks providing nationwide wireless coverage. The GSM application layer MAP protocol to provide mobility management, radio system management, and call processing is incompatible with the ANSI-41 MAP protocol providing those functions. The radio technologies are also quite different; although GSM is a type of time division multiple access, it is incompatible with North American TDMA.

PCS service providers deploying GSM technology currently have limited geographic coverage of their systems because of the time and cost of deploying equipment nationwide. Because of this lack of nationwide coverage, GSM service providers are finding it desirable to interwork with ANSI-41 networks to provide full North American coverage for their subscribers (see Figures 17.17 and 17.18).

Introduction to Interworking

There are two separate areas of system interworking that apply to GSM and ANSI-41 interoperation: radio technology interworking and network interworking. *Radio technology interworking* refers to the capability of a mobile station (MS) to interoperate between GSM radio technology and

Figure 17.17
An interworking
scenario where a
GSM MS roams into
an ANSI-41 network
service area.
Interworking is
required to support
the intersystem
signaling operations
between the two
networks to support
MS service
qualification, location
management, and
many other
functions.

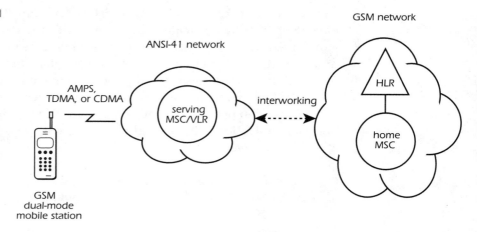

Figure 17.18
An interworking
scenario where an
ANSI-41 MS roams
into a GSM network
service area.
Interworking
analogous to that
shown in Figure
17.17 is required.

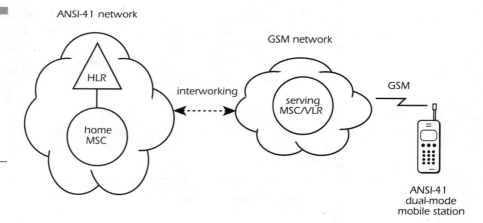

one or more of the North American radio technologies (i.e., AMPS, TDMA, or CDMA) to access either a GSM-based or ANSI-41–based network to receive service. This interworking is not feasible; however, a reasonable solution is a dual-mode MS similar to dual-mode AMPS/TDMA mobile stations used today.

The term *dual-mode* implies a "split personality" MS that would be capable of operating in either of two modes according to the type of network access required. A dual-mode GSM/AMPS MS would enable a single MS to access GSM networks in those service areas and ANSI-41 networks where GSM is not available. In fact, even multimode MSs have been studied to provide access for GSM, AMPS, and either TDMA or CDMA. Multiple radio technologies can coexist (if not interwork), provided that the network technology adopted can simultaneously support the many network access technologies.

Network interworking introduces an entirely different set of problems to be solved. If two networking technologies are highly incompatible, it is difficult for a network service provider to offer the primary service needed for its subscribers: seamless automatic roaming.

It seems obvious that the best solution to incompatible networking protocols is to require no interworking at all! A single MAP protocol supporting the many network access protocols (i.e., radio technologies) is the optimal solution to seamless automatic roaming. However, business issues often outweigh the optimal engineering solutions, and these solutions are not always the most desirable in a fast-moving, high-growth industry like wireless telecommunications.

There are many examples of network protocol interworking, such as TCP/IP-to-X.25 and TCP/IP-to-SS7. The problem in wireless networks is the complexity of the interworking; it is not simply the mapping of data packets and addresses to provide transport, although this is part of it. The primary problem is the interworking of the application-layer MAP protocols, along with the handling of the mobility management, radio system management, and call processing functions.

The Interworking Problem

The best method for approaching the interworking problem is to break it down into the following three steps:

1. Define interworking.
2. Identify the functions requiring interworking.
3. Design the methods to provide interworking for these functions.

Definition of Interworking

The initial step to providing interworking between GSM-based and ANSI-41–based systems is to define interworking and the types of interworking to be supported. There is no standard definition for the term *interworking*. Instead of attempting to define such an all-encompassing term, defining what an interworking solution needs to provide for this specific problem is more appropriate.

We will use the following assumptions to address the MAP interworking problem:

1. *Interworking implies the successful communication between GSM and ANSI-41 MAP protocols.* A more-specific definition might limit a feasible solution to be called interworking.
2. *Interworking provides a degree of seamlessness to subscribers capable of transiting between the GSM and ANSI-41 networks.* Subscribers capable of roaming between the two networks should be able to access basic service seamlessly (i.e., originate and receive calls as usual). Since many call features are not compatible, their use may not be supported while roaming.
3. *It is not feasible to provide an interworking solution enabling full compatibility between GSM and ANSI-41 networks.* Since ANSI-41 and GSM networks are so different, it is not reasonable to provide interworking of all the functions of both networks to each other.
4. *Transmission facilities between GSM and ANSI-41 networks are compatible.* This is a reasonable assumption based on North American transmission standards.
5. *Signaling protocols providing transport between GSM and ANSI-41 networks are compatible.* This is a reasonable assumption based on North American signaling standards.
6. *Dual-mode GSM/AMPS mobile stations are used to access GSM and ANSI-41 networks.* Without a dual-mode MS, it is not feasible to provide access to both network types with a single compatible interface.
7. *The subscriber has a single subscription allowing access to both networks (rather than having two subscriptions, one to each network, which requires no interworking).* From a subscriber perspective, it is unreasonable to support two different subscriptions from two different networks.

These assumptions simplify the basic problem and permit a basic solution to be implemented, which can evolve so that these limitations can be removed.

Functions Requiring Interworking

The next step toward solving the interworking problem involves identifying the set of operational functions to be provided across the two networks. The most necessary functions fall within the scope of mobility management, radio system management, and call processing. These functions include:

- Mobile station service qualification
- Mobile station location management
- Mobile station state management
- Authentication
- Intersystem handoff
- Call origination
- Call delivery
- Call features and control

Since GSM and ANSI-41 are such different systems, and interworking to provide some functions is so complex, an approach that prioritizes these functions into deployment phases can be used. Nearly all these functions require primarily HLR procedures to be performed. One of the functions that requires more processing and greater interaction between functional entities is intersystem handoff. This is due to the complexity of handing off a call between two different radio technologies and the changes required to support those technologies at the anchor and target MSCs. If a dual-mode MS is used, each mode can operate independently to provide the listed functions, except for intersystem handoff.

In high-population wireless coverage areas, it is likely that both GSM and ANSI-41 networks will be deployed. Since an intersystem handoff between the two networks would be an uncommon event (i.e., only where the coverage areas of the two networks do not intersect), intersystem handoff can be considered a lower-priority function.

The other low-priority functions are the call features that GSM and ANSI-41 do not support in common. Implementation of these features also involves radical changes in many functional entities. The features that GSM and ANSI-41 do support in common are call forwarding (i.e., unconditional, busy, and no answer), call waiting, and three-way calling.

Although total compatibility between GSM and ANSI-41 is lacking, there is still benefit to providing basic wireless functionality, through interworking for subscribers who can only receive nationwide coverage, at least in the short term.

Methods to Provide Interworking

To provide interworking between GSM and ANSI-41 MAP protocols, three areas need to be addressed: protocol conversion, database mapping, and transaction management. Protocol conversion provides the translation of messages and parameters from one protocol to the other. Database mapping provides the translation and management of information elements that allow each of the application protocols to provide user service (e.g., subscriber identification, location, and status information). Transaction management enables the completion of queries between the two networks (e.g., reoriginating and maintaining queries and responses from one network to the other).

Interworking Solutions

The basic model of a solution is a logical functional entity that performs the appropriate protocol conversion, database mapping, and transaction management to support the mobility management, call origination, and call delivery functions. This interworking function (IWF) provides the logical interface between GSM and ANSI-41 networks (see Figure 17.19).

Many implementations can be derived from this logical reference model. Note that this model does not necessarily imply the placement of the IWF outside both networks (see Figure 17.20). The IWF can support interworking between any network elements where there is desired interworking functionality. All physical implementations of the IWF need to provide the three basic functions of protocol conversion, database mapping, and transaction management.

Figure 17.19
A logical
interworking function
(IWF) can support
intersystem
operations required
between the ANSI-41
and GSM networks.

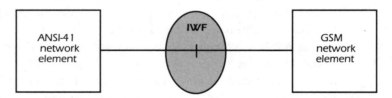

Interworking between GSM and ANSI-41 networks to provide the identified mobility management, call origination, and call delivery functions involves the serving MSC/VLR and the HLR (see Figure 17.21). Signaling is required between the HLR in one network and the MSC/VLR of the other to support all the functions identified for interworking. Trunk connections are required between the MSC/VLR and the PSTN to support call origination, call delivery, and the call features and control.

Figure 17.20
The IWF can reside in either the GSM or ANSI-41 networks or between the networks.

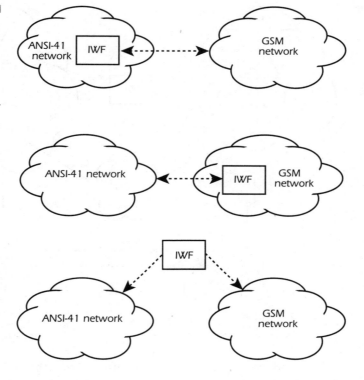

Figure 17.20
The IWF can reside in either the GSM or ANSI-41 networks or between the networks.

The most basic physical implementation of the IWF is known as the *dual-mode HLR*. This network node is essentially a complete ANSI-41–based HLR conjoined with a complete GSM HLR. Each "side" of the HLR adequately serves that "side" of the network. Internally, the protocol conversion, database mapping, and transaction management functions can be performed to support roaming between the two networks. However, the dual-mode HLR solution may be considered "overqualified" as an IWF. It is essentially the implementation of a GSM HLR inside an ANSI-41 HLR or vice versa. Although conceptually this seems to be a solution that intrudes the least on both networks, it is a very expensive implementation and can take many years to design, develop, test, and deploy (see Figure 17.22). It also requires the replacement of existing HLRs in each network with this new dual-mode design.

An alternative—and in some cases preferable—solution is to implement protocol conversion, database mapping, and transaction management for all of the functions supported between the networks as a single separate network element, as shown in Figure 17.21. It provides a basic IWF solution that requires no changes to existing ANSI-41 or GSM net-

Figure 17.21
The IWF involves signaling connections to the serving MSC/VLR and the HLR of both networks to interwork the intersystem operations. Trunk connections to the PSTN are required for call origination and call delivery.

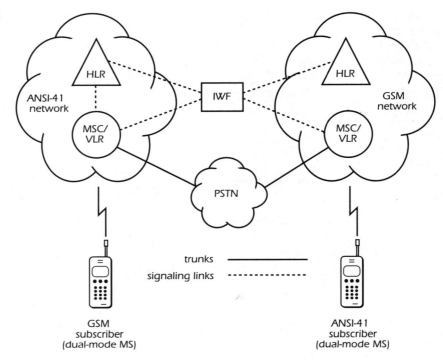

GSM subscriber (dual-mode MS)

ANSI-41 subscriber (dual-mode MS)

works and can reside external or internal to either of the networks. The IWF, in this case, appears as the HLR to the network serving the roaming subscriber, and it also appears as the serving VLR to the subscriber's home network. Intersystem operations can then be routed to and translated by the IWF and sent to the "real" elements in the other network.

Generally, interworking between GSM and ANSI-41 networks does not provide the best of what wireless technology potentially has to offer. There is limited functionality that can reasonably be performed between the two networks. However, it is a method of supporting nationwide roaming to subscribers of GSM networks until those networks provide nationwide coverage themselves.

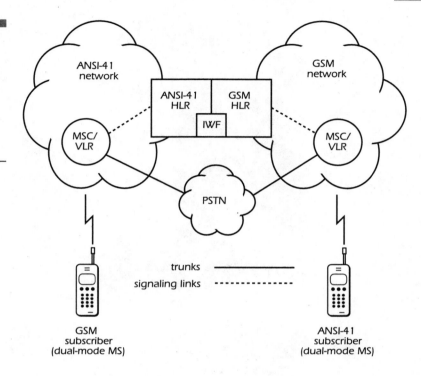

Figure 17.22
The IWF can reside internally to a dual-mode HLR. This solution tends to be prohibitively expensive and can take many years to develop.

CHAPTER 18

ANSI-41
Implementations

The ANSI-41 standard does not specify physical implementations. It does, however, provide the intersystem operations that make many implementations possible. These implementations are generally based on particular network architectures and the means for routing signaling messages between the network elements. Although a variety of implementations are possible, this chapter describes the more common and interesting implementations of ANSI-41 to provide wireless telecommunications services to subscribers.

Use of the Mobile Identification Number

A common misunderstanding about the MIN is that it must also be the directory number (DN), or phone number of the MS. The ANSI-41 standard evolved from many concepts applied to wire-line networks. For wire line networks, the DN of a subscriber is also representative of the line number used to identify the line from the central office to the telephone. The wireless network in North America evolved from the same principle; thus the DN of the subscriber was implemented to represent the unique number used by the network to identify the subscriber. In practice today, the MIN and the DN are typically the same number; however, ANSI-41 never specifies the MIN as a DN. Implementations that provide for a DN separate from the MIN are not at all precluded by the ANSI-41 standard.

Directory numbers are phone numbers that can be dialed from anywhere. They follow the North American Numbering Plan (NANP), which is a subset of the international *ITU-T E.164* numbering plan. The DN is a 10-digit decimal number following the format NXX-NXX-XXXX, where N is any number from 2 to 9 and X is any number from 0 to 9. This format also indicates a geographic address. The first three digits (NXX) indicate the area code, the next three digits (NXX) indicate the exchange or office code, and the last four digits (XXXX) specify an individual subscriber.

DNs are used to call wireless mobile stations, but they can have other uses. If the DN is separate from the MIN, it is used in call delivery by the originating MSC to route a *LocationRequest* message to the correct HLR. The HLR can then find the subscriber's MIN based on the DN (i.e., the DN is a key field in the subscriber's service profile record) and perform subsequent operations based on the MIN rather than the DN.

The MIN is also a 10-digit decimal number, a programmable value entered into the MS by the subscriber or the service provider. The programmable MIN and the permanent electronic serial number (ESN) together represent a subscription to the network. The MIN-ESN pair also represents a key field that uniquely identifies the subscriber's service profile record in the HLR. The MIN identifies an MS over the radio interface and is used to route ANSI-41 messages from the serving MSC to the HLR and vice versa (e.g., RegistrationNotification).

In public wireless implementations, wireless service providers obtain blocks of directory numbers from the NANP administrator. If the DNs are used as MINs, they are distributed to wireless subscribers when they become authorized for service. There are two primary advantages when the MIN and DN are identical:

- The DN does not need to be allocated and managed as a separate information element in the HLR.
- Subscribers need only be aware of a single number.

If the MIN and DN are different numbers, there are many other advantages, most due to the fact that the MIN is not necessarily representative of a geographic location:

- Multiple DNs can be assigned to a MIN, allowing different treatment based on the called number.
- Multiple MINs can be assigned to a DN, allowing multiple MSs to be called with the same number.
- The directory number can be portable among service providers; that is, the phone number doesn't have to change if the subscriber receives a new MIN with a different service provider.
- Directory numbers are a limited resource. Separate MINs allow for numbers that are not limited to the NANP format, providing billions of additional numbers that are distinct from NANP numbers (i.e., numbers where the first and fourth digits can be 0 or 1 in the NXX-NXX-XXXX format).
- Since MINs would not be as limited a resource as DNs, multiple MINs can be implemented in a single MS. Separate MINs could make it easier to route ANSI-41 messages from the MS to different locations based on different routing of the MIN (e.g., short messages routed directly to a message center).
- Mobile phones could easily be preprogrammed with the MINs and sold off the shelf in retail stores in any location. This preprogram-

ming eliminates the geographic tie of a programmed phone to the location indicated by a directory number.

Some wireless service providers today provide private network services using nondialable MIN values programmed into the MSs. However, a directory number must still be assigned to support call delivery to these MSs. Special procedures need to be implemented in the HLR to translate the DN to the MIN to enable the ANSI-41 intersystem operations to perform correctly.

ANSI-41 provides a parameter known as the *mobile directory number* (MDN). This parameter is designed for implementations where the MIN does not also represent the directory number. There are no procedures in ANSI-41 that prescribe a use for the MDN. It was added to the standard in support of implementations where the DN and MIN are separate values.

Use of Signaling System No. 7 (SS7)

SS7 is the primary signaling protocol used to support the ANSI-41 mobile application part (MAP). It provides the transaction capabilities, message transport, and reliable network route management capabilities to support the ANSI-41 intersystem operations.

ANSI-41 specifies the use of the following SS7 protocol levels (see Figure 18.1):

- Message Transfer Part (MTP) Level 1
- Message Transfer Part (MTP) Level 2
- Message Transfer Part (MTP) Level 3
- Signaling Connection Control Part (SCCP) Class 0 service
- Transaction Capabilities Application Part (TCAP)

MTP Level 1 provides the physical transmission layer of the protocol, based on DS0 channels. MTP Level 2 provides the data-link layer of the protocol, providing point-to-point link reliability between SS7 signaling points. MTP Level 3 provides the network address and route management layer of the protocol. SCCP Class 0 service provides basic connectionless transport service, along with an enhanced addressing technique known as *global title addressing*. TCAP provides the application-layer communications protocol supporting transaction management and remote operation capabilities between two ANSI-41 functional entities.

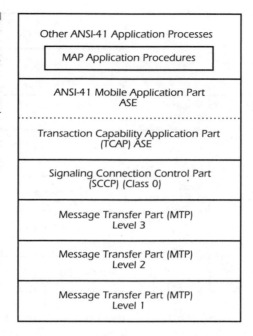

Figure 18.1
Protocol stack diagram showing the SS7 protocols used for ANSI-41 applications.

An ANSI-41 message is encapsulated within a TCAP operation message either to invoke a remote mobile operation or to return some result of an operation. The TCAP operation message is encapsulated within an SCCP unit data (UDT) message for transport. The UDT message is encapsulated within an MTP Level 3 message signal unit (MSU), the basic SS7 packet routed through the network via physical node addresses (see Figure 18.2).

Although SCCP provides three other classes of service, Class 0 is the only one specified for use. It provides unsequenced connectionless data packets routed through the SS7 network independently of one another.

SS7 supports multiple TCAP component operations per transaction (i.e., per message). Although ANSI-41 does not explicitly prohibit this practice, existing implementations generally support only one component per transaction. This is due to simpler designs and because the use of multiple components could exceed the maximum message length for an MSU based on potential ANSI-41 message sizes.

SS7 Routing and Addressing

In ANSI-41 networks, the MSC is an end signaling point (SP) in the SS7 network. The HLR is sometimes considered a service control point (SCP)

Figure 18.2
Diagram showing
the successive
encapsulation of an
ANSI-41 operation
into levels of the SS7
protocol.

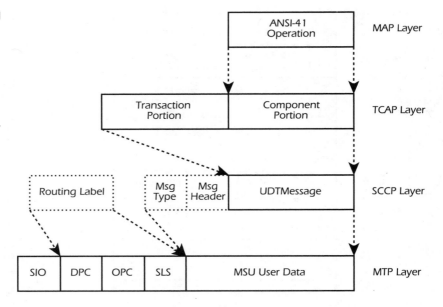

Figure 18.2
Diagram showing
the successive
encapsulation of an
ANSI-41 operation
into levels of the SS7
protocol.

(see Figure 18.3). SS7 routing is based on *point codes*. Each node in the SS7 network has a unique point code address. MSUs sent between SS7 signaling points contain an originating point code (OPC) address and a destination point code (DPC) address. Routing of MSUs through the SS7 network based on point code addresses only is known as *point code routing*.

Figure 18.3
Network architecture
showing the
connectivity of the
serving MSC/VLR to a
subscriber's HLR via
the SS7 network. The
MSC is an end SP,
and the HLR is an
SCP.

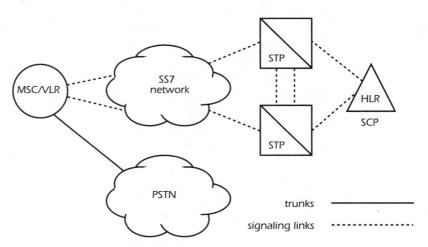

ANSI-41 supports point code routing as well as *global title routing*. A global title (GT) is an address that, by itself, does not explicitly allow routing in the network. This GT address needs to be translated somewhere in the network (usually by STPs) into a point code, *subsystem number* (SSN), or both. The SSN is an address of an SS7 application residing at a signaling point represented by a point code. This is known as *global title translation* (GTT). This addressing mechanism makes network routing of MSUs to and from end signaling points much simpler than point code routing. The end SPs do not need to maintain tables of each point code to route messages properly. They can supply a GT address, and the network will perform the GTT function to route the MSU to the correct SP.

Routing from the MSC to the HLR

The MSC/VLR currently serving the subscriber routes ANSI-41 signaling messages to the HLR based on the subscriber's MIN. The serving MSC/VLR must contain tables, which enable the routing of the messages to the proper STP mated pair that is serving the HLR.

A group of related nodes, such as an STP pair, can be identified by a single point code called an *alias*. The alias point code address is used by network nodes that send MSUs to the related nodes via a combined link set (i.e., A links). The nodes are assigned the alias point code in addition to their individual point codes.

If the serving MSC and HLR reside within an SS7 network that does not support the GTT function, the MSC must perform point code routing of ANSI-41 signaling messages destined to the HLR. The MTP routing tables within the MSC are required to maintain the point code address of the HLR. The serving MSC (which has a roaming agreement with the HLR service provider) maintains a table mapping ranges of MIN values to the point code of the HLR serving those subscribers. If the HLR is deployed as a mated-pair configuration for redundancy, the ranges of MIN values are mapped to an alias point code representing either of the HLRs. In this configuration, one HLR is considered active and the other is standby. The serving MSC is never aware of which one is active, and it has no need to know. Figures 18.4 and 18.5 show the point code routing of ANSI-41 messages (within MSUs) to the HLR.

MINs are mapped to the alias SS7 point code and SSN of the serving HLR. The serving MSC enters the MIN into the called-party address field of the SCCP UDT message. The MSC also enters the alias point code of the HLR mated pair into the DPC of the message. When the MSU is delivered to either of the STPs of the mated pair, the STP maps the actual point code of the active HLR into the DPC field of the MSU and delivers the message across the appropriate link set. In the case of failure in one of the HLRs, all ANSI-41 messages are routed to the other HLR.

Figure 18.4
Network architecture
showing point code
routing of IS-41
messages from the
serving MSC to the
subscriber's HLR.

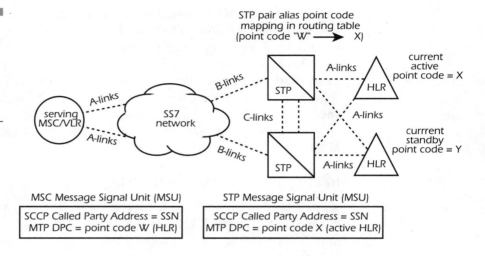

Figure 18.5
Protocol model of
point code routing
from the serving MSC
to the subscriber's
HLR.

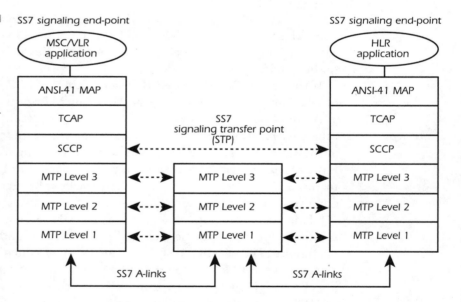

If the serving MSC and HLR reside within an SS7 network that supports the GTT function, a translation is performed on the MIN to route the messages to the HLR. The serving MSC (which has a roaming agreement with the HLR) maintains a table mapping ranges of MIN values to the alias point code of the mated STP pair. Figures 18.6 and 18.7 show the GTT routing of ANSI-41 messages to the HLR.

Figure 18.6
Network architecture
showing global title
routing of ANSI-41
messages from the
serving MSC to the
subscriber's HLR.

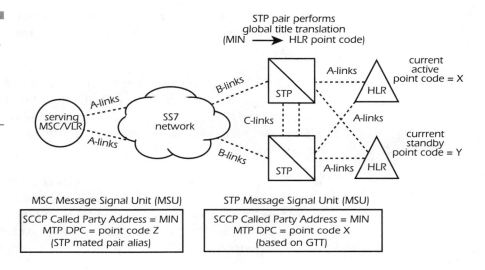

Figure 18.7
Protocol model of
global title routing
from the serving MSC
to the subscriber's
HLR.

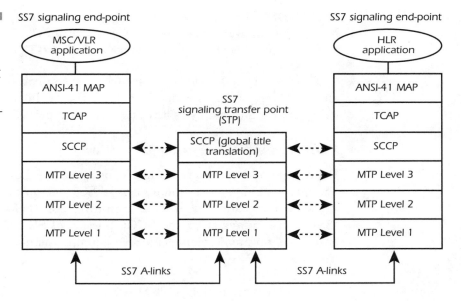

MINs are mapped to the alias SS7 point code of the STP mated pair serving the HLR. The serving MSC enters the MIN into the called-party address field of the SCCP UDT message. The MSC also enters the alias point code of the STP mated pair into the DPC of the message. When the MSU is delivered to either of the STPs of the mated pair, the STP performs GTT and routes the MSU to the appropriate HLR.

The serving MSCs are required to contain routing table information to support routing of ANSI-41 transactions to the subscriber's HLR. Each MSC requires two databases:

- The *number-range database* maps ranges of subscriber numbers (MINs in the form of NPA-NXX codes) onto MSC IDs (i.e., the identification of the HLR which serves the associated subscriber).
- The *current network address database* identifies the DPC for the associated MSC ID.

If GTT is used, the current network address database can contain merely the DPC for the first STP that performs GTT, as opposed to "hard-coding" HLR network addresses into each MSC.

In most cases, the STP provides SS7 message screening, based on the following parameters:

- MTP OPC
- MTP DPC
- SCCP calling-party address (before and after GTT)
- SCCP called-party address (before and after GTT)

This screening provides the security mechanism to prevent unauthorized network access to the HLR. It further validates the identity of a calling-party address by restricting it to preapproved link sets.

Point Code/SSN as System Identifier

Point code addresses are managed by the MTP Level 3 portion of the SS7 protocol. They are unique across SS7 networks and represent the physical node address of each signaling point. Subsystem numbers (SSNs) are managed by the SCCP portion of SS7. They represent the address of an application at an SS7 signaling point. The following SSNs are used for ANSI-41:

- 5—mobile application part (MAP)
- 6—home location register (HLR)
- 7—visitor location register (VLR)
- 8—mobile switching center (MSC)
- 9—equipment identity register (EIR)
- 10—authentication center (AC)
- 11—short-message service (SMS)
- 12—over-the-air service provisioning function (OTAF)

An SSN value of 5, referring to the MAP as an application of the ANSI-41 protocol as a whole, is an artifact of older revisions of ANSI-41. IS-41 Revision A specified a single SSN, representing the entire MAP application. IS-41 Revision B added other SSNs when the need to differentiate ANSI-41 applications (e.g., HLR, MSC, AC) was discovered. ANSI-41 functional entities should be able to accept messages containing the SSN value for the MAP so that they are backward compatible with functional entities that have not implemented the more specific SSN values.

Generally, the MSCID parameter is sent in ANSI-41 application messages to indicate the identity of the serving MSC (or VLR) originating the message. This MSCID parameter is validated by the HLR when messages are received to ensure that the subscriber's HLR has a roaming agreement with that MSC (or VLR).

Many of the ANSI-41 intersystem operations contain the PC_SSN parameter as an option. This parameter is used as an alternative address to the MSCID for the ANSI-41 application layer. This PC_SSN parameter is used to more clearly identify a serving MSC. That is, there are situations that can occur where the MSCID value may not always be indicative of an exact point code address. The MSCID identifies a system and market, including all the cell sites associated with that system. A point code address identifies the actual physical node that sends or receives SS7 signaling messages.

Of course, there already are the separate SS7 originating point code (OPC) and SSN addresses sent with the ANSI-41 messages for use at the MTP and SCCP portions of the SS7 protocol. But the PC_SSN parameter for use in ANSI-41 applications is indicative of the originating address of the transaction, which may not be the same as the OPC and SSN sent along with an individual MSU containing the ANSI-41 message. This can occur, for instance, when the PC_SSN parameter represents a serving MSC, but the message is forwarded to the HLR by a separate VLR, also an SS7 signaling point. In this case, the OPC and SSN values together indicate the VLR and not the MSC originating the message.

Alternate MAP Addressing

Although ANSI-41 specifies the use of SS7 and its addressing techniques, there are other network addressing techniques afforded by ANSI-41. These techniques are based upon the needs of individual service providers. The two additional address types specified by ANSI-41 are:

- MSCIdentificationNumber parameter (not to be confused with MSCID)
- SenderIdentificationNumber parameter

These two parameters represent addresses that can be used in place of lower-layer address parameters, such as PC_SSN, if they are supported by the network. In contrast, the MSCID parameter identifies a system and market including all the cell sites associated with that system. The MSCIdentificationNumber identifies an address that can be used to represent an MSC system in a manner divorced from physical considerations.

The SenderIdentificationNumber identifies an address of a functional entity such as a VLR or HLR similar to the way that a MIN identifies a particular mobile station. The use of these addresses facilitates routing of ANSI-41 messages over X.25 networks or between SS7 and X.25 networks. The format of these identifiers is defined so that they can be supported by global title translation functions.

Procedures for the use of these parameters are not defined in ANSI-41. To know exactly what they identify, their use would have to be supported by systems receiving these addresses. Note that use of these addresses could also facilitate routing between GSM and ANSI-41 systems, since GSM network elements are all identified by unique *ITU-T E.164* addresses. An E.164 address for an ANSI-41 network element can be supported via the MSCIdentificationNumber and SenderIdentificationNumber address parameters.

Short-message Service Architecture

SMS is a service that can have multiple implementation options. Aside from a standardized bearer service capability and some teleservice options, few other aspects of SMS have been standardized in ANSI-41. This allows a variety of implementation-dependent options, which can be made compatible based on the bearer service and standardized addressing techniques.

With the growth of Internet use, it can be most advantageous to implement SMS based on the Internet protocol (IP) as the networking and transport mechanism to deliver short messages. Also, the ANSI-41 network reference model supports connectivity between message centers (MCs) and short-message entities (SMEs). Figure 18.8 shows a sample generic network architecture using IP as the transport method for providing short-message service.

The MC in the ANSI-41 network can provide protocol conversion between IP and SS7 for short messages to and from the serving MSC. If the MSC provides an interworking function supporting IP data transfer, short messages can be transferred directly between the MSC and the MC using IP. The architecture allows ANSI-41 network access to external IP applications (i.e., SMEs) as well as inexpensive IP access between MCs belonging to different networks. This allows for delivery of a short message to a local MC where a subscriber is roaming, or between MCs supporting mobile-to-mobile short messages.

Figure 18.8
A generic SMS architecture based on the internet protocol (IP). IP allows inexpensive access to applications (SMEs) and is an easy protocol to interconnect with other message centers.

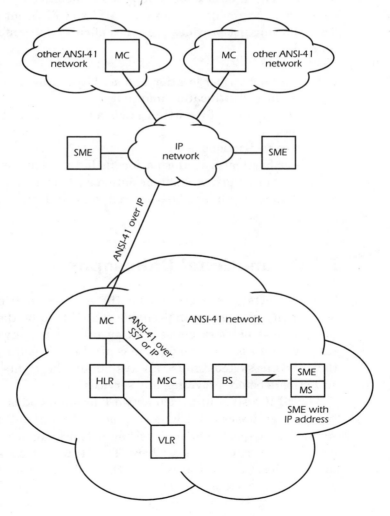

Home Location Register

The HLR is the primary repository for information about a given set of subscribers. In contrast, the VLR is a temporary storage function for those subscribers currently visiting the MSC or MSCs served by that VLR. When a subscriber is served by an MSC, the subscriber's characteristics are retrieved from the HLR and stored in the VLR. The MSC retrieves those characteristics from the VLR as needed based upon subscriber call activity.

The HLR stores both semipermanent and transient data about subscribers. These data enable subscriber mobility between MSCs of the same service provider or between MSCs of different service providers. The following list provides the types of subscriber information stored at an HLR:

- Current subscriber location and activity status
- Subscriber identification information
- Subscription information such as the subscribed features and privileges
- Service restrictions .
- Subscriber feature information including feature activation status and feature registration information such as forward-to numbers
- Data enabling calls to be delivered to subscribers

HLR Functional Philosophy

The HLR acts as a controller for the ANSI-41 network. ANSI-41 provides many procedures that enable the HLR to be the primary controlling wireless network element in an intelligent network–like architecture. The procedures enable the control of all features at the HLR so that serving systems must always obtain the rights and privileges of subscribers before providing any services.

The HLR resides in a fixed network location and is queried to find the location of subscribers, who can move throughout and between networks. These queries enable visiting networks to acquire the rights and privileges of roaming subscribers. The HLR controls a subscriber's features so that those features follow that subscriber while roaming among MSC serving systems.

As a centralized node for subscriber signaling traffic, the HLR is able to control certain features for a subscriber on a networkwide basis. All feature control requests from throughout the network are directed to the specific HLR serving a particular subscriber. A wireless subscriber will be served by only one HLR.

The HLR is essentially a transaction processor handling network queries and operations. It is administered and controlled by the service provider with whom the subscribers have a *service agreement*. The service provider can retain control over the following functionality:

- Subscriber validation
- Subscriber features to maintain a uniform subscriber interface
- Signaling network access
- Delivery of incoming calls to a mobile subscriber regardless of location
- Subscriber roaming
- Strategic market information, such as subscriber activity and number of database accesses
- Fraud protection, investigation, and subscriber shutdown (based on authentication)
- Service offerings

The HLR is the most important network element in the wireless telecommunications network. It can be implemented as a standalone function, such as a service control point (SCP) application, or it can be implemented as an integrated function within an MSC complex.

HLR Databases

The HLR holds all subscriber service profile records, which define the features that have been authorized for each subscriber. It is responsible for validating a subscriber to allow it to incur charges while roaming on another system. It also tracks the location of all its subscribers so that it can direct incoming calls to the MSC currently serving the subscriber. The HLR application is able to perform these functions because of the information it stores in its databases (see Figure 18.9).

The HLR supports the following database tables:

- Subscriber service profile record table
- Subscriber feature option table

■ Served MIN table
■ Roamer agreement table

The HLR supports other tables, but these include the most necessary information for the HLR application functionality.

Figure 18.9
The HLR application has access to a large database. The interface to the database is proprietary.

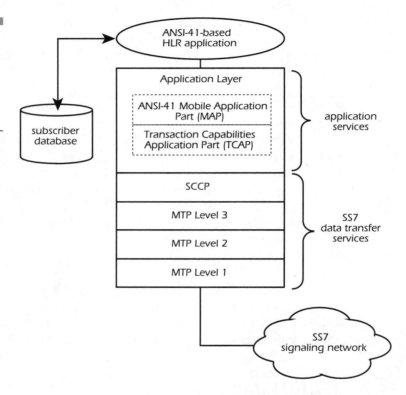

Subscriber Service Profile Record Table

The subscriber service profile record table is the HLR database containing the records established for each subscriber. This table typically includes the following subscriber information:

1. Identification data:
   (a) MIN
   (b) ESN
   (c) MDN (if applicable)
2. Call profile data:
   (a) Credit status (e.g., paid-up, delinquent account)

    (b) Call origination indicator (e.g., local calls only, single NPA-NXX-XXXX only)

    (c) Call restriction digits (e.g., by NPA only, NPA-NXX only, or NPA-NXX-XXXX)

    (d) Call termination restriction (e.g., none, all)

    (e) Preferred interexchange carrier

3. Qualification data (closely related to the authentication procedures)

    (a) Authorization period

    (b) Call history count

    (c) Carrier-requested temporary disconnect (e.g., for delinquent accounts)

    (d) Subscriber-requested temporary disconnect (e.g., for vacation)

4. Feature activations (for each feature available)

    (a) Feature authorized/activated

    (b) Feature states

    (c) Feature timers

5. Feature registrations (one example)

    (a) Call forwarding forward-to-number

6. Location and activity status

    (a) VLR identification information (e.g., MSCID, PC_SSN)

    (b) Registration status (i.e., active, inactive)

7. Fraud protection (e.g., subscriber behavior information)

Subscriber Feature Option Table

The subscriber feature option table maintains the configurations required for certain features to operate. This database information includes:

- Allowed feature code values
- Number of subscriber modifications
- Feature timer values (e.g., message waiting notification)
- Short-message service options

Served MIN Table

The served MIN table defines MIN ranges assigned to a particular HLR. The MINs are defined in terms of office codes, thousands groups, or hundreds groups. If numbers are allocated individually, a subscriber record is usually created for each number in the list and marked as an "unassigned" MIN if it is not yet allocated. The basic structure of the served MIN table is as follows:

- MIN block specification, one of the following:
  - Full office code
  - Thousands group
  - Hundreds group
- Allocated individually (i.e., true or false)

Roamer Agreement Table

The roamer agreement table defines other systems (i.e., MSCs/VLRs) that have a service agreement with the current HLR. The table reflects individual privileges granted to each system. The basic structure of the roamer agreement table is as follows:

- MSCID (primary key)
- Point code (alternate key)
- Subsystem number
- Node type (e.g., HLR, VLR, MSC, AC)
- Allowed screening type:
  - Match point code
  - Match subsystem number
  - Match SID
- Messages allowed from specified subsystem, node, or network
- HLR revision level of node:
  - IS-41 Revision 0
  - IS-41 Revision A
  - IS-41 Revision B
  - IS-41 Revision C
  - ANSI-41 Revision D
  - ANSI-41 Revision E

Visitor Location Register

The VLR is the temporary storage function for those subscribers currently visiting the MSC or MSCs served by that VLR. Subscribers are dynamically added and removed from the VLR database based upon their registration status.

In most cases, the serving VLR is physically integrated with the MSC currently serving a given subscriber. This is primarily due to the close functional relationship required between the MSC and VLR. However, standalone VLR configurations are not precluded by the ANSI-41 protocol.

An interesting implementation option is that a subscriber currently being served in the home area can be served by the VLR associated with the serving MSC. Although the subscriber is not roaming, signaling transactions can be reduced by not serving the subscriber directly from the HLR.

Authentication Center

The authentication function is very closely related to the individual subscriber functions provided by the HLR. The AC can be a separate network element or integrated as an application function within the HLR; the latter is more common. The HLR interface to the AC is based on ANSI-41 as an application-layer interface. Lower-layer protocol interfaces between them are specified as SS7 by ANSI-41, but may be implementation-dependent, of course.

The AC provides the function of executing the authentication algorithm, based on information provided by the HLR about a subscriber. However, the AC can optionally maintain authentication information about subscribers that is separate from the HLR.

ANSI-41 does not preclude the implementation of a centralized AC as a separate and distinct network node; however, there is no separate SS7 global title translation (GTT) type to address the AC as an SS7 node separate from the HLR. All ANSI-41 AC transactions are required to pass through the HLR. ANSI-41 does, however, support a separate subsystem number (SSN) for the AC.

A-key Management

A primary issue with the implementation of ANSI-41-based authentication is A-key management, which consists of control and distribution of the A-key. The basic requirements of A-key management are the following:

- The A-key must remain unknown to the customer and the distributor of the MS.

- The HLR/AC must be very secure.
- The A-key must not be readable by any party except the manufacturer.
- Updates to the A-key in the MS and HLR/AC must be performed securely.

The service provider typically controls distribution of the initial and subsequently reprogrammed A-key. There is currently no standard method to distribute the A-keys for MSs; however, TIA/EIA TSB50 specifies a user programming procedure to enter the A-key into the MS. This specification is not currently considered by the industry to be an adequate method of installing the A-key into the MS.

Authentication is a function that has not, until recently, been widely implemented. The importance of the function did not become pressing to wireless service providers until enormous revenue losses became more important than the implementation costs of a non–revenue-generating function. The service providers deploying digital radio systems (i.e., TDMA and CDMA) perform authentication. The service providers currently providing AMPS-based service are beginning to deploy authentication today. The problem is that most analog MSs currently in use are not authentication capable.

A Interface

The A interface has little impact on the functioning of the ANSI-41 network. It is interesting, though, to see why there are different standardized A interface protocols and the potential impact of these protocols on the rest of the wireless network.

Most A interface protocols deployed today are proprietary. This stems from the original concept in early systems where wireless switches were sold along with the radio systems as a single bundled system. The equipment manufacturers had no need to support a standard A interface since their base station systems were sold with their MSCs. With the tremendous growth of wireless systems, service providers began to desire open standards, allowing them to deploy switches that fit their needs with other vendors' radio systems that fit their needs.

There are three standardized A interface protocols: SS7, ISDN, and frame relay. The use of SS7 for the A interface is a concept originally specified by GSM-based systems. SS7 allows reliable transport of the signaling messages between the base station and the MSC. It is interesting to note that SS7 is not a protocol designed for point-to-point transport (as is the case with the A interface). Most of the functions that SS7 provides are based on packet-switched adaptive routing techniques and the ability to transfer message to different links for rerouting. It is a commonly available protocol, though, and the extraneous functionality can be removed for the A interface implementation.

ISDN was chosen to be standardized specifically for PCS systems. It provides a good user-network interface that can connect to many wire-line telecommunications switches. In fact, almost no MSCs support ISDN. This protocol was desired primarily by service providers implementing the original Bellcore Wireless Access Communications System (WACS) and its cousin, the PCS Access Communications System (PACS). These systems were designed to allow wire-line ISDN-based switches to provide network services to small PCS service providers who did not have their own network. ISDN allows these small PCS service providers to use the intelligent network infrastructure provided by the large wire-line carriers.

Frame relay is the other A interface protocol standardized for use in wireless networks. It enables fast packet-switching technology over very reliable transmission facilities. Frame relay has certain interesting implications for the way that radio systems access the wireless network (see Figure 18.10). It allows for a network of radio systems, which can communicate with each other and access one or more MSCs. With this configuration, radio systems can be added at very little cost since they only need access to an available frame relay network. Handoffs between radio systems can also be performed without MSC intervention. The radio systems can easily be assigned to different MSC serving areas depending on the changing traffic patterns and capacity of adjacent MSCs.

Positive and Negative SID Lists

System IDs (SIDs) are values assigned by the Federal Communications Commission (FCC) and are allocated to wireless service providers. SIDs are unique and identify specific wireless service providers in specific geographic areas where wireless (cellular and PCS) licenses were obtained. A parameter known as the home SID is programmed into the number assignment module (NAM) in each mobile station. This *home SID* represents the system ID of the home service area of the subscriber.

Many wireless service providers offer service in numerous geographic areas because they have multiple licenses, through acquiring or merging

Figure 18.10
The first figure shows
a traditional A
interface architecture
connecting radio
systems to the MSC.
The second figure
shows the
implications of a
frame relay–based A
interface.

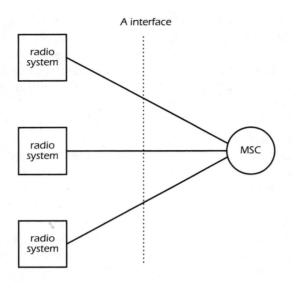

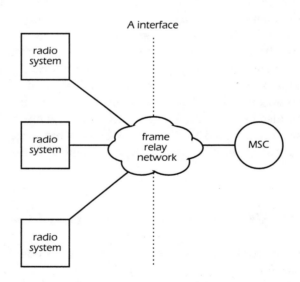

with other service providers. Many offer nationwide service. These carriers generally want to offer service to subscribers in a way that treats the subscribers as if they are being served in the home area (i.e., treat them as if they weren't really roaming). Also, service providers do not want their subscribers to roam onto a competitive service provider's network when their service is provided in the same geographic area. Since the wireless systems in each geographic licensed area broadcast a different

SID value, subscribers will think they are roaming in these other areas (because of the home SID stored in the MS), when they are not really roaming from a service and billing perspective.

To treat these subscribers as if they are always on the home system, although they may be receiving service in a different geographic area, wireless service providers requested that the mobile station (MS) manufacturers implement the ability to add multiple SID values in the mobile stations. This prevents a roam indicator from appearing on the MS, and the subscriber, for all intents and purposes, is not considered roaming. These SID values are known as the *positive SID list* because they enable physical roaming in multiple geographic areas of the same service provider network. Subscribers are never aware of this physical roaming and they are treated as if they are still in the home geographic area for billing purposes. Conversely, many mobile stations support *negative SID lists*. These prevent the mobile station from receiving service from a competing network that offers service in the same geographic area as the subscriber's home network.

Many service providers own licenses in different geographic areas, which provides service in a different frequency band than in the subscribers' home networks. The positive SID list can be mapped to a frequency range within the mobile station, enabling the MS to change frequency bands in those areas. This prevents roaming onto a competitor's network in any geographic area desired. The negative SID list explicitly denies service in certain competitive areas where the competitor's network has a significantly stronger signal that the subscriber's home service provider. Many mobile stations have default functionality that will support service on any network (assuming there is a roaming agreement) if that network's signal strength is significantly stronger than the subscriber's network service provider in that area.

Positive and negative SID lists are primarily used for analog mobile stations since digital MSs typically support more sophisticated mechanisms for roaming. Note that there is no standard for the size or number of entries in these lists and they vary among mobile station manufacturers, makes, and models.

System Operator Codes

System operator codes (SOCs) are identifiers used in TDMA (ANSI/TIA/EIA-136) systems. These 12-bit unique codes are stored in the mobile station NAM and are assigned to wireless service providers by TIA Subcommittee TR-45.3. The SOC identifies a specific wireless service provider and is used by a mobile station to acquire different types of services offered by that service provider. Typically, these services are proprietary to each wireless service provider. SOCs are broadcast from the base station to inform the mobile station of the availability of these services. The most common application of SOCs is to prevent the use of the mobile station on another service provider's network. This is especially important when the mobile station equipment is subsidized by the wireless service provider. The SOC prevents use of the equipment on any other network when an account becomes delinquent or when subscription fraud is perpetrated.

Intelligent Roaming Database (IRDB)

Intelligent roaming databases (IRDBs) are the primary mechanism enabling mobile-directed intelligent roaming in TDMA (ANSI/TIA/EIA-136) systems. Mobile-directed intelligent roaming is a type of intelligent roaming where the mobile station contains all the information necessary to determine the frequency band on which to operate in any given wireless licensed area. Intelligent roaming became necessary due to the deployment of six PCS frequency bands, in addition to the two cellular bands. Hence, a service provider needed to have roaming agreements with a variety of new service providers using these different frequency bands. The service providers also needed the ability to enable subscribers to roam automatically into these networks seamlessly. This intelligent roaming is more sophisticated than the simple use of positive and negative SID lists and the individual use of SOCs.

The mobile station stores the necessary information to perform this automatic intelligent roaming in an IRDB. The IRDB contains lists of SIDs and SOCs in priority order, which identifies different categories of service providers (e.g., home, partner, favored, neutral, and forbidden). When a subscriber roams outside the home service area, the mobile station searches for systems (that are not forbidden) based on the priority order in the IRDB. If the mobile station finds a system broadcasting a SID or SOC not in the IRDB, the service provider is classified as neutral, and service is obtained. However, the MS periodically scans for surrounding higher-priority systems that are contained in the IRDB.

The IRDB can be updated over the air as roaming agreements change, based on the over-the-air programming teleservice (OPTS) portion of the over-the-air parameter administration (OTAPA) function.

Note that there is no standard for the number of entries in the IRDB and they vary among mobile station manufacturers, makes, and models.

Preferred Roaming List (PRL)

A preferred roaming list (PRL) is a CDMA (ANSI/TIA/EIA-95) mechanism using a positive SID list to perform mobile-detected intelligent roaming. A list of SIDs is stored in the NAM of the mobile station. While roaming, the MS searches for a SID contained in the PRL and will provide service from that system if it is found. Typically, the PRL contains SIDs belonging to a single wireless service provider that owns licenses in many different geographic areas serving many different frequency bands. Over-the-air service provisioning (OTASP) and over-the-air parameter administration (OTAPA) are used to download the PRL into the mobile station.

If the subscriber is roaming in a geographic area not contained in the PRL, then roaming will occur as usual, assuming a roaming agreement exists between the home service provider and the visited system.

Note that there is no standard for the size or number of entries in the PRL and they vary among mobile station manufacturers, makes, and models.

ANSI-41 Deployment

ANSI-41 has been deployed by many service providers in North America for many years. As ANSI-41 has been revised and becomes more sophisticated, more service providers have used the protocol to provide enhanced services to subscribers. Examples of the services provided by ANSI-41 that are now taken for granted include automatic call delivery, automatic roaming, and features such as call forwarding and call waiting.

The Cellular Telecommunications & Internet Association (CTIA) tracks the implementation and deployment of ANSI-41 networks. The status has been changing so rapidly over the past few years that it is not worth listing statistics in this book, since by the time of publication those statistics will be outdated. Needless to say, ANSI-41 is currently deployed in *most* networks across the United States and Canada.

In the rest of the world, a large number of countries have wireless systems based on the ANSI-41 protocol and the list is growing. The International Telecommunications Union–Radio (ITU-R) group has officially approved ANSI-41 as an international network standard for support of CDMA-based radio systems.

Wherever AMPS, TDMA, or CDMA radio technologies are chosen for use, ANSI-41 is the networking technology to support them. By the end of the year 2000, there were approximately 30 million AMPS subscribers, 60 million TDMA subscribers, and 80 million CDMA subscribers worldwide using ANSI-41 technology.

CHAPTER 19

Other Technologies Related to ANSI-41

There are many new technologies closely related to ANSI-41 that are in the process of being standardized, or have recently been standardized. In this chapter, we discuss a few of the more interesting technologies related to ANSI-41 networks. Some of these are U.S. government mandates (i.e., lawfully authorized electronic surveillance, enhanced wireless 911, and number portability) and some simply represent the evolution and enhancement of ANSI-41–based networks. We do not go into the detailed protocol information of these technologies since they are not of core importance to ANSI-41.

Authorization and Call Routing Equipment (ACRE)

ACRE is a logical functional entity employing ANSI-41 to provide an interface from a personal base station to the public wireless network. A personal base station is a small radio access system that allows wireless telecommunications in a small geographic area (such as an office building), which may or may not interface with the public wireless network. These personal base stations sometimes interface to a PBX that can act as a local MSC for the small radio systems (see Figure 19.1).

Figure 19.1
Network architecture diagram showing the authorization and call routing equipment (ACRE) and its relationship to wire line and wireless networks.

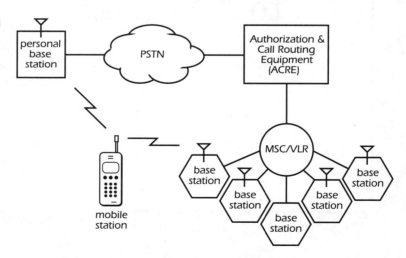

ACRE communicates with personal base stations through the PSTN to authorize, authenticate, and support call routing to mobile stations accessing those personal base stations. It also communicates with the wireless network via ANSI-41 to inform the HLR that the subscriber is now being served by the personal base station. The technique requires no special agreements between the PSTN and the wireless network since the personal base station simply dials through the PSTN to reach the wireless network. The combination of ACRE and the personal base station enables a standard mobile station to access the wireless network as well as any areas covered by the personal base stations.

When subscribers move from the ANSI-41 public wireless network to an area covered by a personal base station, the ACRE system informs the HLR that the subscriber is now being served by the personal base station. In this sense, ACRE acts as a "modified VLR." Subscribers can then receive wireless calls in the environment covered by the personal base station, whether it be an office environment or a home.

Calling Party Pays (CPP)

Calling party pays (CPP) is a method of measuring and billing wireless calls characterized by the calling party's paying for the call and associated airtime when the call is terminated to a mobile station. In North America today, the party receiving the call pays for airtime usage whether or not it initiated the call. This is not true in most other places in the world (e.g., Europe uses the GSM system), where the calling party does pay for the entire call, including airtime, as with wire line-to-wire line calls in North America today.

Calling party pays could be a revolution in the way wireless telecommunications is perceived by subscribers. This is because:

- The cost of owning a wireless phone would be considerably lower, since subscribers would not have to pay for airtime for received calls.
- The number of minutes of use on a wireless phone would increase, since subscribers would be less reluctant to receive calls from unknown parties.
- Anyone making a phone call, regardless of whether the call is made from a wire-line network or a wireless network, would become a customer of the wireless service provider, since they would pay the wireless service provider for the airtime of a called wireless subscriber.

If there are so many benefits to calling party pays, why is this method of charging callers not deployed? The reason is that CPP is a very expensive and complex architectural problem to solve. Wireless networks in North America currently are not implemented to employ such a technology. Today, there are over 2,000 wire-line service providers and over 700 wireless service providers in the U.S. Any wireless service provider offering CPP nationwide would require agreements with many of these networks. Also, standards have not yet been developed to support CPP. For instance, how is the calling party notified that a call is being made to a wireless subscriber? Based on numbering plan deployments today, it may be impossible for a caller to determine that a directory number is a wireless number. In fact, with the deployment of wireless number portability, it will be impossible for a caller to determine this difference. The network would need to perform additional signaling transactions to perform database queries. Also, how will the calling party know how much the airtime costs for the call being made? If a subscriber is roaming across the country, the calling party would pay for the long-distance legs of the call, as well as for the airtime in the network visited by the subscriber. These issues are the most basic, but there are many more complexities such as legal inter-carrier tariff and public education issues.

In July 1999, the FCC issued a Notice of Proposed Rule Making on calling party pays. This notice is not a federal mandate; rather, it is a starting point to inform the wireless industry that it should begin researching solutions to a number of issues that may eventually become federally required for CPP. Typically, the FCC is interested in ensuring a competitive service as well as a reasonable and viable one. These issues include:

- Implied contracts between someone calling a wireless subscriber with CPP and the wireless service provider serving that phone, so that the wireless service provider can collect money from any other carrier network that delivers the call.
- Notification to callers of a wireless subscriber indicating that they are paying for the call, an estimate of the per-minute charges, the name of the wireless service provider, and provision for an opportunity to disconnect the call before charges are incurred.
- Access to the calling party's billing information so that the airtime costs for a call to a wireless subscriber can be properly charged back. One of two competing technologies may be used: ISUP signaling or a query to an alternate billing system database.
- Whether or not CPP should be an optional feature offered to a wireless subscriber. This could be a difficult technical problem since CPP

could not be implemented based on distinctive digits of the wireless called party.
- Solutions to handle calls from non-billable directory numbers, such as pay phones.

In January 1998, the Cellular Telecommunications Industry Association (CTIA) developed a standards requirement document based on the issues raised in the FCC Notice of Proposed Rule Making on CPP. The CTIA then submitted this document to the TIA standardization groups to begin development of a standard technology for CPP. Some standards work began, but eventually came to a halt due to disputes between the CTIA and other industry trade associations on the original proposed standards requirements. Since mid-1999, no standards work on calling party pays has been done.

Because of political, business, and, of course, industry standards issues, CPP may never become a reality. This technology requires widespread industry support from the wireless service providers, industry trade groups, and standards-making bodies.

Communications Assistance for Law Enforcement Act (CALEA)

The Communications Assistance for Law Enforcement Act (CALEA) was enacted and signed into law (Public Law 103-414) by the U.S. Congress in October 1994. This law mandated that wireless network service providers be required to meet law enforcement "wireless wiretap" surveillance requirements within four years (i.e., by October, 1998). Any network infrastructure equipment deployed before January 1, 1995 could be retrofitted with CALEA functionality with the costs (up to $500 million per year for up to five years) reimbursed by the U.S. government. Any network infrastructure deployed after this date would be required to meet the law enforcement requirements at a cost to the wireless service provider. Unfortunately, true estimates of the cost calculated by wireless service providers were as high as 10 times the cost estimated by the government. This has been the seed of bitter political, legal, and business debates within the standards bodies over the past several years, and deployment dates for CALEA functionality have been pushed back. The FBI was the primary law enforcement group providing and supporting technical standards requirements based on their interpretation of CALEA.

The overall functionality specified by the FBI to support CALEA that was the subject of debate within the TIA standards bodies included the following:

- Law enforcement agencies require access to all electronic communications transmitted from the terminal or directory number of the intercept subject. This also includes signaling information such as call identifying information necessary to determine the calling and called parties.
- Law enforcement agencies require real-time and full-time monitoring capability for an intercept subject.
- Law enforcement agencies require provisions for accessing a number of simultaneous intercepts.
- Law enforcement agencies require information from the wireless service providers to verify the association of the intercepted communications with the intercept subject and information on the features and services subscribed to by the intercept subject.
- Law enforcement agencies require wireless service providers to transmit intercepted communications to an external monitoring facility designated by a law enforcement agency.
- Law enforcement agencies require that the reliability and performance standards of the lawfully authorized intercept services be equal to the reliability and performance standards of the telecommunications services provided to the intercept subject.
- Law enforcement agencies require the lawfully authorized interception to be transparent to all parties except the investigative agency requesting the interception and specific individuals involved in implementing the intercept capability.

The CALEA legislation references the development of a "publicly available" standard that fulfills these requirements for wireless law enforcement functionality. This standard is informally referred to as "safe harbor," meaning that if a standard exists, implementation of that standard would put wireless service providers in automatic compliance with CALEA. Noncompliance with CALEA could potentially result in fines to a wireless service provider of up to $10,000 per day while it is noncompliant.

The TIA standards-making bodies battled with law enforcement authorities for five years before the *Lawfully Authorized Electronic Surveillance (J-STD-025)* standard was finally developed. Most of the debate surrounded the following issues:

- *Cost recovery*. Compliance with CALEA is a very expensive proposition for both equipment manufacturers and wireless service providers. Since CALEA is a non–revenue-generating function for the service providers, it is unclear how they would recover the costs of implementation and deployment.
- *Capacity*. The standards requirements put forth from the FBI for the number of simultaneous intercepts allowed is considered to be unrealistic. In some cases, the requirements specified 1 percent of an MSC's engineered capacity. For a switch that supports 100,000 lines, 1,000 simultaneous intercepts was considered to be implausible since, historically, wiretap rates were far fewer. Also, the cost of this additional engineering for each MSC in the network was considered to be enormous.
- *Privacy concerns*. Although wireless intercepts would still require a warrant to be performed, the amount and types of access to wireless calls required by the FBI standards was construed to be unrealistic and a potential infringement on personal freedom.
- *FBI requirements exceeding CALEA requirements*. The standards requirements document promulgated by the FBI specified requirements interpreted as being excessive when compared to the CALEA law itself.

Finally, in 2000, the TIA published the standard Lawfully Authorized Electronic Surveillance (J-STD-025), nearly six years after CALEA was first enacted. The FCC has recognized this standard as a safe harbor. However, government law enforcement agencies are still not satisfied and the U.S. Congress and the FCC were petitioned to review the standard. Because of this petition and the delayed publication of the standard, CALEA has yet to be implemented and fines have not been levied against wireless service providers.

Capabilities that law enforcement insisted on, but were not added by the wireless industry to the J-STD-025 standard include:

- Access to all conference call parties, including the identity of all parties, and access to conversations occurring when the intercept subject is not connected.
- Transmission of intercept information whenever a party is added, deleted, or placed on hold on a conference call.
- Access to user-controlled signaling, for example DTMF tones that could be used to access long-distance service, enter credit information, or access voice-mail.

- Network signaling that cannot be derived from call content, for example signaling information on the air-interface only, such as power-up registrations and power-down deregistrations.
- Precise time correlation between signaling information and call content (e.g., voice).
- Regular reports of all active intercepts and surveillances.
- Reports of feature status, such as the type of call forwarding activated and the forward-to number.
- A continuity check function to verify that call content information can be delivered on all available call channels.
- A standardized interface at all protocol layers; J-STD-025 defines only the application protocol layer.

The FCC will be issuing a new Notice of Proposed Rule Making addressing the concern of law enforcement as they pertain to J-STD-025. Important open questions still requiring resolution include:

- Privacy issues associated with new location technology.
- The retrieval and distinction of both signaling information and user content that can be transmitted transparently through the wireless network as generic packet data. Short-message service is an example of this.
- The interception of multiparty calls when a party, other than the intercept subject, adds another party to a call.
- In-call dialed digit extraction and the difficulty of the network's (other than the equipment actually receiving the signals) to distinguish among call identifying information, information services (e.g., credit information), and extraneous meaningless digits.

At best, CALEA mandates highly controversial wireless functionality. At present, the jury is still out on the final technology to be implemented and deployed by wireless service providers. The FCC has recently ruled that service providers must deploy the final agreed-upon functionality by September 30, 2001. One thing is guaranteed though; either the law enforcement community or the wireless industry will be disappointed by the final outcome.

Emergency Services Calls (Enhanced 911)

It is estimated that 50 million 911 emergency calls were made in the U.S. from mobile phones in 2000. This number has been growing with increasing subscriber rates for many years. Because of the tremendous need for public emergency services, enhanced 911 emergency service (E-911) was first mandated by the FCC in June 1996. This mandate stated that by October 1997, wireless service providers were required to support 911 emergency service calls from all mobile stations regardless of whether the MS was associated with a valid subscription or not. This means that a wireless service provider was required to allow a 911 call to be delivered to a public safety answering point (PSAP), even if the caller had no subscription to any service provider. This is known as "basic 911" service.

The FCC mandated Phase I of E-911 service to be implemented by April 1998. Phase I E-911 required a wireless service provider to pass the caller's directory number (e.g., the MIN or MDN), if known, as well as an approximate location of the MS to the PSAP. The approximate location could simply be the cell site identification or a sector of a cell site. This location information might not be very accurate, as cell site radii can vary from several hundred feet to several miles.

The FCC mandated Phase II of E-911 service to be implemented by October 2001. Phase II E-911 requires the wireless service provider to pass location information in the form of latitude and longitude coordinates to the PSAP. These coordinates are required to be accurate to within 125 meters of the mobile station at least 67 percent of the time.

Note that there are several business issues and technical problems associated with basic 911, Phase I E-911, and Phase II E-911.

Basic 911 Emergency Service

Basic 911 emergency service requires invalid mobile stations (i.e., mobile stations not associated with any wireless service subscription) to make 911 emergency calls. Invalid mobile stations cannot register, although they can be powered on and display radio signal strength. If a normal call is made from an invalid mobile station, the network will not complete the call since validation and service qualification cannot be obtained. Typical call treatment for these invalid calls consists of voice recordings indicating the call cannot be completed or a series of audible tones.

There are many problems with the basic 911 mandate, although the concept has good intentions:

- Invalid mobile stations have no identity and no subscription associated with an HLR. This allows nuisance 911 calls to be made with no indication of who the calling party is.
- Invalid mobile stations do not receive a bill for wireless service. This allows people to avoid paying emergency service fees associated with wireless service. These fees are used to subsidize the infrastructure to provide wireless emergency services and are usually included in the subscriber's bill.
- Wireless service providers are required to complete emergency service calls to invalid mobile stations with no way of recouping revenue for the call delivery and resources used in the network.
- Callback information is unavailable to invalid mobile stations making 911 emergency calls. In wire-line networks, the PSAP retains supervision over 911 calls, and can seize the line associated with the directory number regardless of whether or not the calling party disconnects. The wireless service provider has no way of calling back the invalid mobile station if the call is dropped or otherwise disconnected.

Phase I E-911 Emergency Service

Phase I E-911 emergency services is standardized in J-STD-034 *Wireless Enhanced Emergency Services*. Phase I of these emergency services requires the following capabilities:

- Callback from the PSAP to the mobile station originating the 911 call.
- Location information in the form of cell or cell-sector identification.
- Selective routing of the 911 call to the appropriate PSAP.
- Automatic reconnection to the mobile station calling 911 when the mobile station disconnects due to loss of radio coverage or other radio propagation anomalies.
- Three-way 911 calls, if the three-way calling feature is subscribed to.

Callback is required so that a PSAP operator can call back a mobile phone previously used to make a 911 emergency call. This is necessary if the 911 calling party has disconnected before enough information was obtained, if there is a problem locating the emergency, or if investigations are required following an emergency. The callback feature requires a valid mobile station that has transmitted the MIN or MDN to the serving system during call establishment.

Location information is provided in the form of cell-site or cell-sector identification. This identification is transmitted in the form of a unique directory number belonging to the emergency services network. This number can be associated with the cell site or sector and the PSAP to which to deliver the call.

Selective call routing of E-911 calls is necessary to associate the 911 call with the correct PSAP. This can be done by using the unique directory number employed to identify the cell site or sector. Such a number can be used to associate the calling party's location with the correct PSAP to obtain emergency services. This number is known as the *emergency services routing digits* (ESRD). Figure 19.2 shows E-911 emergency call selective routing with callback while a subscriber is roaming.

Figure 19.2
E-911 selective call routing to a PSAP with callback while roaming. The serving MSC queries the HLR for the subscriber's profile and transmits both the MDN and ESRD to a selective router. This router routes the call to a primary PSAP. If the PSAP determines that the call should be taken by another PSAP, then the call is transferred via the selective router to the appropriate PSAP.

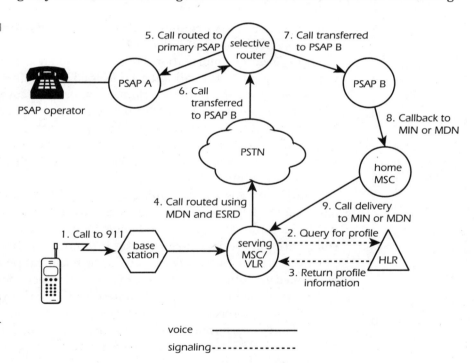

Automatic reconnection of an E-911 call may be required if the call is disconnected due to radio problems. The wireless system can then re-page the mobile station to attempt to reconnect the call. In the case of intersystem handoff, the current serving MSC would need to know that the call being reconnected is an emergency call. This mechanism is not guaranteed, and there may be many instances where the call cannot be reconnected.

Three-way 911 calls are made the same way a normal three-way call is made. However, the call supervision is required to be different. After a third party is connected using the standard three-way call feature, an additional flash request (i.e., pressing SEND again) would disconnect the third party E-911 call. The J-STD-034 standard makes a modification to the ANSI/TIA/EIA-664 standard, which specifies wireless features. This change requires subsequent flashes to be ignored once an E-911 call is connected. This means that the three-way call remains in effect until a party explicitly disconnects the call.

Phase II E-911 Emergency Service

Phase II E-911 emergency services is standardized in J-STD-036 *Enhanced Wireless E-911 Phase II*. This standard requires much more accurate location information for an emergency caller. The mobile station making an E-911 call must be located to within 125 meters 67 percent of the time. Wireless carriers must provide coverage over 50 percent of their service area within 6 months of the Phase II deadline and 100 percent of their service area within 18 months.

Many advanced technologies used to determine the position of a mobile station are much more accurate than cell-site and cell-sector identification. The more common technologies use "smart" antennas or the satellite-based Global Positioning System (GPS). Smart antenna technology used for determining position generally falls into two categories: time difference of arrival (TDOA) and angle of arrival (AOA). TDOA uses multiple antennas to measure minute differences in the arrival time of a mobile station's radio signal. A form of mathematical triangulation is performed to obtain a fairly accurate measurement for the origin of the radio signal. AOA uses multiple antennas to measure the differences in the angle of the received mobile station's radio signal. A form of triangulation is also performed to obtain a measurement for the origin of the radio signal. Both these technologies use very sophisticated digital signal processing techniques to derive the precise location of the mobile station.

These location-determining technologies are beyond the scope of ANSI-41. However, the J-STD-036 standard defines additional functional entities to manage subscriber location information for E-911. The *position determining element* (PDE) is a functional entity that provides the interface from the equipment deriving the location of a mobile station and the ANSI-41 wireless network. The PDE interfaces with a base station and specialized location equipment to pass the location information to a PSAP. The location equipment is proprietary and can be any technology that can derive location information from a mobile station. The *mobile position coordinator* (MPC) is a functional entity that provides the interface between the PDE and the PSAP as well as the interface between the serving MSC and the PSAP. The MPC is used to correlate the location information with the identity of the mobile station making an E-911 emergency call. Figure 19.3 shows the basic wireless E-911 network architecture incorporating the PDE and MPC.

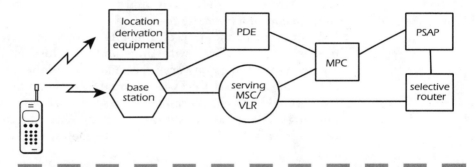

Figure 19.3 Basic Phase II E-911 network architecture. The location derivation equipment obtains the position of the mobile station making the 911 call. This position is passed to the PDE for collection and interconnection to the wireless network. The MPC correlates the position information with the subscriber identity associated with the 911 call. The selective router routes the call to the PSAP, which can correlate the position information and subscriber identity with the 911 call.

There are three basic methods for determining location and position information for a mobile station: network-based, mobile station–based, and a hybrid of these.

Network-based solutions can work for all mobile stations regardless of technology (i.e., AMPS, TDMA, and CDMA). These solutions are specific to the smart antenna technologies, since no additional capabilities need exist in the mobile station. The MS originates a 911 call and the network can determine the position based on radio signal propagation.

Mobile station–based solutions require the MSs to support special technology so that they can communicate their position to the network. An example of this is a GPS receiver embedded in the mobile station equipment. However, subscribers would require new, upgraded mobile stations with this embedded technology and there is currently no national standard for mobile station–based positioning. These solutions are required by the FCC to be more accurate than network-based solutions. Mobile station–based solutions for positioning are required to be accurate to within 50 meters of the caller 67 percent of the time and within 150 meters of the caller 95 percent of the time.

Hybrid solutions also require some enhancement to the mobile station equipment, but all the logic and signal analysis can reside in the network. An example of this solution can be found in ANSI/TIA/EIA-95 for CDMA systems. This standard specifies a special power-up function that the network can use to command the mobile station to raise its transmit signal strength high enough, and for a long enough time, for the network to obtain a fix on its position.

Position determination of a mobile station is a complex and difficult problem to solve. Even with the standard solutions, it may be difficult for the network to get a very precise location of the mobile station. Many of the solutions do not provide adequate location information (e.g., deriving vertical location information when an E-911 call is made from a tall building). The industry is aware of this; hence, the requirement only to determine a subscriber's position to 125 meters 67 percent of the time.

General Packet Radio Service (GPRS)

The general packet radio service (GPRS) is a technology first defined for GSM-based wireless networks in the GSM Phase 2+ version of the standard. It is generally considered to be a 2.5G technology; that is, the system is designed to provide wireless high-speed data rates close to broadband rates, but not actually considered a third-generation (3G) technology; it does not require an overhaul to existing network deployments. GPRS defines a high-speed packet-switching standard for GSM technology. However, because GSM uses a TDMA-based radio technology different from, but similar to, ANSI/TIA/EIA-136, the North American TDMA service providers who use ANSI/TIA/EIA-136 also chose GPRS as a standard technology to provide high-speed data services.

GPRS supports both X.25 and IP protocols, although IP-based technologies are preferred today to provide wireless Internet access. The design of GPRS is elegant because it is a relatively simple technology to add onto existing wireless telecommunications networks. Although both

GSM and TDMA networks support several types of data transfer, GPRS offers a standard means of providing high-speed Internet access that can operate with little or no change to the basic network infrastructure. The GPRS system differs from the other wireless data technologies standardized for GSM and TDMA because it bypasses the circuit-switch–based voice network and provides direct access to public packet data networks (PPDNs), like the Internet.

The GPRS technology incorporates two new functional entities into the wireless network infrastructure: Serving GPRS support node (SGSN) and gateway GPRS support node (GGSN).

These support nodes, along with some modifications to the radio base stations and new GPRS-capable mobile stations, are required to provide high-speed Internet access. The subscriber profile in the HLR would need to support the GPRS data feature as well as routing information. Two GPRS services are supported for a subscriber: point-to-point and point-to-multipoint (i.e., broadcast) service.

Data rates vary for GPRS, depending on the type of radio channel coding used and the number of time slots allocated for a subscriber. In the rawest form, data rates can technically be as high as 170kb/s, but figures of 115kb/s are more commonly accepted. Many systems are configured for a maximum of 64kb/s. These data rates are derived from the aggregate concatenation of time slots supported by the basic TDMA-based radio technology. GPRS technology uses the basic physical TDMA structure of the existing radio protocol, but supports concatenation of the time slots to achieve higher data rates than those available on individual time slots used for voice.

The SGSN functions as the delivery node in the network. It "serves" the mobile station by tracking its location (via the HLR) and delivering the data to the base station serving the MS. The GGSN functions as the gateway node, enabling data traffic to and from external packet data networks. Both these functional entities are based on IP data routing equipment, which is commonly available and much less expensive than circuit-switched telecommunications network equipment. Figure 19.4 shows the basic network architecture of GPRS.

GPRS makes very efficient use of the air interface. Radio channels are allocated only when data are sent or received and the forward and reverse traffic channels are assigned separately. This is different from voice-based circuit switching, where the traffic channels are held in both directions for the duration of the call. GPRS provides capacity-on-demand and an always-on data connection, enabling an Internet session to be held continuously with no penalty, since resources are used only when data are being transmitted. Normal voice calls can still be originated and terminated. GPRS-based mobile stations are specified in the standard as providing GPRS only, or as both a voice and a data terminal.

Figure 19.4
Basic GPRS network architecture. The serving GPRS support node (SGSN) passes high-speed packet data to and from the base station serving the subscriber. The gateway GPRS support node (GGSN) provides the interfaces to external packet-data networks such as the Internet.

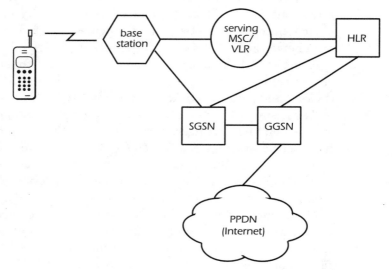

Wireless Intelligent Network (WIN)

ANSI-41 wireless networking technology can be considered an intelligent network (IN) in the broadest definition of the term. It employs the concepts of IN, as characterized by distributed functional entities that provide, process, and interpret information locally at the functional entities. However, WIN is typically characterized by much more sophisticated functionality designed to make the implementation and deployment of new enhanced services fast and efficient.

The WIN concept employs a call processing model with "triggers" that enable call treatment to be efficiently distributed to specialized nodes designed for specific services. This allows greater network intelligence and removes processing of multiple features and services from switching platforms to these special network nodes. A call model is a generic model used to treat any type of call. The call model has specific event points and trigger points that enable calls requiring special treatment to be processed if they fulfill the criteria of those events and triggers. A trigger is any event, action, or occurrence of a condition that is used to initiate a call-processing function. During the processing of a call, a trigger point may be reached that requires special call treatment. Signaling messages can then be sent to the appropriate intelligent node to perform special actions, such as database queries or process invocations, which provide the requested service that needs to be applied to the call.

When WIN functionality is properly implemented, the MSC supports a full suite of triggers and obviates the need for equipment manufacturers to develop new features and services within each switch. If a trigger is activated by the wireless service provider, a message need be sent only during call processing to the appropriate external node that has the intelligence to treat the call properly. These nodes can be service control points (SCPs), intelligent peripherals (IPs), or service nodes (SNs), which provide the functionality of both an SCP and an IP. The greatest advantage of WIN over basic ANSI-41 technology is the ability for MSCs to communicate with multiple SCPs, IPs, or SNs instead of a single HLR to provide features and services. WIN uses the SS7 backbone network for signaling to the external SCPs, IPs, and SNs and builds on the distributed wireless architecture provided by ANSI-41. Figure 19.5 shows the basic architecture for WIN.

Figure 19.5
Basic WIN architecture. SS7 is used as the signaling transport protocol to pass WIN messages to the HLR, VLR, SCP, IP, and SN. Note that the VLR may not be a separate network node and is typically implemented as a functional entity within the MSC.

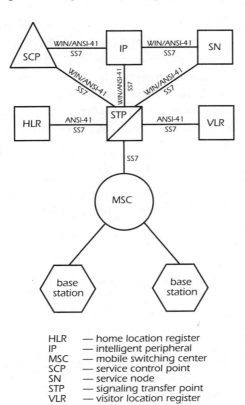

HLR	— home location register
IP	— intelligent peripheral
MSC	— mobile switching center
SCP	— service control point
SN	— service node
STP	— signaling transfer point
VLR	— visitor location register

Service control points (SCPs) are physical network elements in the SS7 signaling network. They are essentially end-signaling points that use transaction capabilities (i.e., the TCAP protocol) for direct access to application databases and other service processing functions. A typical example of SCP functionality is the processing of toll-free 800-number calls. Switches in the network send SS7 TCAP queries to SCPs, which access a database to translate the 800-number to the actual directory number of the called party. This directory number is sent back to the requesting switch as a result of the TCAP query. The switch can then route the call properly via the actual directory number provided. Intelligent peripherals (IPs) are functional entities particular to intelligent networks. IPs are typically responsible for providing resource service support functionality. Examples of IP services are dialed-digit analysis and interactive voice response systems. IPs are typically used for interactive voice-controlled services and in-call interactions with a subscriber; however, they are certainly not limited to this functionality. Service nodes (SNs) are network elements that combine the functions of an SCP and an IP.

One of the most desired aspects of WIN is the concept of a service creation environment (SCE). The SCE is an intelligent network feature that enables a service provider to design, implement, and deploy new features by using a standard set of triggers and responses to those triggers. A properly implemented SCE enables a wireless service provider to implement inexpensive, custom features and services for subscribers. The WIN concept, along with an SCE, provides the following advantages to wireless service providers:

- Ability to integrate wireless and wire-line functions
- Simplified individual functional entities and their connectivity
- Optimized MSCs for basic switching rather than intelligent call processing
- Faster feature development
- Custom feature development
- Rapid introduction of advanced services
- A flexible service creation environment
- Uniform features and services in a multivendor network infrastructure environment

Wireless intelligent network functionality is specified in TIA/EIA/IS-771, *Wireless Intelligent Network*. This standard specifies the following primary functionality:

- Addition of the SCP, IP, and SN functional entities
- Use of existing ANSI-41 and SS7 signaling
- Use of existing voice channel interconnections
- Enhancement of existing ANSI-41 intersystem operations for interfaces to new functional entities (i.e., SCP, IP, and SN)
- New intersystem operations for service logic and service data access
- Originating call model
- Terminating call model
- Origination triggers
- Termination triggers
- Service-independent triggers for distribution of service logic

Examples of the services and features that can be provided with WIN include:

- Voice-controlled services, such as:
  - speech-to-text conversion
  - voice-based user identification
  - voice-controlled dialing
  - voice-controlled feature control
- Single-number service
- Calling name presentation
- Incoming call screening
- Virtual private networking

In addition, number portability and calling-party-pays can both be implemented using WIN.

Wireless Local Number Portability (WNP)

Wireless local number portability is a service mandated by the FCC. This mandate is designed to promote more competition among local wireless and wire-line service providers. The impetus for this mandate is the belief that customers are less likely to change to a new service provider if they cannot keep their existing directory number. Changing directory numbers generally results in missed calls and large expenditures incurred by subscribers (e.g., business marketing and collateral materials, reprinting stationery, and efforts to communicate the number change to all concerned parties).

In July 1996, the FCC issued its first order defining Phase I and Phase II of number portability. Phase I requires wireless service providers offering service in the top 100 markets in the U.S. to provide the ability to route calls to ported wire-line numbers by December 31, 1998. This is known simply as *local number portability*. Phase II requires wireless service providers to support ported wireless numbers by June 30, 1999. This is known as *wireless number portability*. This date, however, was extended to March 31, 2000. After much lobbying by the Cellular Telecommunications & Internet Association (CTIA), the FCC agreed to delay the mandated deployment of WNP to November 24, 2002. The reasoning put forth for this delay was to allow new wireless PCS providers to build out their networks to a more mature state.

Number portability is standardized in TIA/EIA/IS-756, *Enhancements for Wireless Number Portability*. This standard provides specifications for both Phase I and Phase II. The basic requirement for Phase I, local number portability, is to correctly terminate calls to ported wire line numbers. An example of this is when a business or residence chooses to change local exchange carriers for a wire-line phone, but wishes to keep the same phone number. The network is required to deliver the call made to the phone number to the correct network so that it can be received by the called party. Historically, local exchange carriers *owned* large blocks of numbers that they obtained from the North American Numbering Plan Administrator. With local number portability, a number is owned by the service provider determined by the customer. The original standard for Phase I local number portability was published as *TIA/EIA/IS-756* in April, 1998. This standard was revised as *TIA/EIA/IS-756 Revision A* to incorporate the standards for Phase II. IS-756 is considered an addition to the ANSI-41 standard.

Phase I Local Number Portability

The basic method used to fulfill the requirements of Phase I is to use a *local reference number* (LRN) to identify the new switch used to terminate a call to the ported number. The original switch that terminated calls to the phone number prior to the port is not involved in the call. LRNs are stored in the *number portability database* (NPDB), a new functional entity that has been added to the ANSI-41 network reference model. The first six digits of the ported directory number (i.e., NPA-

NXX) of the entire 10-digit directory number are used to route the call to the proper terminating switch. Figure 19.6 shows the routing scenario for Phase I local number portability when delivering local mobile-to-wire line calls. Figure 19.7 shows the routing scenario for Phase I local number portability when delivering interexchange mobile-to-wire line calls.

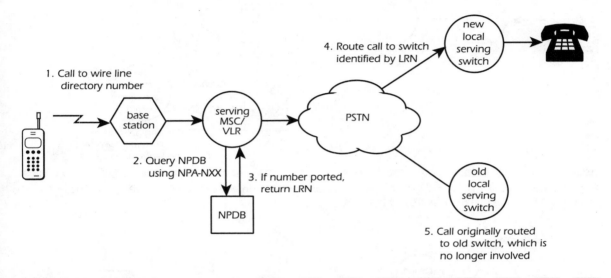

Figure 19.6 A call to a ported wire-line directory number. The serving MSC queries the number portability database (NPDB) for an LRN. The LRN identifies the correct local switch (known as the recipient switch) to route the call to. If no LRN is returned for the query, the call is routed to the original switch (known as the donor switch).

The MSC uses the NumberPortabilityRequest (NPREQ) message to query the NPDB. SS7 ISUP signaling is used to establish the call to the correct local switch. The LRN returned from the NPDB is used as a parameter within the ISUP InitialAddressMessage (IAM) to establish the call via the appropriate trunk group.

Phase II Wireless Number Portability

Phase II WNP allows termination of wire-line or wireless calls to ported wireless directory numbers. It also allows wireless subscribers to retain their directory numbers even when they change wireless service providers. Local reference numbers (LRNs) are also used for Phase II in the same way they are used for Phase I (see Figure 19.8).

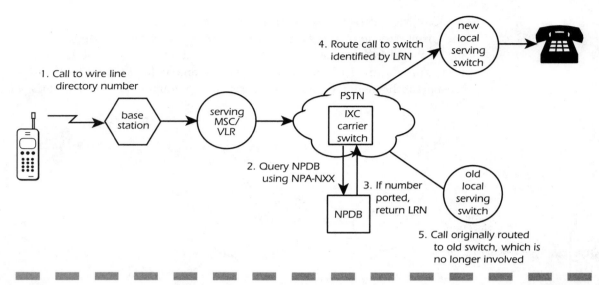

Figure 19.7 A call to a ported wire-line directory number for an interexchange carrier (IXC) call. The IXC is responsible for obtaining the LRN from the NPDB.

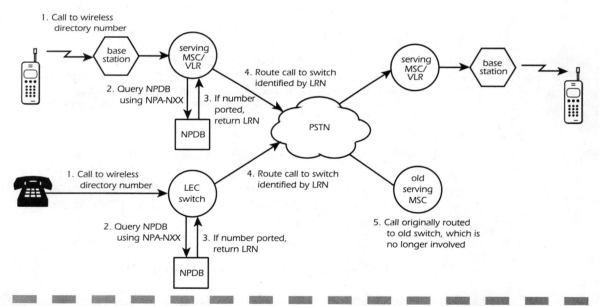

Figure 19.8 A call to a ported wireless directory number. For mobile-originated calls, the serving MSC queries the number portability database (NPDB) for an LRN. The LRN identifies the correct MSC (known as the recipient switch) to route the call to. If no LRN is returned for the query, the call is routed to the original MSC (known as the donor switch). For wire line–originated calls, the local exchange carrier (LEC) switch queries the NPDB for an LRN. For interexchange carrier calls, the scenario is the same as in Figure 19.7.

Number portability for wireless subscribers is a bit more complex. In most cases today, a subscriber's mobile identification number (MIN) also serves as the dialable wireless directory number. However, this is certainly not a requirement and this implementation is not specified in ANSI-41. In the case of wireless directory number portability, separation of the MIN and the mobile directory number (MDN) is required since MINs are the primary information element used for mobility management signaling in the network. If the MIN were portable, number portability queries would need to be performed for all mobility management signaling queries. When the MIN acts as the MDN, it is associated with an MSC that is the terminating switch for that MIN. However, the MIN is also associated with a home location register (HLR), authentication center (AC), message center (MC), etc. of a specific wireless service provider. Queries to each of these functional entities would require database queries that would significantly increase call processing time and add unreasonable amounts of traffic to the signaling network.

Separation of the MIN and MDN optimizes ANSI-41 signaling and requires no changes to network routing based on the MIN. However, this requires a new MIN to be assigned to the wireless subscriber and reprogrammed into the mobile station whenever a wireless phone is ported; a positive tradeoff when compared to the network processing required if this were not the case. Note that the international mobile station identifier (IMSI) is completely unaffected by number portability since the IMSI never acts as the mobile directory number.

Phase III Wireless Number Portability

Phase III wireless number portability is not a mandate of the FCC. This phase of WNP is being worked on by the TIA independently to address number portability issues that need to be resolved even though they are not mandated. Because Phase II of WNP was delayed until 2002, no work is being performed on Phase III, although many issues have already been recognized by the standards organizations, including:

- Number portability for short-message service (SMS) routing.
- Number portability for SS7 global title routing (GT) to perform proper network routing for each service that uses GT.

APPENDIX A

STANDARDS AND BULLETINS RELATED TO ANSI-41

This appendix contains the standards, bulletins, and specifications most closely related to ANSI-41 technology. The latest versions listed are valid as of the publication date of this book and are subject to change.

Note that there are many other standards and bulletins not listed here. These standards pertain to other aspects of wireless systems not directly related to ANSI-41, such as the ancillary air-interface standards that belong to the AMPS, TDMA, and CDMA families of standards.

Name: ANSI/TIA/EIA-41
Title: Cellular Radio-Telecommunications Intersystem Operations
Summary: Provides standard intersystem operations among MSCs, VLRs, and HLRs for wireless networks to support subscriber mobility.
Latest Version: ANSI/TIA/EIA-41 Revision E (publication scheduled for 2001)

Name: ANSI/TIA/EIA-93
Title: Wireless Telecommunications A_i–D_i Interfaces Standard
Summary: Provides the standard protocol interfaces from a wireless MSC to wireline switches in the PSTN and ISDN.
Latest Version: ANSI/TIA/EIA-93 Revision A (1998)

Name: ANSI/TIA/EIA-95
Title: Mobile Station–Base Station Compatibility Standard for Dual-Mode Wideband Spread Spectrum Cellular Systems
Summary: Provides the standard dual-mode AMPS–CDMA air-interface protocol supported by ANSI-41. Note that there are other standards and TSBs closely associated with this air-interface standard that do not affect ANSI-41.
Latest Version: ANSI/TIA/EIA-95 Revision B (1999)

Name: ANSI/TIA/EIA-124
Title: Wireless Radio Telecommunications Intersystem Non-Signalling Data Communication DMH (Data Message Handler)
Summary: Provides the standard protocol for the transport of near-real-time call detail records in support of billing and fraud management for ANSI-41 networks.
Latest Version: ANSI/TIA/EIA-124 Revision B (1999)

Name: ANSI/TIA/EIA-136
Title: TDMA Cellular PCS
Summary: Provides the standard dual-mode AMPS–TDMA air-interface protocol supported by ANSI-41. Note that there are other standards and TSBs closely associated with this air-interface standard that do not impact ANSI-41.
Latest Version: ANSI/TIA/EIA-136 Revision B (2000)

Name: ANSI/TIA/EIA-634
Title: MSC-BS Interface for Public Wireless Communications Systems
Summary: Provides the standard A interface transport protocol for wireless systems based on both SS7 and frame relay and supported by ANSI-41.
Latest Version: ANSI/TIA/EIA-634 Revision B (1999)

Name: TIA/EIA/IS-653
Title: ISDN-Based A-Interface (Radio System–PCSC) for 1800 MHz Personal Communications Systems
Summary: Provides the standard A interface transport protocol for PCS systems based on ISDN and supported by ANSI-41.
Latest Version: TIA/EIA/IS-653 Revision A (1996)

Name: ANSI/TIA/EIA-660
Title: Uniform Dialing Procedures and Call Processing Treatment for Use in Cellular Radio Telecommunications
Summary: Provides the standard subscriber dialing plan for wireless networks and the network treatment of the call types dialed.
Latest Version: ANSI/TIA/EIA-660 (1996)

Name: ANSI/TIA/EIA-664
Title: Cellular Features Description
Summary: Provides the standard stage 1 descriptions (from the subscriber's perspective) of the wireless supplementary services supported by ANSI-41.
Latest Version: ANSI/TIA/EIA-664-A (2000)

Name: ANSI/TIA/EIA-691

Title: Mobile Station–Land Station Compatibility Standard for 800 MHz Analog Cellular

Summary: Provides the standard analog AMPS air-interface specification supported by ANSI-41 networks. Note that there are other standards and TSBs closely associated with this air-interface standard, but they do not affect ANSI-41.

Latest Version: ANSI/TIA/EIA-691 (1999)

Name: TIA/EIA/IS-129

Title: Interworking/interoperability Between DCS 1900 and IS-41 Based MAPs for 1800 MHz Personal Communications Systems–Phase

Summary: Provides the interim standard for interoperability between a GSM network system and an IS-41 network system to enable roaming between these systems.

Latest Version: TIA/EIA/IS-129 (1996)

Name: TIA/EIA/IS-680

Title: Personal Base Station–Authorization and Call Routing Equipment Compatibility Standard

Summary: Provides the interim standard for communication between a personal base station and authorization and call routing equipment (ACRE).

Latest Version: TIA/EIA/IS-680 (1996)

Name: TIA/EIA/IS-725

Title: Cellular Radiotelecommunications Intersystem Operations— Over-the-Air Service Provisioning (OTASP) and Parameter Administration (OTAPA)

Summary: Provides the interim standard for the OTASP subscriber feature description and intersystem operations for the CDMA and TDMA air interfaces.

Latest Version: TIA/EIA/IS-725-A (1999)

Name: TIA/EIA/IS-728

Title: Intersystem Link Protocol

Summary: Provides the interim standard for an intersystem link protocol (ISLP) for circuit-mode data services. These data services include asynchronous data and group-3 fax. The ISLP adapts between air-interface data rates and higher-speed intersystem rates. The ISLP may be used between a serving system and an anchor system, possibly through one or more tandem systems.

Latest Version: TIA/EIA/IS-728 (1998)

Name: TIA/EIA/IS-730

Title: Intersystem Operations Support for the IS-136 Digital Control Channel

Summary: Provides the interim standard changes and additions to ANSI/TIA/EIA-41 to support basic roaming and handoff scenarios for wireless systems following the IS-136 air-interface technology.

Latest Version: TIA/EIA/IS-730 (1997)

Name: TIA/EIA/IS-735

Title: Enhancements to TIA/EIA-41-D & TIA/EIA-664 for Advanced Features in Wideband Spread Spectrum Systems

Summary: Provides the interim standard modifications and additions to TIA/EIA-41 Revision D and TIA/EIA-664 required to support advanced features for CDMA. These features include network-directed system selection and subscriber confidentiality supported by the temporary mobile station identity.

Latest Version: TIA/EIA/IS-735 (1998)

Name: TIA/EIA/IS-737

Title: Enhancements for Circuit Mode Services

Summary: Provides the interim standard service allowing TDMA wireless subscribers to send and receive asynchronous data.

Latest Version: TIA/EIA/IS-737 (1998)

Name: TIA/EIA/IS-751

Title: TIA/EIA-41-D Modifications to Support IMSI

Summary: Provides the interim standard modifications to support the International Mobile Station Identity (IMSI) and should be used in parallel with ANSI/TIA/EIA-41.

Latest Version: TIA/EIA/IS-751 (1998)

Name: TIA/EIA/IS-756

Title: TIA/EIA-41-D Enhancements for Wireless Number Portability

Summary: Provides the interim standard specifying the network impact of number portability, specifically service provider portability, as mandated by the Federal Communications Commission in docket No. 95-116. Service provider portability means subscribers can retain the same telephone numbers whenever they change from one service provider to another.

Latest Version: TIA/EIA/IS-756 Revision A (1998)

Name: TIA/EIA/IS-764

Title: TIA/EIA-41-D Enhancements for Wireless Calling Name Feature Descriptions

Summary: Provides the interim standard describing the CNAP and CNAR features, and specifies the operation of CNAP and CNAR so that a roaming wireless subscriber can use these features in a seamless manner.
Latest Version: TIA/EIA/IS-764 (1998)

Name: TIA/EIA/IS-771
Title: Wireless Intelligent Network
Summary: Provides the interim standard features, descriptions, and operational procedures for WIN.
Latest Version: TIA/EIA/IS-771 (1999)

Name: TIA/EIA/IS-778
Title: Wireless Authentication Enhancements Descriptions
Summary: Provides the interim standard identifying authentication enhancements so that a roaming wireless subscriber will be authenticated in a seamless manner.
Latest Version: TIA/EIA/IS-778 (1999)

Name: TIA/EIA/IS-807
Title: TIA/EIA-41-D Enhancements for Internationalization
Summary: Provides the interim standard enhancements necessary to support international intersystem operations.
Latest Version: TIA/EIA/IS-807 (1999) & TIA/EIA/IS-807-1 Addendum 1 (2000)

Name: TIA/EIA/IS-808
Title: Network Based Enhancements for User Identity Module (UIM)
Summary: Provides the interim standard for stage 1, stage 2, and stage 3 enhancements to support a mobile station equipped with a user identity module (UIM). The UIM can be in one of two forms, either integrated within the mobile station or removable, permitting the UIM to be inserted or removed from the mobile equipment. This document addresses the removable UIM (R-UIM) only.
Latest Version: TIA/EIA/IS-808 (2000)

Name: TIA/EIA/IS-812
Title: TIA/EIA-41-D Message Segmentation
Summary: Provides the interim standard enhancements to enable a wireless system to determine whether lower-layer message segmentation is supported in communication with another system.
Latest Version: TIA/EIA/IS-812 (1999)

Name: TIA/EIA/IS-820
Title: Removable User Identity Module (R-UIM) for TIA/EIA Spread Spectrum Standards
Summary: Provides the interim standard containing the requirements for (R-UIM), smart card-based technology for CDMA systems.
Latest Version: TIA/EIA/IS-820 (2000)

Name: J-STD-025
Title: Lawfully Authorized Electronic Surveillance
Summary: Provides the standard for the interfaces between a telecommunications service provider and a law enforcement agency to assist law enforcement in conducting lawfully authorized electronic surveillance.
Latest Version: J-STD-025-A (2000)

Name: J-STD-034
Title: Wireless Enhanced Emergency Services
Summary: Provides the standard for the Phase I wireless enhanced emergency services capabilities. These capabilities include provision of base station, cell site, or sector identification information; subscriber identification; callback and reconnect features.
Latest Version: J-STD-034 (1997)

Name: J-STD-036
Title: Enhanced Wireless E-911 Phase II
Summary: Provides the standard for the Phase II wireless enhanced emergency services capabilities. These capabilities primarily include advanced position and location information.
Latest Version: J-STD-036 (2000)

Name: TIA/EIA TSB29
Title: International Implementation of Wireless Telecommunication Systems Compliant with ANSI/TIA/EIA-41
Summary: Provides recommended guidelines for the implementation, assignment, and administration of MIN and SID values for international implementations of AMPS and ANSI-41.
Latest Version: TIA/EIA TSB29 Revision C (1999)

Name: TIA/EIA TSB50
Title: User Interface for Authentication Key Entry
Summary: Provides recommended user procedures for programming the A-key required for authentication into the mobile station.
Latest Version: TIA/EIA TSB50 (1993)

Name: TIA/EIA TSB76

Title: ANSI-41 Enhancements for PCS Multi-Band Support

Summary: Provides modifications to ANSI-41 messages and procedures to enable interworking between cellular and PCS systems.

Latest Version: TIA/EIA TSB76 (1996)

Name: TIA/EIA TSB100

Title: Wireless Network Reference Model

Summary: Provides the specification for the basic wireless network reference model including network entities and reference points.

Latest Version: TIA/EIA TSB100 (1998)

Figure A.1

Relationship of the TIA/EIA standards to the interfaces using those standards. (Reproduced under written permission of Cellular Networking Perspectives)

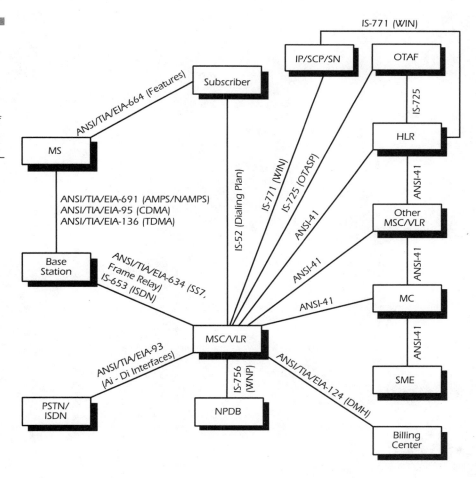

APPENDIX B

THE STANDARDS CREATION PROCESS

Telecommunications Industry Association (TIA) Interim Standards and specifications are created to define and promote compatibility among telecommunications systems equipment.

The process begins when a representative of a member company introduces a contribution to an engineering committee or subcommittee (i.e., a standard-formulating group) requesting the creation of a new standard specification for a particular technology. If a large enough consensus is reached within the formulating group to develop the specification, a project request letter along with a project initiation notice form is submitted to the TIA Technical Standards Subcommittee (TSSC) for approval.

If the TSSC approves the project request, the TIA Standards Secretariat will assign a project number (PN) to the specification and inform the formulating body of the approval. The standards-formulating group then solicits written contributions to begin creating the technical standard. The formulating group assigns an editor to develop a draft of the standard based on the contributions, or portions of the contributions, that have been accepted by consensus. Typically, contributions continue to be added to the draft over a period of several months or years. The draft specification is continually revised and rewritten based on accepted contributions. Eventually, when enough core material has been agreed upon, the draft specification becomes baseline text. The advance to baseline text is an affirmation that the material in the draft is in a relatively stable state.

Written contributions continue to be added to the baseline text until it is decided by consensus that the technical material in the specification is complete, accurate, and adequate for the standard. The baseline text is then frozen and no more new technical changes are allowed to be made to the text, unless, of course, an error has been found. Editorial changes are always allowed, up until publication, so that clarifications as well as grammatical corrections can be made.

After the technical information is frozen, the specification may be submitted to a process known as verification and validation (V&V). The V&V process is meticulous, since the baseline text is carefully analyzed, line by line, for correctness and accuracy. Once the V&V process is complete, the specification can be submitted to the TSSC to become a standards proposal (SP). This SP and a ballot are distributed to the TIA committee membership for vote. Ballots returned with a negative vote or with technical comments must be resolved before the SP can be published as a standard. If enough significant technical changes are required to the SP, it may need to be reballoted. Once all the comments and negative votes have been resolved, the SP can be submitted to the TIA for publication as a standard.

TIA interim standards are considered satisfactory for a maximum of five years. Within this time, the standards-formulating group must review the standard and take one of the following three actions:

- Reaffirm the standard; that is, determine that the technical content is still valid and requires no changes.
- Revise the standard; that is, determine that the technical content is still valid, but requires changes.
- Rescind the standard; that is, determine that the technical content no longer has value and declare the standard to be no longer in force.

The ANSI-41 standard is one of the few TIA or ANSI standards to be officially revised at least six times for technical modifications and enhancements.

Figure B.1
Flow chart of the standards creation process.

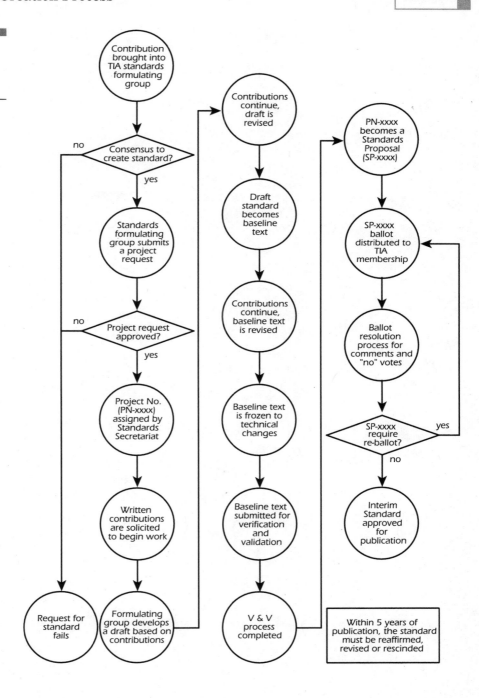

APPENDIX C

GLOSSARY

A-band—One of the two sets of frequency spectra licensed to cellular service providers and generally characterized by mobile station transmit-frequencies between 824 and 836 MHz and receive-frequencies between 869 and 880 MHz. Also, one of the sets of frequency spectra licensed to PCS providers generally characterized by mobile station transmit frequencies between 1850 and 1865 MHz and receive-frequencies between 1930 and 1945 MHz.

access signaling—A type of signaling used to manage communications between a network user and an access point into the network.

adaptive routing—In packet-switched networks, the ability of traffic to be dynamically routed around congested or inaccessible switches and trunks, enabling communications to continue over alternate network paths.

advanced intelligent network (AIN)—A network specified by Bellcore (now Telcordia) and defined as a service-independent network architecture that allows an exchange carrier or its agent quickly and economically to create, modify, and support telecommunications services and their operations to meet the needs of subscribers.

advanced mobile phone system (AMPS)—The analog cellular telecommunications system used in North America. It is characterized by analog radio frequencies between 800 and 900 MHz.

alerting—The function that provides an indication to the MS that a mobile-terminated call is being delivered or that some other feature has been invoked. Alerts are typically in the form of audible tones, visual signals, or vibrations.

alternate procedures—These are optional substitute procedures that may be used under certain circumstances for feature operation. They may be used in addition to the normal specified procedures, but not in place of them.

American National Standards Institute (ANSI)—The primary United States national public standards-making group responsible for creating and accrediting standards related to telecommunications systems and services.

anchor MSC—The role that an MSC assumes when it is the initial serving MSC for the call; the MSC that controls the base station that is first to assign an air-interface traffic channel to an MS-originated or MS-terminated call. It is called the *anchor* MSC because it serves as the *fixed* MSC—with special processing responsibilities—for the duration of the call, even if the serving MSC changes as a result of call handoff.

ANSI-41 network—A wireless network based on the ANSI/TIA/EIA-41 standard. This term is generic and describes any network that uses ANSI-41 intersystem operations to provide mobility to subscribers.

antenna system—A system that converts electrical signals from a radio transmitter into electromagnetic waves that comprise the radio transmission signals or vice versa.

application service element (ASE)—The portion of the application layer (layer 7) of the OSI reference model that performs a specific task for the application, such as specialized communications and transaction-based operations. Many ASEs can serve a single application.

application service interface (ASI)—The interface between the MAP application processes and the application service elements (ASEs) serving the MAP.

authentication—A procedure used to validate a mobile station's or subscriber's identity.

authentication center (AC)—A functional entity in the ANSI-41 network that manages the authentication information related to a mobile station or subscriber.

authentication controller—This term is not used in ANSI-41; we use it to refer to the entity controlling various authentication processes. This can be either the VLR or the AC, depending on whether or not SSD is shared with the serving system.

authentication key (A-key)—The primary private 64-bit key used in the CAVE algorithm. Private-key encryption means that the same key is used to encrypt and decrypt the information; therefore, the key must be shared by at least two network elements, the authentication center (AC) and the mobile station (MS). The A-key is only used to generate the temporary private key, called the shared secret data (SSD).

automatic number identification (ANI)—The number that provides the charge number of the line or trunk that originated a call. It is a number used to support exchange access and billing. This number may or may not be identical to the calling party number.

automatic roaming—The ability of a visited wireless network to support roaming of a mobile subscriber automatically, without additional actions being taken by the subscriber or a party calling the subscriber.

autonomous registration—Another name for timer-based registration, whereby the mobile station periodically registers with the network without the subscriber's taking any actions.

B-band—One of the two sets of frequency spectra licensed to cellular service providers generally characterized by mobile station transmit-frequencies between 837 and 849 MHz and receive-frequencies between 881 and 894 MHz. Also, one of the sets of frequency spectra licensed to PCS providers generally characterized by mobile station transmit-frequencies between 1870 and 1885 MHz and receive-frequencies between 1950 and 1965 MHz.

bandwidth—The specific range of frequencies that a given signal can occupy.

base station (BS)—The functional entity that represents all the functions that terminate radio communications at the network side of the MS interface to the wireless network. It controls radio resources and manages network information required to provide telecommunications services to the MS. The BS essentially consists of the radio transceiver and radio transceiver controller equipment serving one or more cells.

basic trading area (BTA)—One of the 493 personal communications services (PCS) areas defined by Rand McNally & Company and adopted by the FCC. The BTA overlays part of the area covered by a major trading area (MTA) and is significantly smaller than an MTA.

bearer information—The necessary information that enables communications between end-users of a bearer service to occur.

bearer service—A communications service defined by the basic capabilities of the transmission medium that the communications traverse. In SMS, the transport protocol, along with basic addressing and identification functions, define the bearer service.

Bell System—The name of the pre-divestiture conglomerate of AT&T, Western Electric, Bell Laboratories, and the Regional Bell Operating Companies.

border MSC—A term synonymous with neighbor MSC, although its use is generally limited to discussions of *problems* that occur between neighboring systems.

border system—This term generally refers to an MSC (or MSC and associated VLR) that serves cells in close proximity to those of the current serving system; either adjacent to or sufficiently close so that signals intended for one system are overheard by the other. The use of this term is limited to discussions of problems that occur between neighboring systems.

C-band—One of the sets of frequency spectra licensed to PCS providers generally characterized by mobile station transmit-frequencies between 1895 and 1910 MHz and receive-frequencies between 1975 and 1990 MHz.

call—A temporary communication between telecommunications end-users for the purpose of exchanging information. A call includes the sequence of events that allocates and assigns resources and signaling channels required to establish a communications connection.

call control—*See* "call processing."

call history count—The value of a 6-bit counter maintained by both the AC (and the serving system if SSD is shared) and the MS. The MS counter is increased under command of the AC (or serving system) and is used to detect the existence of an illegal MS.

call processing—A set of functions that establishes, maintains, and releases calls to and from the mobile subscriber.

called party—The intended recipient of a telephone call. For mobile-terminated calls, the MS is the called party. For mobile-originated calls, any valid telecommunications termination point can be the called party.

calling party—The party originating a call. For mobile-originated calls, the MS is the calling party. For mobile-terminated calls, any valid telecommunications terminal can be the calling party.

calling party number—A set of digits and related indications that provide numbering information related to a calling party.

calling party pays (CPP)—A method of billing where the calling party rather than the called party is charged for the wireless call.

candidate MSC—A neighbor MSC, which the current serving MSC is considering for the role of the new serving MSC before intersystem handoff occurs.

cell—An individual geographic area that is the topological component of a cellular or personal communications services system. The area is defined by the telecommunications coverage of the radio equipment located at the cell site.

cell sector—A geographic portion of a cell (typically a third) that is served by directional antennas. The aggregate of the cell sectors that form an entire cell increases capacity by providing frequency reuse among the sectors.

cell site—The physical location of a cell's radio equipment and supporting systems. This term also refers to the equipment located at the cell site.

cellular authentication and voice encryption (CAVE)—The common term for the collection of algorithms specified for the authentication processes. The CAVE algorithms make use of the MIN, ESN, A-key, and SSD values to generate authentication results that determine the legitimacy of a mobile station.

cellular digital packet data (CDPD)—A technology that provides digital packet-switched data services through standard analog cellular channels. CDPD consists of a radio portion (the air interface) and a network that employs standard OSI protocols to support end-to-end transfer of packetized messages.

cellular geographic service area (CGSA)—A cellular market coverage area defined by the United States Government. Two cellular network operators are licensed in each of these areas (A-band and B-band). A CGSA is further defined as either a metropolitan statistical area (MSA) or a rural service area (RSA). There are 733 CGSAs.

Cellular Intercarrier Billing Exchange Roamer (CIBER)—The call detail record format defined by CIBERNET Corp. that is specified for wireless calls.

cellular subscriber station (CSS)—*See* "mobile station."

cellular system—A type of wireless system characterized by high capacity, frequency reuse, and mobility management that controls communications to and from specific geographic "cells."

Cellular Telecommunications & Internet Association (CTIA)—Formerly, the Cellular Telecommunications *Industry* Association, this is an international organization of the wireless communications industry, including both wireless carriers and manufacturers, formed in 1984. The membership of the association includes all commercial mobile radio service providers, including cellular, personal communications services, enhanced specialized mobile radio, and mobile satellite services.

central office—An exchange carrier switching system that terminates station loops and connects the loops to each other and to trunks. Sometimes called "end office."

channel—A single unidirectional or bidirectional path for transmitting or receiving electrical signals between two entities.

channel associated signaling—A type of signaling in which the signals necessary for the traffic carried by a single channel are transmitted in the channel itself or in a channel permanently associated with it.

charge number—*See* "automatic number identification."

circuit—An electromagnetic path between two or more points capable of providing a number of channels to support communications.

circuit-switched connection—A circuit connection established and maintained, usually on demand, between two or more stations to allow the exclusive use of the circuit until the connection is released.

code division multiple access (CDMA)—A direct sequence spread spectrum radio technique that digitally codes a base signal and employs different signal encoding patterns for each channel to redistribute that base signal across a broad bandwidth. Channels for the signal are distinguished from one another through the use of unique code sequences.

common carrier—A company that provides public telecommunications facilities at nondiscriminatory rates. Common carriers are regulated by the FCC.

common channel signaling—A type of signaling in which signaling information relating to many circuits or channels is conveyed over a single channel.

conversation—The portion of a call in which the calling and called parties may engage in a spoken exchange.

customer service center (CSC)—A functional entity in the ANSI-41 wireless network reference model required for over-the-air service provisioning. The CSC is responsible for handling all customer issues, specifically initial activation, subsequent mobile station programming, and service requests. The CSC is typically a proprietary function incorporating proprietary communications interfaces.

D-band—One of the sets of frequency spectra licensed to PCS providers generally characterized by mobile station transmit-frequencies between 1865 and 1870 MHz and receive-frequencies between 1945 and 1950 MHz.

data circuit-terminating equipment (DCE)—The interface equipment that provides all the functions that are required to establish, maintain, and terminate a connection. It couples the data terminal equipment to a transmission circuit and vice versa.

data terminal equipment (DTE)—The equipment comprising the data source, data termination, or both. It converts user information into data signals for transmission or reconverts received data signals into user information.

digital AMPS (D-AMPS)—An informal term for the dual-mode TDMA/AMPS technology standardized by the Telecommunications Industry Association in North America (i.e., ANSI/TIA/EIA-136).

digital signal level 0 (DS0)—A 64-kb/s digital pulse-code-modulated signal (or channel) derived from a 4 kHz voice-band analog signal.

digital signal level 1 (DS1)—A time division multiplexed digital signal comprised of 24 DS0 level signals (or channels) at a composite rate of 1.544 Mb/s.

directory number (DN)—The dialable number used to call a mobile station or telecommunications terminal. It is the actual number used by a calling party to reach a called party.

distributed network—A type of network characterized by network resources, such as switching equipment and databases, which are distributed throughout the geographical area being served by the network.

dual tone multifrequency (DTMF)—A method for signaling digits in-band over a voice channel by adding two multifrequency signal tones together.

E-1 carrier—A digital transmission system used in Europe, which multiplexes 32 separate 64 kb/s digital channels onto one circuit at a composite rate of 2.048 Mb/s. Only 30 of the channels are used for voice and data. The other two are used for frame synchronization and signaling.

E-band—One of the sets of frequency spectra licensed to PCS providers generally characterized by mobile station transmit-frequencies between 1885 and 1890 MHz and receive-frequencies between 1965 and 1970 MHz.

electronic serial number (ESN)—A 32-bit number permanently stored in North American mobile stations, uniquely identifies the mobile station equipment.

emergency services—A generic term for the ability of a mobile subscriber to make an emergency service call (e.g., calling 911) from the mobile station, have an operator respond to that call, and receive appropriate emergency assistance. Emergency services are typically associated with enhanced 911 service (*see* "enhanced 911").

emergency services routing digits (ESRD)—These are digits of a unique directory number associating the cell-site or cell-sector identification information with the appropriate PSAP to properly deliver E-911 emergency calls.

end-user—The user of or subscriber to wireless network services. The end-user is either the calling party for mobile-originated calls, or the called party for mobile-terminated calls.

enhanced 911 (E-911)—A 911 emergency call placed from a mobile station that supports the transmission of location information to a public safety answering point so that emergency service can be provided at the calling party's location.

equipment identity register (EIR)—The functional entity in the network that provides the database repository for mobile equipment–related data. Information such as normal, stolen, or unknown, pertaining to individual mobile station equipment status is stored.

F-band—One of the sets of frequency spectra licensed to PCS providers generally characterized by mobile station transmit-frequencies between 1890 and 1895 MHz and receive-frequencies between 1970 and 1975 MHz.

feature—A bearer service, teleservice, or supplementary service provided to a subscriber, such as call waiting, call forwarding, or short-message services.

feature activation—An action taken by the service provider, subscriber, or network to enable a feature operation to be invoked. While active, a feature may be either "quiescent" or "operative" according to whether or not the system is actually providing the feature at a given time. For example, call forwarding can be active and quiescent while no mobile-terminated calls are made. It becomes operative when a call is actually made and the feature is subsequently invoked.

feature authorization—The administrative process of making the feature available to a subscriber, involving the subscriber and the service provider. Subscription options may be associated with the authorization. A feature can be made available to all subscribers or to an individual subscriber by prearrangement with the service provider.

feature code—A special sequence of digits entered by a subscriber into the MS to perform some action within the wireless network.

feature interactions—The rules to follow when multiple features are invoked simultaneously.

feature invocation—An action or event that triggers the feature to be performed. It may be an action taken by the subscriber (e.g., by pressing a specific key), automatically by the network, or automatically by the terminal as a result of a particular condition that occurs.

feature registration—The process of programming, by the subscriber or service provider, information related to the feature (e.g., *registering* the forward-to number for call forwarding).

Federal Communications Commission (FCC)—The United States regulatory body consisting of a board of seven commissioners appointed by the President. The FCC has the power to regulate all interstate and foreign electrical communications systems originating in the United States, including radio, television, facsimile, telephone, and cable systems.

flash—A message sent on a voice channel from an MS to the MSC used to invoke some special processing. Typically, the pressing of the "SEND" key by the subscriber during a call is considered a flash from the MS.

frequency modulation (FM)—The modulation of the frequency of an electromagnetic carrier wave, with another wave serving as the modulating signal, such that the frequency excursions of the carrier wave are proportional to the modulating signal bearing the intelligence to be transmitted.

frequency reuse—A method of simultaneously transmitting the same channel frequency at many cells at specific power levels to expand the capacity of the system (provided that the distance between cell sites using the same frequency is great enough to avoid interference).

functional entity (FE)—A logical entity in the network consisting of one or more logical functions. The relationship between logical functional entities and physical network nodes can be one-to-one or many-to-one.

gateway GPRS support node (GGSN)—A functional entity required for the general packet radio service (GPRS), which acts as a logical interface to external packet data networks.

gateway MSC—This term is not used in ANSI-41 (except for a very brief description in ANSI-41.1) but is usually acknowledged to mean an MSC assigned the MS's directory number, but does not provide radio access services (i.e., does not control any base stations).

general packet radio service (GPRS)—A packet data technology for TDMA-based wireless networks (originally GSM and then adopted by ANSI/TIA/EIA-136). GPRS is based on capacity-on-demand technology and provides packet data at near-broadband rates (i.e., above 100 kb/s) by transmitting data over aggregated TDMA channels. GPRS requires two additional functional entities: the gateway GPRS support node (GGSN) and the serving GPRS support node (SGSN).

handoff—The seamless transfer of a wireless call from one base station to another as a mobile unit crosses cell boundaries. (*Also see* "intersystem handoff," "intrasystem handoff," "mobile station-controlled handoff," "mobile station-assisted handoff," "network-controlled handoff," "hard handoff," "soft handoff," *and* "softer handoff.")

handoff chain—As a call is handed off from the first serving MSC (the anchor MSC) to another serving MSC and so on, a handoff chain develops: The sequence of MSCs, from the anchor to the current serving MSC, actively involved in the call at a given point in time.

hard handoff—A "break-before-make" form of call handoff between radio channels, whereby the MS temporarily disconnects from the network as it changes channels. This is the only type of handoff that requires ANSI-41 signaling support (as of ANSI-41); specifically, ANSI-

41 intersystem operations are required when the handoff is between radio channels served by different MSCs. AMPS and TDMA MSs are limited to hard handoff, whereas CDMA MSs can perform soft handoff.

home location register (HLR)—The functional entity in the network that provides the primary database repository of subscriber information used to provide control and intelligence in wireless networks. The HLR contains a record for each home subscriber, which includes location information, subscriber status, subscribed features, and directory numbers.

home message center—The message center belonging to the home service provider that stores and forwards short messages for a particular subscriber.

home MSC—The MSC associated with an MS that broadcasts the system identifier (SID) that matches one programmed into the MS's memory. The home MSC is assigned the MS's directory number (i.e., calls to the MS's directory number will terminate on the home MSC).

home system—The home MSC and its associated home location register (HLR). These two functional entities are sometimes combined in the same physical equipment; however, this need not be the case.

identification—A process to identify a network functional entity, telecommunications end-user, telecommunications terminal, or telecommunications service provider.

implementation dependent—A specification of functionality in the network that is determined or influenced by a particular method of implementation and is thus limited to that implementation.

implementation independent—A specification of functionality in the network that is not influenced by a particular method of implementation and is thus independent of that implementation.

in-band signaling—A type of signaling in which the frequencies or time slots used to carry the signals are within the bandwidth of the information channel.

incoming call—A call described from the mobile station perspective that is mobile terminated and originated from some calling party.

initiating MSC—The MSC that initiates the release of an inter-MSC trunk.

Integrated Services Digital Network (ISDN)—An integrated digital network in which the same time division multiplexing switches and transmission routes are used to establish connections for different types of services, such as telephone, data, electronic mail, or facsimile services.

Integrated Services Digital Network User Part (ISUP)—The functional part of the SS7 protocol that is responsible for providing call control signaling functions required to support basic bearer services and supplementary services for voice and nonvoice applications in an ISDN.

intelligent network (IN)—A type of distributed network characterized by the capabilities of the network nodes to provide and interpret information via local processing and functionality at those nodes.

intelligent peripheral (IP)—A functional entity in an intelligent network architecture that is responsible for providing telecommunications resource service support functionality and communicates with an MSC. Examples of IP services are dialed-digit analysis and interactive voice response systems.

intelligent roaming—A method of automatic roaming where either the mobile station or the network determines which frequency band and system the mobile station should use to send and receive calls without subscriber intervention.

intelligent roaming database (IRDB)—A database contained within the NAM of the mobile station that comprises a list of SIDs and SOCs in priority order. This list identifies the service providers the mobile station uses to access the network while roaming.

inter-MSC trunks—The dedicated voice circuits between MSCs required for call-handoff purposes.

interexchange carrier (IXC)—A carrier authorized to provide interexchange telecommunications services using the North American Numbering Plan. IXCs typically carry traffic between local exchange areas.

interface—The boundary or point of connection between two adjacent physical or logical entities. This boundary establishes the technical interface, the test points, and the points of operational responsibility of the two entities. The boundary is defined in communications systems by functional, interconnection, and signaling characteristics.

Interim Standard (IS)—The name for a standard that has been approved by the Telecommunications Industry Association membership, but not yet by the entire ANSI membership as a full national standard.

Interim standards are not considered a permanent standard; rather they are an interim step toward a permanent national ANSI standard.

international mobile station identity (IMSI)—A unique identifier standardized in *CCITT Recommendation E.212*. It is stored in the mobile station or on a removable smart card (*see* "subscriber identity module" and "removable user identity module") and identifies a subscriber and the home network of that subscriber.

international roaming—The ability for mobile stations to roam between two or more countries. This type of roaming is usually difficult to achieve seamlessly due to differences in protocol implementations in different countries.

internet protocol (IP)—A packet-switched layer 3 data protocol.

intersystem handoff—A type of handoff where the MS is moving between two cells subtending two different MSCs.

intersystem operations—Operations performed between wireless systems that enable subscriber mobility in and between those systems.

interworking function (IWF)—A functional entity that provides interoperability between two other functional entities or networks. Examples of this interoperability are protocol conversion and transaction management.

intrasystem handoff—A type of handoff in which the MS is moving between two cells subtending the same MSC.

local exchange carrier (LEC)—Any exchange carrier that provides telecommunication exchange and exchange access service. An LEC serves subscribers directly and provides local service inside its service area.

local reference number (LRN)—A number used to identify the correct local switch of a ported directory number. LRNs are held in the number portability database (NPDB) for local number portability.

logical channel—A means of two-way simultaneous transmission across a communications link comprised of associated transmit and receive channels. A number of logical channels may be derived from a common physical channel by multiplexing the logical channels onto the physical channel.

local number portability (LNP)—A type of number portability where the subscriber's directory number is essentially owned by the subscriber within the local geographic area. LNP enables the directory number to be portable among local wireless and wire-line service providers within the geographic area specified by the numbering plan. With LNP, a subscriber can change service providers and retain the same directory number.

major trading area (MTA)—One of the 51 personal communications services (PCS) areas, defined by Rand McNally & Company and adopted by the FCC, which cover large geographic portions of the United States.

Message Transfer Part (MTP)—The portion of the Signaling System No. 7 protocol that is responsible for the transfer of signaling messages throughout the SS7 network as required by the application user parts. The MTP comprises Levels 1, 2, and 3 of the SS7 protocol.

message center (MC)—The functional entity in the network that provides the store-and-forward function for short-message services (SMS).

metropolitan statistical area (MSA)—One of the 305 cellular geographic service areas (CGSAs), defined by the U.S. government, which cover the most populated urban cellular markets.

mobile application part (MAP)—An application-layer signaling protocol (Layer 7) mainly providing the mobility management function supporting wireless network operations.

mobile country code (MCC)—A portion of the IMSI specified by ITU Recommendation E.212. It is a three-digit number identifying the country of the service provider for a wireless network subscription.

mobile directory number (MDN)—The mobile station's dialable directory number, which may be different from its mobile identification number (MIN).

mobile identification number (MIN)—The 10-digit number that represents a mobile station's identity, directory number, or both.

mobile network code (MNC)—A portion of the IMSI specified by ITU Recommendation E.212. It is a two- or three-digit number identifying the network (i.e., service provider) within a country for a wireless network subscription.

mobile position coordinator (MPC)—A functional entity in the wireless network primarily used for enhanced 911 emergency calls. The MPC can route, validate, filter, and transfer messages between the wireless network and the emergency services network.

mobile radio-telephone—*See* "mobile station."

mobile station (MS)—The mobile or portable subscriber radio-telephone equipment.

mobile station-assisted handoff (MAHO)—A type of handoff where the network requires the MS to measure signals from surrounding cells and report those measurements back to the network. The network then uses these measurements to determine where a handoff is required and to which channel it should be made.

mobile station-controlled handoff—A type of handoff where the MS continuously monitors the radio signal strength and quality. When predefined criteria are met, the MS checks the best candidate cell for an available traffic channel and requests that the handoff occur.

mobile station identity (MSID)—An identifier that can take on the value of either a MIN or an IMSI. This identifier is required during the transition from MIN to IMSI as the primary unique identifier of a subscriber.

mobile station identification number (MSIN)—A portion of the IMSI specified by ITU Recommendation E.212. It is a nine- or ten-digit number identifying the individual subscriber within a network within a country for a wireless network subscription.

mobile switching center (MSC)—The functional entity in the network that provides the wireless radio-telephone switching function.

mobile telephone switching office (MTSO)—The wireless switching office including the mobile switching equipment, hardware interfaces, and real estate where the switching system resides.

mobility management—A general set of network application functions that enables the mobility of wireless subscribers in a wireless network.

Modification of the Final Judgement (MFJ)—The 1984 settlement (associated with the 1982 Consent Decree) between AT&T and the U.S. government involving the separation of the Regional Bell Operating Companies, Western Electric, AT&T Long Lines, and Bell Laboratories.

multifrequency (MF)—A trunk signaling method in which a combination of audio frequencies is used to convey address and control signaling.

narrowband advanced mobile phone system (NAMPS)—A narrowband version of the analog cellular telecommunications system used in North America and patented by Motorola, Inc. It is characterized by dividing the 30 kHz analog radio channels used for AMPS into three 10-kHz channels, thereby increasing capacity three times.

national mobile station identity (NMSI)—A portion of the IMSI specified by ITU Recommendation E.212. It is the concatenation of the mobile network code (MNC) and the mobile station identification number (MSIN) for a wireless network subscription.

neighbor MSC—An MSC that serves cells that are adjacent to those of the current serving MSC. *See* "border MSC."

network—The telecommunications equipment that has any part in processing a call or a supplementary service for the subscriber referred to. It may include local exchanges and transit exchanges, but does not include the mobile station and is not limited to the "public network" or any other particular set of equipment.

network architecture—The design principles, physical structure, functional organization, data formats, operational procedures, and other features used as the basis for the design, development, and operation of a network.

network operator—*See* "wireless service provider."

network reference model—A diagram that depicts the entities of a network and the interfaces between those entities. The model is used as a graphical representation of the mobile telecommunications system as a whole and does not depict a specific network implementation.

network signaling—A type of signaling used between network nodes to operate, manage, and control the network to support certain types of functionality (e.g., mobility, voice traffic, etc.).

network-controlled handoff—A type of handoff where the radio base station, MSC, or both monitor the radio signal of the MS. When the signal strength and quality deteriorate below a predefined threshold, the network arranges for a handoff to another channel.

North American Numbering Plan (NANP)—A plan for the allocation of unique 10-digit directory numbers. The numbers consist of a 3-digit area (numbering plan area) code, a 3-digit office code, and a 4-digit line number. The plan also extends to format variations (e.g., 3-digit and 7-digit numbers), prefixes (e.g., 1, 0, 01, and 011) and special code applications (e.g., service access codes like 800-numbers).

number assignment module (NAM)—The mobile station's electronic memory module where the MIN and other subscriber specific parameters are stored. MSs that have multi-NAM features offer subscribers the option of obtaining multiple subscriptions with different service providers simultaneously with the same MS.

number portability database (NPDB)—A functional entity required for number portability. It is a repository that supports delivery of calls to ported directory numbers.

operations, administration, and maintenance (OA&M)—A set of functions that enable the network operator to monitor and control the network. OA&M functions are primarily used to observe and record operational characteristics of the network.

originating MSC—The MSC that initiates the mobile-terminated call delivery procedures for an MS. Generally, this would be the home system for the MS, but in some network architectures, a gateway MSC provides this function.

out-of-band signaling—A type of signaling in which the frequencies or timeslots used to carry the signals are outside the bandwidth of the information channel.

outgoing call—A call described from the mobile station perspective that is mobile originated and destined toward some called party.

over-the-air activation teleservice (OATS)—A teleservice provided by the over-the-air function (OTAF) used to provide over-the-air service provisioning (OTASP).

over-the-air function (OTAF)—A functional entity required for over-the-air service provisioning (OTASP). It interfaces with one or more customer service centers (CSCs) to support service provisioning activities. The OTAF interfaces with the MSC to send mobile station orders necessary to complete service provisioning requests.

over-the-air parameter administration (OTAPA)—A feature that provides updates to the NAM of a mobile station via remote programming of the MS by the service provider, over the air interface. Parameters stored in the NAM of the mobile station that enable network subscription are updated by the service provider remotely.

over-the-air programming teleservice (OPTS)—A teleservice provided by the over-the-air function (OTAF) used to provide over-the-air programming administration (OTAPA).

over-the-air service provisioning (OTASP)—A feature that provides initial activation of a mobile station via remote programming of the MS by the service provider, over the air-interface. Parameters stored in the NAM of the mobile station that enable network subscription are programmed by the service provider remotely.

packet-switched network—A network in which logical connections are established between two or more network nodes, which allows the routing and transfer of data in the form of packets so that a channel is occupied only during the transmission of the packet. Upon completion, the channel is made available for the transmission of other packets.

packet-switched connection—A logical connection established between two or more stations to allow the routing and transfer of data in the form of packets. The channel is occupied only during the transmission of the packet. Upon completion of the transmission, the channel is made available for the transmission of other packets.

paging—The function performed at the cell site to seek an MS for a mobile-terminated call.

party—Any participant in a mobile telephone call.

path minimization—The process of removing the second MSC from the handoff chain when an MS is handed off to a third MSC during the same call.

personal communications services (PCS)—A family of mobile or portable radio communications services that provides services to individuals and businesses and is integrated with a variety of other networks. In North America, PCS is characterized by the use of radio frequencies between 1800 and 2200 MHz.

physical channel—A single unidirectional or bidirectional path for transmitting or receiving electrical signals between two entities using a particular encoding and modulation scheme.

point-to-point—A service designed to convey data between two well-defined entities. In SMS, the short-message entities (SMEs) represent the communication points that can originate and receive short messages.

position determining element (PDE)—A functional entity in the wireless network that collects precise location and position information primarily for enhanced 911 emergency service calls. The PDE determines the location of the mobile station making an E-911 call and passes this information to the wireless network.

preferred roaming list (PRL)—A list of SIDs stored in the NAM of the mobile station, enabling mobile-directed intelligent roaming in CDMA (ANSI/TIA/EIA-95) systems. The PRL can be programmed into the phone via the over-the-air parameter administration (OTAPA) feature.

prepaid wireless—A type of wireless telephony service where the subscriber pays in advance for the service to be used. Prepaid wireless systems require no billing of the subscriber and can be controlled from within the network or from the mobile station.

protocol—A set of rules or conventions governing the interactions of processes, applications, and components in a communications system.

public safety answering point (PSAP)—The E-911 emergency services coordinator, typically a public agency that provides emergency services upon completion of an E- 911 emergency call.

public-switched packet data network (PPDN)—Any public data communications network that provides communication via packet-switched protocols. Examples of a PPDN are the Internet or any IP-based or X.25-based public network.

public-switched telephone network (PSTN)—The telecommunications network, commonly accessed by ordinary telephones, key telephone systems, private branch exchange trunks, and data transmission equipment, which provides service to the general public.

radio frequency (RF) channel—A single allocation of contiguous spectrum (i.e., 30 kHz in AMPS, 10 kHz in NAMPS, 30 kHz in TDMA, and 1.25 MHz in CDMA).

radio frequency (RF)—The frequencies of the electromagnetic spectrum that are normally associated with radio-wave propagation.

radio system management—A set of functions that manages the radio resources, connections, and radio transmission paths between the MS and the network, before, during, and after a call.

radio system—The portion of a mobile telecommunications network that includes antenna systems, transceivers, and transceiver controllers and is responsible for managing and controlling the radio resources in the network.

radio-telephone—*See* "mobile station."

receiving MSC—The MSC that receives the request for inter-MSC trunk release.

registration—A set of messages and functions that enables the network to become aware of a subscriber in a particular location. Registration encompasses the location management, service qualification, and mobile station state management functions.

removable user identity module (R-UIM)—A type of smart card supported by CDMA (ANSI/TIA/EIA-95) mobile stations that can be removed from and reinserted into the MS. The R-UIM contains parameters similar to those stored in a number assignment module (NAM), which enable subscription to the wireless network. The primary advantage of the R-UIM is the ability to use any R-UIM–supported CDMA MS to access the network without service provider intervention. *Also see* "subscriber identity module (SIM)" for use in GSM-based networks.

reseller—A wireless service provider that buys service in bulk from another wireless network operator and "resells" that service to subscribers.

revertive call—A call to one's self. In the case of an MS, it is a call to the MS's own mobile directory number (MDN).

Revision "0"—An informal term used to refer to the initial version of a specification published by the Telecommunications Industry Association. This revision number is not officially listed as an actual revision number by the TIA.

roaming—The act of operating a mobile station in a wireless system or network other than the one from which service was subscribed.

rural service area (RSA)—One of the 428 cellular geographic service areas (CGSAs), defined by the U.S. government, which cover the less-populated rural cellular markets.

seamless roaming—A type of roaming whereby the subscriber obtains mobile telecommunications service in the exact same way that it is provided in the home service area (i.e., no apparent *seams* between network service areas).

service—Any function that can be performed or supported in the wireless network. This term is sometimes used to refer to a "feature," which is a specific type of service.

service creation environment (SCE)—An intelligent network feature that enables a service provider to design, implement, and deploy new features by using a standard set of triggers and responses to those triggers.

service control point (SCP)—A network element in the SS7 network (i.e., an end-signaling point) that uses transaction capabilities (i.e., the TCAP protocol) to directly access application databases.

service node (SN)—A functional entity in an intelligent network that provides both service control point (SCP) and intelligent peripheral (IP) functionality.

service switching point (SSP)—A network element in the SS7 network (usually an end-signaling point) that uses transaction capabilities (i.e., the TCAP protocol) to communicate with an SCP that can directly access application databases.

serving GPRS support node (SGSN)—A functional entity required for the general packet radio service (GPRS) that is responsible for the delivery of packets to mobile stations within the service area.

serving MSC—The MSC that, at any given point in time, is "serving" the mobile station (MS); in the case of an active call, the serving MSC controls the base station through which the connection is made between the MS and the other party to the call.

serving VLR—The visitor location register (VLR) associated with the serving MSC.

serving system—The serving MSC and its associated VLR. In general, the serving system maintains the most current location and status of the MS. These two functional entities are often combined in the same physical equipment; however, this need not be the case.

shared secret data (SSD)—The temporary private 128-bit key calculated and stored in the MS and known by the network. The SSD is a concatenation of two 64-bit subsets: SSD_A, which is used to support the authentication procedures, and SSD_B, which is used to support the voice encryption procedures.

short message—A message, represented as a single packet of data, no more than 200 octets in length, which is the user-traffic sent and received via the short-message service (SMS).

short-message entity (SME)—A functional entity in the network that can originate short messages, terminate short messages, or both.

short-message service (SMS)—A packet-switched messaging service that provides store-and-forward functions for the handling of short messages destined to or originating from wireless service subscribers.

signal—The intelligence, operation, message, or control function to be conveyed over a communication system.

signaling—The interchange of signals for the purpose of operating, controlling, managing, supervising, or maintaining a communication system.

Signaling Connection Control Part (SCCP)—The portion of the SS7 protocol responsible for connectionless and connection-oriented network data transfer including the Global Title Translation function.

Signaling System No. 7 (SS7)—An out-of-band common-channel signaling protocol standard designed to be used over a variety of digital telecommunication switching networks. It is optimized to provide a reliable means for information transfer for call control, remote network management, and maintenance.

signaling link—In SS7, a bidirectional transmission path for signaling, consisting of two data channels operating together in opposite directions at the same data rate. A signaling link consists of a physical signaling data link together with its transfer control functions. The signaling link provides reliable transfer of signaling messages between nodes in a signaling network.

signaling point (SP)—A node in an SS7 signaling network that either originates or receives signaling messages, or transfers signaling messages from one signaling link to another, or both.

signaling transfer point (STP)—A signaling point at which a message is received on a signaling link and is transferred to another link. It is neither the source nor the final destination of the message.

soft handoff—A "make-before-break" form of call handoff between radio channels, whereby the MS temporarily communicates with both the serving cell and one or more target cells before being directed to release all but the selected target radio channel. Soft handoff generally refers to the case where all the cells—serving and target—are controlled by a single MSC; therefore, this type of handoff does not require ANSI-41 signaling. However, future revisions of ANSI-41 may support soft handoff *between* MSCs. Currently, soft handoff is supported only for handoffs based on the CDMA air interface.

softer handoff—A soft handoff between two different parts or sectors of a cell controlled by a single base station. This type of handoff does not require ANSI-41 signaling and is currently supported by CDMA mobile stations only.

Specification and Description Language (SDL)—An internationally standardized flowcharting language used to specify the behavior and general parameters of a telecommunications system.

spectrum—A continuous range of frequencies of electromagnetic waves having common characteristics.

spread spectrum—A communication technique in which an information signal is transmitted across an entire bandwidth of frequencies based upon a special coding algorithm.

SSD sharing—Network signaling traffic between the serving system and the AC can be reduced if some authentication tasks are off-loaded from the AC to the serving system. To do this, the AC must *share* the SSD with the serving system.

stage 1—This stage is part of the overall method used to characterize telecommunication services. Stage 1 defines the service aspects of a capability. Specifically, stage 1 provides a service description of a telecommunication service from the subscriber perspective.

stage 2—This stage is part of the overall method used to characterize telecommunication services. Stage 2 defines the functional network aspects of a capability. Specifically, stage 2 provides a description of the functions at the user-network interface and inside the network between network elements.

stage 3—This stage is part of the overall method used to characterize telecommunication services. Stage 3 defines the network implementation aspects of a capability. Specifically, stage 3 provides a description of the actual protocols and formats used to develop the telecommunication service.

standard—A specification that establishes engineering and technical requirements for processes, procedures, practices, and methods that have been decreed by authority or adopted by consensus.

store-and-forward—A method of transmitting data to its destination recipient and the storage of that data if the recipient is unavailable to receive it. The data are stored until sending is convenient.

subscriber—A person authorized for a wireless feature or service.

subscriber identity module (SIM)—A type of removable smart card used in GSM-based networks. A primary feature of the SIM is the ability to perform "SIM roaming," enabling the use of any GSM MS to provide service to a subscriber. The subscriber inserts the SIM into any GSM MS and the network can identify that subscriber and apply all subscribed features.

switch—A function that transfers connection, contact, or signal continuity from one circuit or channel to another.

switching system—An electronic system that performs the selection and switching of telecommunication circuits.

system identifier (SID)—A 15-bit value used for unique identification of a licensed wireless network operator within a cellular geographic service area (for cellular systems) and a trading area (for PCS systems).

system operator code (SOC)—A type of identifier used in TDMA (ANSI/TIA/EIA-136) systems. One or more SOCs are stored in the NAM of the MS, providing mobile-directed intelligent roaming. The SOC limits the MS to receiving service from those wireless service providers identified by the SOCs.

T-1 carrier—A digital transmission system used in North America that multiplexes 24 separate 64 kb/s digital channels onto one circuit at a composite rate of 1.544 Mb/s.

tandem MSC—An MSC in the handoff chain that is neither the anchor MSC nor the current serving MSC.

target MSC—The candidate MSC chosen by the serving MSC as the most appropriate system to take on the serving MSC responsibilities.

Telecommunications Industry Association (TIA)—A U.S. public standards-making group responsible for accrediting other standards groups and approving standards for use in the United States.

Telecommunications Systems Bulletin (TSB)—The name for a publication that is not a standard but has significant value to the wireless telecommunications industry. TSBs are used as addenda to existing standards or to publicize material of industry importance.

teleservice—A type of telecommunication service that provides the complete capability, including terminal equipment functions, for communication between end-users according to protocols established between administrations. In SMS, the applications that use the bearer service are considered the teleservices. These include application protocol elements such as supported character sets, priorities, and time stamps.

terminal mobility—The ability of the terminal to access telecommunication services from different locations while in motion, and the capability of the network to identify and locate that terminal.

time division multiple access (TDMA)—The use of time interleaving to provide multiple, and apparently simultaneous, transmission in a single radio transceiver with a minimum of interference.

traffic—The aggregate of all user requests and communications being serviced by the network.

transaction—A controlled exchange of information between two functional entities in a network.

Transaction Capabilities (TC)—A set of functions that enables transaction processing between two functional entities in a network.

Transaction Capabilities Application Part (TCAP)—The portion of the SS7 protocol responsible for information transfer between two or more nodes in the signaling network. The TCAP consists of transaction capabilities that manage remote operations via SCCP message transfer.

transcoding—The conversion of digital signals from one coding plan to another.

transmission facilities management—A set of functions that manages the physical means of providing voice and data communications services to wireless subscribers. These functions manage the trunk and line types and support the physical rates and digital encoding of the trunks and lines associated with the switching system.

trigger—An event, action, or occurrence of a condition that is used to initiate a call processing function.

trunk—A transmission channel connected between two network elements in a telecommunications system.

trunk group—A group of trunks connected between two elements in a telecommunications system.

trunk maintenance—A set of functions that controls and maintains individual trunks.

user identity module (UIM)—A generic term for a type of smart card supported by mobile stations. The UIM contains parameters similar to those stored in a number assignment module (NAM), which enable subscription to the wireless network.

user information—Information transferred across the functional interface from a source user to a communication system for ultimate delivery to a destination user.

user part—The portion of the SS7 protocol that transfers signaling messages via the Message Transfer Part (MTP). Different types of user parts exist, each of which specifies a particular use of the signaling system.

virtual circuit—A logical circuit or channel established between source and destination packet switches and usually requiring some form of setup prior to data transfer. This type of circuit may not be visible or obvious to a user.

visited system—A serving system for an MS that is broadcasting a SID different from any programmed into the MS's memory. Since ANSI-41's primary application is in the case where the serving system is a visited system (not the home system), the terms *visited system* and *serving system* can be used interchangeably.

visitor location register (VLR)—A local database function that maintains temporary records associated with individual network subscribers. The VLR contains subscriber location and service information. The local network switch accesses the VLR to retrieve information for the handling of calls to and from subscribers.

wireless carrier—*See* "wireless service provider."

wireless intelligent network (WIN)—A type of distributed intelligent wireless network characterized by the capabilities of the network nodes to provide and interpret information via local processing and functionality at those nodes. The WIN incorporates service control points (SCPs), intelligent peripherals (IPs), service nodes (SNs) to support a call model, and triggers, which can provide enhanced and easily modifiable services to a subscriber.

wireless number portability (WNP)—A type of number portability that applies specifically to wireless ported numbers. WNP allows a wireless subscriber to port the directory number to another wireless service provider or to a wire-line service provider.

wireless service provider—A company, organization, business, etc. authorized to provide wireless radio-telephone service. It is the organization that sells, administers, maintains, and charges for the wireless service. The service provider may or may not be the provider and operator of the network.

wireless system—Any network system accessed via the propagation of electromagnetic waves through the air (i.e., instead of wires), supporting system communications.

wire-line system—Any network system accessed via the propagation of electromagnetic waves through a physical wire path to provide connections supporting system communications.

world zone 1 (WZ1)—The group of countries in the World Numbering Plan that are identified by the single-digit country code 1. WZ1 is defined in ITU-T Recommendation E.164. WZ1 generally includes the U.S., Canada, and the Bahamas.

X.25—An ITU-T specification for a packet-switched protocol specified for use on public data networks between data terminal equipment (DTE) and data circuit-terminating equipment (DCE).

APPENDIX D

GLOSSARY OF ACRONYMS

3G	third-generation wireless technology
3WC	three-way calling
AC	authentication center
ACG	automatic code gapping
ACRE	authorization and call routing equipment
ADS	asynchronous data service
AFREPORT	ANSI-41 Authentication Failure Report operation
AH	answer hold
AHAG	TR45.2 ad hoc authentication group
AIN	advanced intelligent network
A-key	authentication key
AMPS	Advanced Mobile Phone System
ANI	automatic number identification
ANSI	American National Standards Institute
AOA	angle of arrival
AOC	advice of charge
APDU	application protocol data unit
ASE	application service element
ASI	application service interface
ASP	application service part
ASREPORT	ANSI-41 AuthenticationStatusReport operation
ATIS	Alliance for Telecommunications Industry Solutions

AUTHDIR	ANSI-41 AuthenticationDirective operation
AUTHDIRFWD	ANSI-41 AuthenticationDirectiveForward operation
AUTHREQ	ANSI-41 AuthenticationRequest operation
BCD	binary coded decimal
Bellcore	Bell Communications Research, Inc.
BID	billing identification
BLK	ANSI-41 Blocking operation
BS	base station
BSC	base station controller
BSCHALL	ANSI-41 BaseStationChallenge operation
BTA	basic trading area
BTS	base transceiver system
BTTC	broadcast teleservice transport capability
BULKDEREG	ANSI-41 BulkDeregistration operation
CALEA	Communications Assistance for Law Enforcement Act
CAVE	cellular authentication voice encryption
CC	conference calling or country code
CCI	conference calling indicator
CCITT	International Telegraph and Telephone Consultative Committee (currently ITU-T)
CD	call delivery
CDMA	code division multiple access
CFB	call forwarding—busy
CFD	call forwarding—default
CFNA	call forwarding—no answer
CFU	call forwarding—unconditional
CGSA	cellular geographic service area
CIBER	Cellular Intercarrier Billing Exchange Roamer
CMEA	cellular message encryption algorithm

CMRS	commercial mobile radio service
CMT	cellular messaging teleservice
CNAP	calling name presentation
CNAR	calling name restriction
CNI	calling number identification
CNIP	calling number identification presentation
CNIR	calling number identification restriction
COUNTREQ	ANSI-41 CountRequest operation
CPN	calling party number
CPP	calling party pays
CPT	cellular paging teleservice
CSC	customer service center
CSS	cellular subscriber station (MS preferred)
CT	call transfer
CTIA	Cellular Telecommunications Industry Association
CW	call waiting
D-AMPS	Digital AMPS
DCE	data circuit-terminating equipment
DECT	Digital European Cordless Telephone
DMH	data message handler
DN	directory number
DND	do not disturb
DP	data privacy
DS0	digital signal level 0
DS1	digital signal level 1
DTE	data terminal equipment
DTMF	dual-tone multifrequency
E-911	enhanced 911
ECSA	Exchange Carrier Standards Association
EIA	Electronic Industries Association

EIR	equipment identity register
EPE	enhanced privacy and encryption
ESN	electronic serial number
ESRD	emergency services routing digits
ETSI	European Telecommunications Standards Institute
FA	flexible alerting
FACDIR	ANSI-41 FacilitiesDirective operation
FACDIR2	ANSI-41 FacilitiesDirective2 operation
FACREL	ANSI-41 FacilitiesRelease operation
FC	feature code
FCC	Federal Communications Commission
FCS	feature code string
FE	functional entity
FEATREQ	ANSI-41 FeatureRequest operation
FLASHREQ	ANSI-41 FlashRequest operation
FM	frequency modulation
GGSN	gateway GPRS support node
GPRS	general packet radio service
GPS	global positioning system
HANDBACK	ANSI-41 HandoffBack operation
HANDBACK2	ANSI-41 HandoffBack2 operation
HANDMREQ	ANSI-41 HandoffMeasurementRequest operation
HANDMREQ2	ANSI-41 HandoffMeasurementRequest2 operation
HANDTHIRD	ANSI-41 HandoffToThird operation
HANDTHIRD2	ANSI-41 HandoffToThird2 operation
HLR	home location register
IA5	International Alphabet 5
ID	identification or identifier
IEC	International Electrotechnical Commission
IEEE	Institute of Electrical and Electronic Engineers

IMSI	international mobile station identity
IN	intelligent network
INFOBACK	ANSI-41 InformationBackward operation
INFODIR	ANSI-41 InformationDirective operation
INFOFWD	ANSI-41 InformationForward operation
IP	intelligent peripheral or Internet protocol
IRDB	intelligent roaming database
IS	interim standard
ISANSWER	ANSI-41 InterSystemAnswer operation
ISDN	Integrated Services Digital Network
ISO	International Organization for Standardization
ISP	intermediate service part
ISPAGE	ANSI-41 InterSystemPage operation
ISPAGE2	ANSI-41 InterSystemPage2 operation
ISSETUP	ANSI-41 InterSystemSetup operation
ISUP	ISDN User Part
ITU	International Telecommunications Union
ITU-T	International Telecommunications Union—Telecommunications (formerly CCITT)
IVR	interactive voice-response
IWF	interworking function
IXC	interexchange carrier
kb/s	kilobits per second
kHz	kilohertz
LEC	local exchange carrier
LNP	local number portability
LOCREQ	ANSI-41 LocationRequest operation
LRN	local reference number
MAH	mobile access hunting
MAHO	mobile station-assisted handoff

MAP	mobile application part
Mb/s	megabits per second
MC	message center
MCC	mobile country code
MDN	mobile directory number
MF	multifrequency
MFJ	modification of the final judgement
MHz	megahertz
MIN	mobile identification number
MNC	mobile network code
MPC	mobile position coordinator
MS	mobile station
MSA	metropolitan statistical area
MSC	mobile switching center
MSID	mobile station identity
MSIN	mobile station identification number
MSINACT	ANSI-41 MSInactive operation
MSONCH	ANSI-41 MobileOnChannel operation
MTA	major trading area
MTP	message transfer part
MTSO	mobile telephone switching office
MWN	message waiting notification
NAM	number assignment module
NAMPS	Narrowband Advanced Mobile Phone System
NANP	North American Numbering Plan
NANPA	North American Numbering Plan Administration
NDC	national destination code
NDSS	network-directed system selection
NIST	National Institute of Standards and Technology
NIUF	National ISDN Users Forum

NMSI	national mobile station identity
NPA	numbering plan area
NPDB	number portability database
NSN	national significant number
NPREQ	ANSI-41 NumberPortabilityRequest operation
OA&M	operations, administration, and maintenance
OAM&P	operations, administration, maintenance, and provisioning
OATS	over-the-air activation teleservice
OMT	overhead message train
OPTS	over-the-air programming teleservice
ORREQ	ANSI-41 OriginationRequest operation
OSI	Open Systems Interconnection
OTA	over-the-air activation
OTAF	over-the-air function
OTAPA	over-the-air parameter administration
OTASP	over-the-air service provisioning
PACA	priority access and channel assignment
PACS	Personal Access Communications System
PARMREQ	ANSI-41 Parameter Request operation
PBX	private branch exchange
PCA	password call acceptance
PCIA	Personal Communications Industry Association
PCN	personal communications network
PCS	personal communications services
PCW	priority call waiting
PDE	position determining element
PIN	personal identification number
PL	preferred language
PN	project number

POTS	plain old telephone service
PPC	prepaid charging
PPDN	public-switched packet data network
PRL	preferred roaming list
PSAP	public safety answering point
PSTN	public-switched telephone network
QUALDIR	ANSI-41 QualificationDirective operation
QUALREQ	ANSI-41 QualificationRequest operation
RANDREQ	ANSI-41 RandomVariableRequest operation
RBOC	Regional Bell Operating Company
REDDIR	ANSI-41 RedirectionDirective operation
REDREQ	ANSI-41 RedirectionRequest operation
REGCANC	ANSI-41 RegistrationCancellation operation
REGNOT	ANSI-41 RegistrationNotification operation
RESET	ANSI-41 ResetCircuit operation
RF	radio frequency
RFC	remote feature control
ROUTREQ	ANSI-41 RoutingRequest operation
RSA	rural service area
RUIDIR	ANSI-41 RemoteUserInteractionDirective operation
R-UIM	removable user identity module
SCA	selective call acceptance
SCCP	Signaling Connection Control Part
SCE	service creation environment
SCP	service control point
SDL	Specification and Description Language
SERVREQ	ANSI-41 Service Request operation
SGSN	serving GPRS support node
SID	system identifier
SIM	subscriber identity module

SMDBACK	ANSI-41 SMSDeliveryBackward operation
SMDFWD	ANSI-41 SMSDeliveryForward operation
SMDPP	ANSI-41 SMSDeliveryPointToPoint operation
SME	short-message entity or signaling message encryption
SMS	short-message services
SMSNOT	ANSI-41 SMSNotification operation
SMSREQ	ANSI-41 SMSRequest operation
SN	subscriber number or station number or service node
SOC	system operator code
SP	signaling point or standards proposal
SPINA	subscriber PIN access
SPINI	subscriber PIN intercept
SS#7	Signaling System No. 7 (ITU-T Standard)
SS7	Signaling System No. 7 (ANSI Standard)
SSD	shared secret data
SSD_A	shared secret data "A"
SSD_B	shared secret data "B"
STP	signaling transfer point
SWNO	switch number
TC	transaction capabilities
TCAP	Transaction Capabilities Application Part
TDMA	time division multiple access
TDOA	time difference of arrival
TIA	Telecommunications Industry Association
TID	transaction identifier
TLDN	temporary local directory number
TMSI	temporary mobile station identity
TRANUMREQ	ANSI-41 TransferToNumberRequest operation
TSAR	teleservice segmentation and reassembly
TSB	Telecommunications Systems Bulletin

TSSC	Technical Standards Subcommittee
TTEST	ANSI-41 TrunkTest operation
TTESTDISC	ANSI-41 TrunkTestDisconnect operation
UBLK	ANSI-41 Unblocking operation
UG	user group
UIM	user identity module
UNRELDIR	ANSI-41 UnreliableRoamerDataDirective operation
UNSOLRES	ANSI-41 UnsolicitedResponse operation
USCF	user-selective call forwarding
V&V	verification and validation
VLR	visitor location register
VMR	voice message retrieval
VP	voice privacy
WG	working group
WIN	wireless intelligent network
WNP	wireless number portability
WZ1	world zone 1

BIBLIOGRAPHY

American National Standards Institute (ANSI) T1 Committee Standards

American National Standards Institute, Inc., American National Standard for Telecommunications, Signaling System Number 7 (SS7)–General Information, Exchange Carriers Standards Association Committee T1, T1.110, 1992.

American National Standards Institute, Inc., American National Standard for Telecommunications, Signaling System Number 7 (SS7)–Message Transfer Part (MTP), Exchange Carriers Standards Association Committee T1, T1.111, 1992.

American National Standards Institute, Inc., American National Standard for Telecommunications, Signaling System Number 7 (SS7)–Signaling Connection Control Part (SCCP), Exchange Carriers Standards Association Committee T1, T1.112, 1996.

American National Standards Institute, Inc., American National Standard for Telecommunications, Signaling System Number 7 (SS7)–Transaction Capabilities Application Part (TCAP), Exchange Carriers Standards Association Committee T1, T1.114, 1992.

American National Standards Institute, Inc., American National Standard for Telecommunications, Signaling System Number 7 (SS7)–Supplementary Services for Non-ISDN-Subscribers, Exchange Carriers Standards Association Committee T1, T1.611, 1991.

American National Standards Institute, Inc., American National Standard for Telecommunications, Operations, Administration, Maintenance and Provisioning (OAM&P)–Network Tones and Announcements, Exchange Carriers Standards Association Committee T1, T1.209, 1989.

International Organization for Standardization (ISO) Standards

International Organization for Standardization (ISO), ISO-7776, Information Technology–Telecommunication and Information Exchange Between Systems–High Level Data Link Control Procedures–Description of the X.25 LAPB-Compatible DTE Data Link Procedures, September, 1994.

International Organization for Standardization (ISO), ISO-8208, Information Technology–Data Communications–X.25 Packet Layer Protocol for Data Terminal Equipment, April, 1993.

International Organization for Standardization (ISO), ISO-8878, Information Technology–Telecommunication and Information Exchange Between Systems–Use of X.25 to Provide the OSI Connection-Mode Network Service, October, 1992.

International Telegraph and Telephone Consultative Committee (CCITT) Standards and International Telecommunications Union–Telecommunication Sector (ITU-T) Standards

CCITT, Volume II—Fascicle II.2, Telephone Network and ISDN–Operation, Numbering, Routing and Mobile Service, Recommendations E.100–E.333, 1992.

CCITT, Volume III—Fascicle III.7, Integrated Services Digital Network (ISDN) General Structure and Service Capabilities, Recommendations I.110–I.257, 1988.

CCITT, Volume VI—Fascicle VI.1, General Recommendations on Telephone Switching and Signaling, Functions and Information Flows for Services in the ISDN, Supplements, Recommendation Q.65, 1988.

CCITT, Volume VI—Fascicle VI.13, Public Land Mobile Network—Mobile Application Part and Interfaces, Recommendations Q.1051–Q.1063, 1988.

CCITT, Volume VI—Fascicle VI.7, Specifications of Signaling System No. 7, Recommendations Q.700–Q.716, 1996.

CCITT, Volume VI—Fascicle VI.8, Specifications of Signaling System No. 7, Recommendations Q.721–Q.766, 1992.

CCITT, Volume VI—Fascicle VI.9, Specifications of Signaling System No. 7, Recommendations Q.771–Q.795, 1992.

CCITT, Volume VII—Fascicle VII.3, International Reference Alphabet (IRA) (formerly International Alphabet No. 5 [IA5]), Recommendation T.50, 1988.

CCITT, Volume VIII—Fascicle VIII.2, Data Communication Networks: Services and Facilities, Interfaces, Recommendations X.1–X.32, 1988.

CCITT, Volume VIII—Fascicle VIII.3, Data Communication Networks: Transmission, Signaling and Switching, Network Aspects, Maintenance and Administrative Arrangements, Recommendations X.40–X.181, 1988.

CCITT, Volume VIII—Fascicle VIII.4, Data Communication Networks: Open Systems Interconnection (OSI)—Model and Notation, Service Definition, Recommendations X.200–X.219, 1988.

CCITT, Volume VIII—Fascicle VIII.5, Data Communication Networks: Open Systems Interconnection (OSI)—Protocol Specifications, Conformance Testing, Recommendations X.220–X.290, 1988.

ITU-T, Numbering Plan for the ISDN Era, Recommendation E.164, 1991.

Telecommunications Industry Association (TIA) Standards

*To purchase the complete text of any TIA document, log on to **www.tiaonline.org** or call Global Engineering Documents at (800) 854-7179.

ANSI/TIA/EIA-41-D, Cellular Radio-Telecommunications Intersystem Operations, December 1997.

ANSI/TIA/EIA-41-E, Cellular Radio-Telecommunications Intersystem Operations, 2001.

ANSI/TIA/EIA-93-A, Wireless Telecommunications A_i–D_i Interfaces Standard, November 1998.

ANSI/TIA/EIA-95-B, Mobile Station–Base Station Compatibility Standard for Wideband Spread Spectrum Cellular Systems, February 1999.

ANSI/TIA/EIA-124-B, Wireless Radio Telecommunications Intersystem Non-Signaling Data Communication DMH (Data Message Handler), July 1999.

ANSI/TIA/EIA-136-B, TDMA Cellular PCS, March 2000.

ANSI/TIA/EIA-634-B, MSC-BS Interface for Public Wireless Communications Systems, April 1999.

ANSI/TIA/EIA-660, Uniform Dialing Procedures and Call Processing Treatment for Cellular Radio Telecommunications, July 1996.

ANSI/TIA/EIA-664, Cellular Features Description, June 1996.

ANSI/TIA/EIA-691, Mobile Station–Base Station Compatibility Standard for Enhanced 800 MHz Analog Cellular, November 1999.

TIA, 1999 TIA Annual Report, 1999.

TIA, Engineering Manual (online), 2000.

TIA, TR45.AHAG, Common Cryptographic Algorithms, Revision D, March 14, 2000.

TIA, TR45.AHAG, Interface Specification for Common Cryptographic Algorithms, Revision D, March 14, 2000.

TIA/EIA, Interim Standard IS-41, Cellular Radio-Telecommunications Intersystem Operations, February 1988.

TIA/EIA, Interim Standard IS-41-A, Cellular Radio-Telecommunications Intersystem Operations, January 1991.

TIA/EIA, Interim Standard IS-41-B, Cellular Radio-Telecommunications Intersystem Operations, December 1991.

TIA/EIA, Interim Standard IS-41-C, Cellular Radio-Telecommunications Intersystem Operations, February 1996.

TIA/EIA/IS-129, Interworking/interoperability Between DCS 1900 and IS-41-based MAPs for 1800 MHz Personal Communications Systems–Phase I, July 1996.

TIA/EIA/IS-653-A, ISDN-based A-interface (Radio System–PCSC) for 1800 MHz Personal Communications Systems, May 1996.

TIA/EIA/IS-680, Personal Base Station–Authorization and Call Routing Equipment Compatibility Standard, May 1996.

TIA/EIA/IS-725-A, Cellular Radiotelecommunications Intersystem Operations—Over-the-Air Service Provisioning (OTASP) and Parameter Administration (OTAPA), July 1999.

TIA/EIA/IS-728, Intersystem Link Protocol, April 1998.

TIA/EIA/IS-730, Intersystem Operations Support for the IS-136 Digital Control Channel, August 1997.

TIA/EIA/IS-735, Enhancements to TIA/EIA-41-D & TIA/EIA-664 for Advanced Features in Wideband Spread Spectrum Systems, February 1998.

TIA/EIA/IS-737, Enhancements for Circuit Mode Services, April 1998.

TIA/EIA/IS-751, TIA/EIA-41-D Modifications to Support IMSI, February 1998.

TIA/EIA/IS-756-A, TIA/EIA-41-D Enhancements for Wireless Number Portability Phase II, December 1998.

TIA/EIA/IS-764, TIA/EIA-41-D Enhancements for Wireless Calling Name Feature Descriptions, June 1998.

TIA/EIA/IS-771, Wireless Intelligent Network, July 1999.

TIA/EIA/IS-778, Wireless Authentication Enhancements Descriptions, March 1999.

TIA/EIA/IS-807, TIA/EIA-41-D Enhancements for Internationalization, August 1999.

TIA/EIA/IS-807-1, TIA/EIA-41-D Enhancements for Internationalization, Addendum 1, March 2000.

TIA/EIA/IS-808, Network Based Enhancements for User Identity Module (UIM), 2000.

TIA/EIA/IS-812, TIA/EIA-41-D Message Segmentation, August 1999.

TIA/EIA/IS-820, Removable User Identity Module (R-UIM) for TIA/EIA Spread Spectrum Standards, May 2000.

J-STD-025-A, Lawfully Authorized Electronic Surveillance, May 2000.

J-STD-034, Wireless Enhanced Emergency Services, December 1997.

J-STD-036, Enhanced Wireless E-911 Phase II, September 2000.

TIA/EIA TSB29-C, International Implementation of Wireless Telecommunication Systems Compliant with ANSI/TIA/EIA-41, June 1999.

TIA/EIA TSB29-C-1, International Implementation of Wireless Telecommunication Systems Compliant with ANSI/TIA/EIA-41, Addendum 1, December 1999.

TIA/EIA TSB50, User Interface for Authentication Key Entry, March 1993.

TIA/EIA TSB76, ANSI-41 Enhancements for PCS Multi-Band Support, September 1996.

TIA/EIA TSB100, Wireless Network Reference Model, July 1998.

Other Sources

Cellular Networking Perspectives, David Crowe. Calgary, Canada: Cellular Networking Perspectives Ltd., 1992–2000.

Wireless Security Perspectives, Les Owens and David Crowe. Calgary, Canada: Cellular Networking Perspectives Ltd., 1999–2001.

*To purchase a subscription to *Cellular Networking Perspectives* or *Wireless Security Perspectives*, call 800-633-5514 (+1-403-274-4749 if outside USA and Canada) or send a facsimile to +1-403-289-6658.

Coursey, Cameron Kelly. *Understanding Digital PCS—The TDMA Standard*. Boston: Artech House Publishers, 1998.

Telcordia, Compatibility Information for Interconnection of a Wireless Services Provider and a Local Exchange Carrier Network, Generic Requirement GR-145, Issue 2, May 1998.

Suggested Readings

Bates, Bud. *Wireless Networked Communications: Concepts, Technology, and Implementation*. New York: McGraw-Hill, 1994.

Lee, William C.Y. *Mobile Cellular Telecommunications Systems: Analog and Digital Systems*. New York: McGraw-Hill, 1995.

Russell, Travis. *Signaling System #7*. New York: McGraw-Hill, 2000.

■■ ■■ INDEX

ABOUT THE AUTHORS

RANDALL A. SNYDER is Director of Emerging Technologies for Openwave Systems, in Redwood City, California. He has extensive experience in wire line and wireless telecommunications, systems engineering, and network design, as well as development of SS7, ANSI-41, and GSM networks. He also co-authored *Mobile Telecommunications Networking with IS-41*.

MICHAEL D. GALLAGHER is a telecommunications consultant, based in San Jose, California. His experience spans hardware design, software design, and development in wire line and wireless telecommunications. He co-authored *Mobile Telecommunications Networking with IS-41*, the popular predecessor to this book.